Embedded C Programming and the Microchip PIC

THOMSON
DELMAR LEARNING

Australia Canada Mexico Singapore Spain United Kingdom United States

Embedded C Programming and the Microchip PIC

RICHARD BARNETT
LARRY O'CULL
SARAH COX

Embedded C Programming with the Microchip PIC

Richard Barnett, Larry O'Cull, Sarah Cox

Vice President, Technology and Trades SBU:
Alar Elken

Editorial Director:
Sandy Clark

Acquisitions Editor:
David Garza

Development Editor:
(Senior) Michelle Ruelos Cannistraci

Editorial Assistant:
(Senior) Dawn Daugherty

Marketing Director:
Cyndi Eichelman

Channel Manager:
Fair Huntoon

Marketing Coordinator:
Casey Bruno

Production Director:
Mary Ellen Black

Production Manager:
Andrew Crouth

Production Editor:
Sharon Popson

Art/Design Coordinator:
Francis Hogan

Technology Project Manager:
Kevin Smith

Printed in Canada
1 2 3 4 5 XX 06 05 04

For more information contact Delmar Learning
Executive Woods
5 Maxwell Drive, PO Box 8007,
Clifton Park, NY 12065-8007
Or find us on the World Wide Web at
www.delmarlearning.com

Library of Congress Cataloging-in-Publication Data:

ISBN: 1401837484

NOTICE TO THE READER

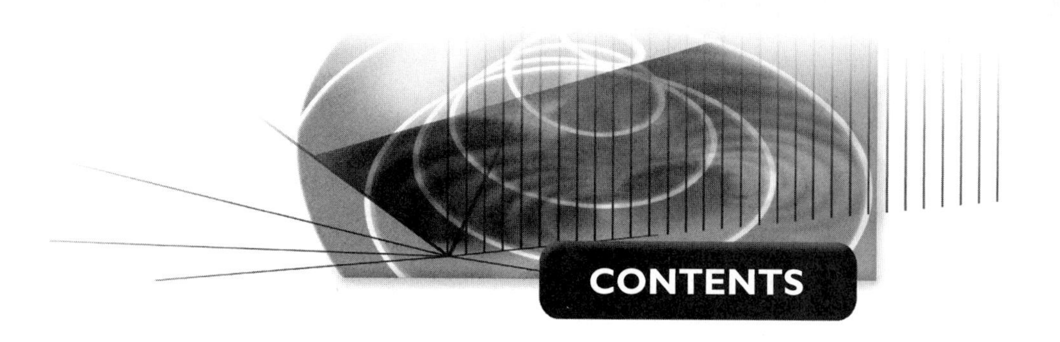

CONTENTS

CHAPTER 3 STANDARD I/O AND PREPROCESSOR DIRECTIVES

CHAPTER 4 THE CCS-PICC C COMPILERAND IDE

CHAPTER 5 PROJECT DEVELOPMENT

PREFACE

Embedded C Programming and the Microchip PIC is designed to teach both C language programming as it applies to embedded microcontrollers and to provide knowledge in the application of the Microchip® family of PIC® RISC microcontrollers.

INTENDED AUDIENCE

This book is designed to serve two diverse audiences:

- Students in Electrical and Computer Engineering, Electronic Engineering, Electrical Engineering Technology, and Computer Engineering Technology curricula. Two scenarios for students fit the book very well:
  - Beginning students who have not yet had a C programming course: The book serves a two-semester or four-quarter sequence in which students learn C language programming and how to apply C to embedded microcontroller designs. Within the same sequence they can advance to the more sophisticated embedded applications, which can all be run on an embedded microcontroller with very little hardware knowledge required. After Chapter 1, Embedded C Language Tutorial, is completed, it will serve as a programming reference for the balance of the courses.
  - Students who have already taken a C programming course can use the book for a one-semester or two-quarter course in embedded microcontrollers. In this instance, the students study only those portions of Chapter 1 that relate to programming for the embedded environment and move quickly to the advanced hardware concepts. Chapter 1 is organized (as are all the chapters in the book) to provide a usable reference for information needed in other courses.
- Practicing engineers, technologists, and technicians who want to add a new microcontroller to their areas of expertise: Chapter 1 can be used as needed (depending on the user's level of programming experience) either to learn needed concepts or as a reference. Chapter 2 in which the Microchip PIC

microcontroller hardware is discussed will lead such an individual through the steps of learning a new microcontroller and serve as a reference for future projects.

PREREQUISITES

Some knowledge of digital systems, number systems, and logic design is required. Preliminary versions of Chapter 1, Embedded C Language Tutorial, have been used successfully in a fundamental microcontrollers course (sophomore-level class—no prerequisite programming) following two semesters of basic digital logic courses. This textbook has also proven to be excellent for an advanced microcontrollers elective course. In many cases, the students have elected to keep the book and use it as a reference through their senior project design courses, and have taken it with them into industry as a useful reference.

ORGANIZATION

The text is organized in logical topic units so that instructors can either follow the text organization, starting with the C language and progressing through the PIC hardware and into more advanced topics, or they can choose the order of the topics to fit their particular needs. Topics are kept separate and identified for easy selection. The chapter exercises and laboratory exercises are also separated by topic to make it easy to select those that apply in any particular instance.

CHAPTER CONTENTS SUMMARY

Chapter 1, Embedded C Language Tutorial

The C language is covered in detail in a step-by-step method as it applies to programming embedded microcontrollers. One or more example programs accompany each programming concept to illustrate its use. At the conclusion of the chapter, students will be able to create C language programs to solve problems. Chapter 1 features the CCS-PICC C Compiler as a model for C compilation as it applies to the Microchip PIC, but the basis of what is demonstrated applies to other PIC C compilers as well.

Chapter 2, PIC Microcontroller Hardware

The PIC RISC processors are covered from basic architecture through use of all of the standard peripheral devices included in the microcontrollers. Example programs are used to demonstrate common uses for each of the peripherals. On completion of Chapters 1 and 2, students will be able to apply PIC RISC processors to solve problems.

Chapter 3, Standard I/O and Preprocessor Directives

Chapter 3 introduces students to the built-in functions available in C and to their use. Again, example programs are used to illustrate how to use the built-in functions. Finishing Chapter 3 prepares students to use the built-in functions to speed their programming and efforts at problem solving.

Chapter 4, The CCS-PICC C Compiler and IDE

This chapter can be used as a handbook for using the CCS-PICC C Compiler and its accompanying integrated development environment (IDE). Students can learn to use the PICC and its IDE effectively to create and debug C programs using the evaluation edition of the compiler that is included on the enclosed CD, as well as Microchip's MPLAB debugging environment available from Microchip's Web site: http://www.microchip.com.

Chapter 5, Project Development

This chapter focuses on the orderly development of a project using microcontrollers. An electronic scooter is developed in its entirety to illustrate the process. Students learn to efficiently develop projects to maximize their successes.

Appendices:

Appendix A, Library Functions Reference. A complete reference to the built-in library functions available at the time of publication.

Appendix B, Programming the PIC Microcontrollers. This is a guide to actually programming the FLASH memory area of the PIC devices, so that students can understand the programming function.

Appendix C, CCS ICD-S Serial In-System Programmer/Debugger. This is an introduction to the CCS ICD-S programming and debugging tool.

Appendix D, Microchip ICD-II Serial In-System Programmer/Debugger. This is an introduction to the Microchip ICD-II programming and debugging tool.

Appendix E, The "FlashPIC-Dev" Development Board. This is an introduction to the FlashPIC-Dev development board by Progressive Resources LLC.

Appendix F, ASCII Table.

Appendix G, PIC16F877 Instruction Set Summary. An assembly code instruction summary for use with the PIC16F877 assembly code programming examples.

Appendix H, PIC18F458 Instruction Set Summary. An assembly code instruction summary for use as a reference.

Appendix I, Answers to Selected Questions (By Chapter).

RATIONALE

The advancing technology surrounding microcontrollers continues to provide amazingly greater amounts of functionality and speed. These increases have led to the almost universal use of high-level languages such as C to program even time-critical tasks that used to require assembly language programs to accomplish. Simultaneously, microcontrollers have become easier and easier to apply, making them excellent vehicles for educational use. Many schools have adopted microcontroller vehicles as target devices for their courses. Also, the price of microcontroller development boards has dropped to the level wherein a number of schools have the students buy the board as a part of their "parts kit" so that all students have their own development board. Some of these courses require C programming as a prerequisite, and others teach C language programming and the application of embedded microcontrollers in an integrated manner.

Embedded C Programming and the Microchip PIC is an answer to the need for a textbook that is usable in courses with and without a C language prerequisite course and that can be used as a useful reference in later course work. The CD-ROM included with this book contains a "Student Edition" compiler and other software so that students with their own development boards have everything they need to work outside of class as well as in the school labs.

HARDWARE USED

Most of the programming application examples in this textbook were developed using a 'FlashPIC-Dev' evaluation board provided by Progressive Resources, LLC (refer to Appendix E for specifics). This board is particularly well suited for educational use and is a good general purpose development board. However, the Microchip PIC microcontrollers are very easy to use and can be run perfectly well by simply plugging them into a prototype board, adding the oscillator crystal, along with two capacitors, and connecting four wires for programming. Students have been very successful with either method.

The PIC16F877 and PIC18F458 microcontrollers have been used to work out the examples for the text. One of the major advantages of the PIC microcontrollers is that they are parallel in their architecture and the programming approach for the devices. This means that the examples shown will work on virtually any PIC microcontroller, provided that it contains the peripherals and other resources to do the job—it is not necessary to make changes to use the code for other members of the PIC family. Consequently, the text is useful with other members of the PIC family as well.

The more common peripherals are covered in this textbook, and the code can be used as a template to apply to more exotic peripherals in some of the other PIC family members.

CD-ROM CONTENTS AND SOFTWARE USED IN THE TEXTBOOK

The software used with the textbook includes the Microchip MPLAB® (which is free by accessing http://www.microchip.com/), and the CCS PICC compiler from Custom Computer Services Inc. (http://www.ccsinfo.com/). The CD-ROM included with this book contains the source code for all of the software examples from the text. These can be used as references or as starting points for specific assignments. All of the programs in the textbook can be compiled using the evaluation version of CCS PICC that is included on the enclosed CD. The included evaluation version is limited to two processor types, the PIC16F877 and the PIC18F458, and will produce a maximum of 2K of output code. Refer to the above-listed Web site information to obtain the latest version of the compiler. More information about purchasing the full version may be found at http://www.prllc.com/.

ACKNOWLEDGMENTS

The material contained in this textbook is not only a compilation of years of experience but of information available from Microchip Technology, Inc., Custom Computer Services, Inc., and Progressive Resources LLC.

Microchip Technology Inc.
2355 West Chandler Blvd.
Chandler, AZ 85224-6199
Telephone: (480) 792-7200
Fax: (480) 792-9210
http://www.microchip.com/

Custom Computer Services, Inc.
P.O. Box 2452
Brookfield, WI 53008
Telephone: (262) 797-0455
Fax: (262) 797-0459
http://www.ccsinfo.com/

Progressive Resources LLC
4105 Vincennes Road
Indianapolis, IN 46468
Telephone: (317) 471-1577
Fax: (317) 471-1580
http://www.prllc.com/

The authors and Delmar Learning would like to extend their thanks to the following reviewers:

Harold Broberg, Purdue University
Shujen Chen, DeVry University, Tinley Park, IL
Gerard Gambs, Pittsburgh Institute of Technology
Richard Helps, Brigham Young University
Ron Krahe, Penn State – Behrend College
Faramarz Mortezaie, DeVry University
James Stewart, DeVry University

AUTHOR-SPECIFIC ACKNOWLEDGMENTS

The support of my family, Gay, Laura, and April, has made this book both a pleasure to work on and a joy to complete. It is also important to acknowledge the motivation supplied by Larry O'Cull and the fantastic pleasure of working with Larry and Sarah on this project.

Richard H. Barnett, PE, Ph.D.
September 2003

It was a great pleasure to work on this project with Dr. Barnett, teacher and mentor, and Sarah Cox, partner and coauthor. They kept this project exciting. This work would not have been possible without the patience and support of my wife, Anna, and children, James, Heather, Max, and Alan, who have been willing to give up some things now to build bigger and better futures.

Larry D. O'Cull
September 2003

This book has been a challenging and exciting endeavor. I have a tremendous amount of respect for Larry O'Cull and Dr. Barnett and have considered it a great privilege to work with them. I must specifically thank Larry for having the vision for this project. I also want to thank my husband, Greg, and daughter, Meredith, for their support throughout the entire process.

Sarah A. Cox
September 2003

AUTHORS

This textbook is definitely a highly collaborative work by the three authors. Each section was largely written by one author and then critically reviewed by the other two, who rewrote chunks as needed. It is not possible to delineate who is responsible for any particular part of the book. The authors:

Richard H. Barnett, PE, Ph.D.
Professor of Electrical Engineering Technology
Purdue University

Dr. Barnett has been instructing in the area of embedded microcontrollers for the past eighteen years, starting with the Intel 8085, progressing to several members of the 8051 family of embedded microcontrollers, and now teaching Advanced Embedded Microcontrollers using the Atmel AVR devices. During his summers and two sabbatical periods, he worked extensively with multiple-processor embedded systems, applying them in a variety of control-oriented applications. In addition, he consults actively in the field. Prior to his tenure at Purdue University, he spent ten years as an engineer in the aerospace electronics industry.

In terms of teaching, Dr. Barnett has won a number of teaching awards, including the Charles B. Murphy Award, as one of the best teachers at Purdue University. He is also listed on Purdue University's Book of Great Teachers, a list of the 225 most influential teachers over Purdue's entire history. This is his third textbook.

He may be contacted with suggestions/comments at Purdue University at (765) 494-7497 or by e-mail at rbarnett@purdue.edu.

Larry D. O'Cull
Senior Operating Member
Progressive Resources LLC

Mr. O'Cull received a Bachelor of Science degree from the School of Electrical Engineering Technology at Purdue University. His career path started in the design of software and control systems for computer numeric controlled (CNC) machine tools. From there he moved to other opportunities in electronics engineering and software development for vision systems, laser-robotic machine tools, medical diagnostic equipment, and commercial and consumer products, and he has been listed as inventor/coinventor on numerous patents.

Mr. O'Cull started Progressive Resources in 1995 after several years of working in electrical and software engineering and engineering management. Progressive Resources LLC (http://www.prllc.com) specializes in innovative commercial, industrial, and consumer product development. Progressive Resources has been a Microchip consultant member since 1995.

He may be contacted with suggestions/comments by e-mail at locull@prllc.com.

Sarah A. Cox
Director of Software Development
Progressive Resources LLC

Ms. Cox has a Bachelor of Science degree in both Computer and Electrical Engineering from Purdue University where she focused her studies on software design.

After a short career for a large consulting firm working on database management systems, she was lured away by the fast pace and the infinite possibilities of microprocessor designs. She worked independently on various pieces of medical test equipment before becoming a partner at Progressive Resources LLC.

At Progressive Resources, Ms. Cox has developed software for projects ranging from small consumer products to industrial products and test equipment. These projects have spanned several fields, among them automotive, medical, entertainment, child development, public safety/education, sound and image compression, and construction. Along the way she has been listed as coinventor on numerous patent applications.

She may be contacted with suggestions/comments by e-mail at sac@prllc.com.

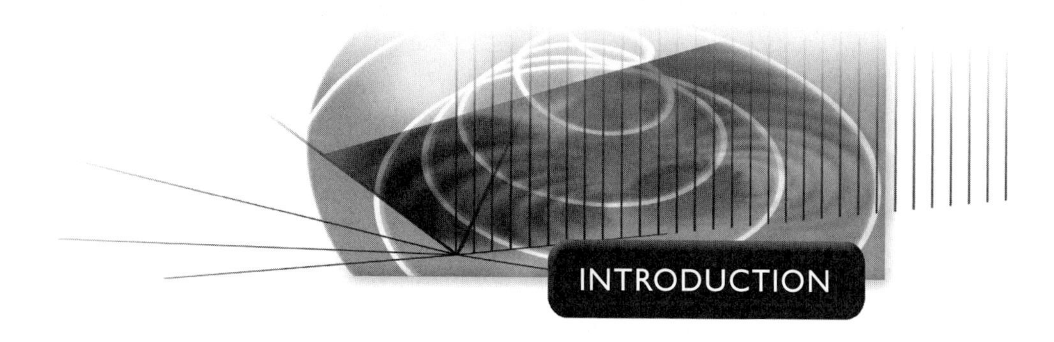

INTRODUCTION

An embedded microcontroller is a microcomputer that contains most of its peripherals and required memory inside a single integrated circuit along with the CPU. It is in actuality "a microcomputer on a chip." Embedded microcontrollers have been in use for more than three decades. The Intel 8051 series was one of the first microcontrollers to integrate the memory, I/O, arithmetic logic unit (ALU), program ROM, as well as some other peripherals all into one very neat little package. These processors are still being designed into new products. Other companies that followed Intel's lead into the embedded microcontroller arena are General Instruments, National Semiconductor, Motorola, Philips/Signetics, Zilog, AMD, Hitachi, Toshiba, Microchip Technology, and Atmel, among others.

In the past decade, Microchip Technology has become a world leader in the development of a reduced instruction set computing (RISC) core architecture that provides for very low-cost, yet amazing solutions. The PIC® processor family has been based on EEPROM and, more recently, FLASH memory technology. FLASH technology is a nonvolatile, yet reprogrammable memory often used in products such as digital cameras, portable audio devices, and PC motherboards. This memory technology has allowed Microchip to push ahead in the microcontroller industry by providing an in-system programmable solution.

The next great step in this high-tech evolution was the implementation of high-level language compilers that are targeted specifically for use with these new microprocessors. The code generation and optimization of the compilers is quite impressive. The C programming language, with its free form, "make your own rules" structure lends itself to this application by its ability to be tailored to a particular target system, while still allowing for code to be portable to other systems. The key benefit of a language like this one is that it creates pools of intellectual property that can be drawn from again and again. This lowers development costs on an ongoing basis by shortening the development cycle with each subsequent design.

One of the finest C language tools developed to date for the Microchip PIC processor family is PIC-C. Created by Custom Computer Services, Inc., this completely *integrated development environment* (IDE) allows editing, compiling, part programming, and debugging to be performed from one PC Windows® application. The motivation that has led to the development of this book is the growing popularity of the PIC and other RISC microcontrollers, the ever-increasing level of integration (more on a chip and fewer chips on a circuit board), and the need for "tuned thinking" when it comes to developing products utilizing this type of technology. You may have experience writing C for a PC, or assembler for a microcontroller. But when it comes to writing C for an embedded microcontroller, the approach must be modified to get the desired final results: small, efficient, reliable, reusable code. This textbook is designed to provide a good baseline for the beginner as well as a helpful reference tool for those experienced in embedded microcontroller design.

Embedded C Language Tutorial

1.1 OBJECTIVES

At the conclusion of this chapter, you should be able to:

- Define, describe, and identify variable and constant types, their scope and uses.
- Construct variable and constant declarations for all sizes of numeric data and for strings.
- Apply enumerations to variable declarations.
- Assign values to variables and constants by means of the assignment operator.
- Evaluate the results of all of the operators used in C.
- Explain the results that each of the control statements has on program flow.
- Create functions that are composed of variables, operators, and control statements to complete tasks.
- Apply pointers, arrays, structures, and unions as function variables.
- Create C programs that complete tasks using the concepts in this chapter.

1.2 INTRODUCTION

This chapter provides a baseline course in the C programming language as it applies to embedded microcontroller applications. The chapter includes extensions to the C language that are a part of the CCS-PICC® C language. You will go from beginning concepts through writing complete programs, with examples that can be implemented on a microcontroller to further reinforce the material.

The information is presented somewhat in the order that is needed by a programmer:

- Declaring variables and constants
- Simple I/O, so that programs can make use of the parallel ports of the microcontroller

- Assigning values to the variables and constants and doing arithmetic operations with the variables
- C constructs and control statements to form complete C programs

The final sections cover the more advanced topics such as pointers, arrays, structures, and unions, and their use in C programs. Advanced concepts such as real-time programming and interrupts complete the chapter.

1.3 BEGINNING CONCEPTS

Writing a C program is, in a sense, like building a brick house: A foundation is laid, sand and cement are used to make bricks, these bricks are arranged in rows to make a course of blocks, and the courses are then stacked to create a building. In an embedded C program, sets of instructions are put together to form functions; these functions are treated as higher-level operations, which are then combined to form a program.

Every C language program must have at least one function, namely, *main()*. The function *main()* is the foundation of a C language program, and it is the starting point when the program code is executed. All functions are invoked by *main()* either directly or indirectly. Although functions can be complete and self-contained, variables and parameters can be used to cement these functions together.

The function *main()* is considered to be the lowest-level task, because it is the first function called from the system starting the program. In many cases, *main()* will contain only a few statements that do nothing more than initialize and steer the operation of the program from one function to another.

An embedded C program in its simplest form appears as follows:

```
void main()
{
        while(1) // do forever..
            ;
}
```

The program shown above will compile and operate perfectly, but you will not know that for sure because there is no indication of activity of any sort. We can embellish the program such that you can actually see life, review its functionality, and begin to study the syntactical elements of the language.

```
#include <stdio.h>

void main()
{
      printf("HELLO WORLD"); /* the classic C test program.. */
      while(1)          // do forever..
          ;
}
```

This program will print the words "HELLO WORLD" to the standard output, which is most likely a serial port. The microcontroller will sit and wait forever or until the microcontroller is reset. This demonstrates one of the major differences between a personal computer program and a program that is designed for an embedded microcontroller; namely, that the embedded microcontroller applications contain an infinite loop. Personal computers have an operating system, and once a program has executed, it returns control to the operating system of the computer. An embedded microcontroller, however, does not have an operating system and cannot be allowed to fall out of the program at any time. Hence every embedded microcontroller application has an infinite loop built into it somewhere, such as the *while(1)* in the example above. This prevents the program from running out of things to do and doing random things that may be undesirable. The **while** construct will be explained in a later section.

The example program also provides an instance of the first of the common preprocessor compiler directives. **#include** tells the compiler to include a file called stdio.h as a part of this program. The function *printf()* is provided for in an external library and is made available to us because its definition is located in the stdio.h file. As we continue, these concepts will come together quickly.

These are some of the elements to take note of in the previous examples:

;	A semicolon is used to indicate the end of an expression. An expression in its simplest form is a semicolon alone.
{ }	Braces "{}" are used to delineate the beginning and the end of the function's contents. Braces are also used to indicate when a series of statements is to be treated as a single block.
"text"	Double quotes are used to mark the beginning and the end of a text string.
// or /* ... */	Slash-slash or slash-star/star-slash are used as comment delimiters.

Comments are just that, a programmer's notes. Comments are critical to the readability of a program. This is true whether the program is to be read by others or by the original programmer at a later time. The comments shown in this text are used to explain the function of each line of the code in the example. The comments should always explain the actual *function* of the line in the program, and not just echo the specific instructions that are used on the line.

The traditional comment delimiters are the slash-star (/*), star-slash (*/) configuration. Slash-star is used to create block comments. Once a slash-star (/*) is encountered, the compiler will ignore the subsequent text, even if it encompasses multiple lines, until a star-slash (*/) is encountered. Refer to the first line of the *main()* function in the previous program example for an example of these delimiters.

The slash-slash (//) delimiter, on the other hand, will cause the compiler to ignore the comment text only until the end of the line is reached. These are used in the second line of the *main()* function of the example program.

As we move into the details, a few syntactical rules and some basic terminology should be remembered:

- An identifier is a variable or function name made up of a letter or underscore (_), followed by a sequence of letters and/or digits and/or underscores.
- Identifiers are generally case sensitive. The **#case** directive is used to control the case sensitivity in the CCS-PICC compiler.
- Identifiers can be any length, but some compilers may recognize only a limited number of characters such as the first thirty-two. So beware!
- Particular words have special meaning to the compiler and are considered reserved words. These reserved words must be entered in lowercase and should never be used as identifiers. Table 1.1 lists reserved words.

auto	default	for	int32	return	union
break	defined	goto	register	short	unsigned
bit	do	if	return	signed	void
byte	double	inline	short	sizeof	volatile
case	else	int	signed	static	while
char	enum	int1	sizeof	struct	
const	extern	int8	long	switch	
continue	float	int16	register	typedef	

Table 1.1 *Reserved Word List*

- Because C is a free-form language, "white space" is ignored unless delineated by quotes. This includes blank (space), tab, and new line (carriage return and/or line feed).

1.4 VARIABLES AND CONSTANTS

It is time to look at data stored in the form of variables and constants. Variables, as in algebra, are values that can be changed. Constants are fixed. Variables and constants come in many forms and sizes; they are stored in the program's memory in a variety of forms that will be examined as we go along.

1.4.1 VARIABLE TYPES

A variable is declared by the reserved word indicating its type and size followed by an identifier:

```
unsigned char Peabody;
int dogs, cats;
long int total_dogs_and_cats;
```

Variables and constants are stored in the limited memory of the microcontroller, and the compiler needs to know how much memory to set aside for each variable without wasting memory space unnecessarily. Consequently, a programmer must declare the variables, specifying both the size of the variable and the type of the variable. It is always important to understand the size of each data type for the compiler you are using because not all compilers are created equally. The potential for math errors will be minimized as long as you understand the size and sign of what you are declaring. Table 1.2 lists the standard variable types and their typically associated sizes.

Standard Type	***CCS-PICC Default Type***	***Size (Bits)***	***Range***
bit	int1, short	1	0, 1
char	int8, int, char, signed char, int, short int	8	−128 to 127
unsigned char	unsigned int, unsigned char	8	0 to 255
signed char	int8, int, char, signed char, int, short int	8	−128 to 127
int	int16, long int	16	−32768 to 32767
short int		16	−32768 to 32767
unsigned int	unsigned long int	16	0 to 65535
signed int		16	−32768 to 32767
long int	int32	32	−2147483648 to 2147483647
unsigned long int	unsigned int32	32	0 to 4294967295
signed long int	signed int32	32	−2147483648 to 2147483647
float	float	32	±1.175e-38 to ±3.402e38
double		32	±1.175e-38 to ±3.402e38

Table 1.2 *Variable Types and Sizes*

The default types for the CCS-PICC compiler for the given sizes listed in the "Size" column of Table 1.2 are shown in the "CCS-PICC Default Type" column of that table. The types **int1, int8, int16,** and **int32** are specific to the CCS-PICC compiler. The sizes in CCS-PICC can be set to the more standardized ones listed in the "Standard Type" column of Table 1.2 by using the #type preprocessor directive.

```
#TYPE  SHORT=8, INT=16, LONG=32
```

More information on the preprocessor can be found in Chapter 3, Standard I/O and Preprocessor Functions. Also, the type **bit** can be created using the **typedef** operator:

typedef int1 bit;

covered in Section 1.10.5, "Typedef Operator."

It is not uncommon to find type/size differences from compiler to compiler. Each compiler writer has his or her own sense of what makes the language efficient for the processors it supports and the mathematical capabilities of those processors in their potential applications.

1.4.2 VARIABLE SCOPE

As noted earlier, constants and variables must be declared prior to their use. The *scope* of a variable is its accessibility within the program. A variable can be declared to have either *local* or *global* scope.

Local Variables

Local variables are memory spaces allocated by the function when the function is entered, typically on the program stack or a compiler-created heap space. These variables are not accessible from other functions, meaning that their scope is limited to the functions in which they are declared. The local variable declaration can be used in multiple functions without conflict, because the compiler sees each of these variables as being part of that function only.

Global Variables

A global or external variable is a memory space that is allocated by the compiler and can be accessed by all the functions in a program (unlimited scope). A global variable can be modified by any function and will retain its value to be used by other functions.

Global variables are typically cleared (set to zero) when *main()* is started. This operation is most often performed by the startup code generated by the compiler, invisible to the programmer.

An example piece of code is shown below to demonstrate the scope of variables:

```
unsigned char globey;      //a global variable
void function_z (void)     //this is a function called from main()
{
        unsigned int tween;         //a local variable

        tween = 12;                 //OK because tween is local
        globey = 47;                //OK because globey is global
        main_loc = 12;      // This line will generate an error
                            // because main_loc is local to main
}

void main()
{
        unsigned char main_loc;     //a variable local to main()
        globey = 34;                //Ok because globey is a global
```

```
        tween = 12;          //will cause an error - tween is local
                             // to function_z
        while(1)             // do forever..
            ;
    }
```

When variables are used within a function, if a local variable has the same name as a global variable, the local will be used by the function. The value of the global variable, in this case, will be inaccessible to the function and will remain untouched.

1.4.3 CONSTANTS

As described earlier in the textbook, constants are fixed values—they may not be modified as the program executes. Constants in many cases are part of the compiled program itself, located in read-only memory (ROM), rather than an allocated area of changeable random access memory (RAM). In the assignment

```
x = 3 + y;
```

the number *3* is a constant and will be coded directly into the addition operation by the compiler. Constants can also be in the form of characters or a string of text:

```
printf("hello world");
            // The text hello world is placed in program memory
            // and never changes.
x = 'B';    // The letter 'B' is permanently set
            // in program memory.
```

You can also declare a constant by using the reserved word const and indicating its type and size. An identifier and a value are required to complete the declaration:

```
const char c = 57;
```

Identifying a variable as a constant will cause that variable to be stored in the program code space rather than in the limited variable storage space in RAM. This helps preserve the limited RAM space.

Numeric Constants

Numeric constants can be declared in many ways by indicating their numeric base and making the program more readable. Integer or long integer constants may be written in the following forms:

- Decimal form without a prefix (such as 1234)
- Binary form with **0b** prefix (such as 0b101001)
- Hexadecimal form with **0x** prefix (such as 0xff)
- Octal form with **0** prefix (such as 0777)

There are also modifiers to better define the intended size and use of the constant:

- Unsigned integer constants can have the suffix **U** (such as 10000U).

- Long integer constants can have the suffix **L** (such as 99L).
- Unsigned long integer constants can have the suffix **UL** (such as 99UL).
- Floating point constants can have the suffix **F** (such as 1.234F).
- Character constants must be enclosed in single quotation marks, **'a'** or **'A'**.

Character Constants

Character constants can be printable (like 0–9 and A–Z) or non-printable characters (such as new line, carriage return, or tab). Printable character constants may be enclosed in single quotation marks (such as 'a'). A backslash followed by the octal or hexadecimal value in single quotes can also represent character constants:

't' can be represented by **'\164'** (octal)

or

't' can be represented by '\x74' (hexadecimal)

Table 1.3 lists some of the "canned" non-printable characters that are recognized by the C language.

Character	*Representation*	*Equivalent Hex Value*
BEL	'\a'	'\x07'
Backspace	'\b'	'\x08'
TAB	'\t'	'\x09'
LF (new line)	'\n'	'\x0a'
VT	'\v'	'\x0b'
FF	'\f'	'\x0c'
CR	'\r'	'\x0d'

Table 1.3 *Non-printable Character Notations*

Backslash (\) and single quote (') characters themselves must be preceded by a backslash to avoid confusing the compiler. For instance, '\' is a single quote character and '\\ ' is a backslash. BEL is the bell character and will make a sound when a computer terminal or terminal emulator receives it.

1.4.4 ENUMERATIONS AND DEFINITIONS

Readability in C programs is very important. Enumerations and definitions are provided so that the programmer can replace numbers with names or other more meaningful phrases.

Enumerations are listed constants. The reserved word **enum** is used to assign sequential integer constant values to a list of identifiers:

```
int num_val;              //declare an integer variable

//declare an enumeration
enum { zero_val, one_val, two_val, three_val );

num_val = two_val;        // the same as:  num_val = 2;
```

The name *zero_val* is assigned a constant value of 0, *one_val* of 1, *two_val* of 2, and so on. An initial value may be forced as in

```
enum { start=10, next1, next2, end_val };
```

which will cause *start* to have a value of 10, and then each subsequent name will be one greater. *next1* is 11, *next2* is 12, and *end_val* is 13.

Enumerations are used to replace pure numbers, which the programmer would have to look up, with the words or phrases that help describe the number's use.

Definitions are used in a manner somewhat similar to the enumerations, in that they will allow substitution of one text string for another. Observe the following example:

```
.
.
enum { red_led_on = 1, green_led_on, both_leds_on };
#define leds PORTB
.
.
PORTB = 0x1;     //means turn the red LED on
leds = red_led_on;  //means the same thing
```

The *'#define leds PORTB'* line causes the compiler to substitute the label *PORTB* wherever it encounters the word *leds*. Note that the **#define** line is not ended with a semicolon and may only have comments that are delimited by /* and */. The enumeration sets the value of *red_led_on* to 1, *green_led_on* to 2, and *both_leds_on* to 3. This might be used in a program to control the red and green LEDs, where outputting 1 turns the red LED on, 2 turns the green LED on, etc. The point is that *'leds = red_led_on'* is much easier to understand in the program's context than *'PORTA = 0x1'*.

#define is a preprocessor directive. Preprocessor directives are not actually part of the C language syntax, but they are accepted as such because of their use and familiarity. The preprocessor is a step separate from the actual compilation of a program, which happens before the actual compilation begins. More information on the preprocessor can be found in Chapter 3, Standard I/O and Preprocessor Functions.

1.4.5 STORAGE CLASSES

Variables can be declared in three storage classes: **auto, static**, and **register**. Auto, or automatic, is the default class, meaning the reserved word **auto** is not necessary.

Automatic

An automatic class local variable is uninitialized when it is allocated, so it is up to the programmer to make sure that it contains valid data before it is used. This memory space is released when the function is exited, meaning the values will be lost and will not be valid if the function is reentered. An automatic class variable declaration would appear as follows:

```
auto int value_1;
     or
int value_1;          // this is the common, default form
```

Static

A static local variable has only the scope of the function in which it is defined (it is not accessible from other functions), but it is allocated in global memory space. The static variable is initialized to zero the first time the function is entered, and it retains its value when the function is exited. This allows the variable to be current and valid each time the function is reentered.

```
static int value_2;
```

Register

A register local variable is similar to an automatic variable in that it is uninitialized and temporary. The difference is that the compiler will try to use an actual machine register in the microprocessor as the variable in order to reduce the number of machine cycles required to access the variable. There are very few registers available compared to the total memory in a typical machine. Therefore, this would be a class used sparingly and with the intent of speeding up a particular process.

```
register char value_3;
```

1.4.6 TYPE CASTING

There are times when a programmer might wish to temporarily force the type and size of the variable. Type casting allows the previously declared type to be overridden for the duration of the operation being performed. The cast, called out in parentheses, applies to the expression it precedes.

Given the following declarations and assignment,

```
int x;          // a signed, 16-bit, integer (-32768 to 32767)
char y;         // a signed, 8-bit character (-128 to +127)

x = 12;         // x is an integer, (but its value will fit
                // into a character)
```

type cast operations on these variables could appear as

```
y = (char)x + 3;    // x is converted to a character and then
                    // 3 is added,
                    // and the value is then placed into y.
x = (int)y;         // y is extended up to an integer, and
                    // assigned to x
```

Type casting is particularly important when arithmetic operations are performed with variables of different sizes. In many cases, the accuracy of the arithmetic will be dictated by the variable of the smallest type. Consider the following:

```
int z;              // declare z
int x = 150;        // declare and initialize x
char y = 63;        // declare and initialize y

z = (y * 10) + x;
```

As the compiler processes the right side of the equation, it will look at the size of *y* and make the assumption that *(y * 10)* is a character (8-bit) multiplication. The result placed on the stack will have exceeded the width of the storage location, one byte or a value of 255. This will truncate the value to 118 (0x76) instead of the correct value of 630 (0x276). In the next phase of the evaluation, the compiler would determine that the size of the operation is integer (16 bits) and 118 would be extended to an integer and then added to *x*. Finally, *z* would be assigned 268 … WRONG!!

Type casting should be used to control the assumptions. Writing the same arithmetic as

```
z = ((int)y * 10) + x;
```

will indicate to the compiler that y is to be treated as an integer (16 bits) for this operation. This will place the integer value of 630 onto the stack as a result of a 16-bit multiplication. The integer value x will then be added to the integer value on the stack to create the integer result of 780 (0x30C); *z* will be assigned the value 780.

C is a very flexible language. It will give you what you ask for. The assumption the compiler will make is that you, the programmer, know what you want. In the preceding example, if the value of *y* were 6 instead of 63, there would have been no errors. When writing expressions, you should always think in terms of the maximum values that could be given to the expression and what the resulting sums and products may be.

A good rule to follow: "When it doubt—cast it out." Always cast the variables unless you are sure you do not need to.

1.5 I/O OPERATIONS

Embedded microcontrollers must interact directly with other hardware. Therefore, many of their input and output operations are accomplished using the built-in parallel ports of the microcontroller.

Most C compilers provide a convenient method of interacting with the parallel ports through a library or header file. CCS-PICC uses the **#byte** compiler command to assign labels to each of the parallel ports as well as other I/O devices. This command will be discussed in a later section, but the example below will serve to demonstrate the use of the parallel ports:

```
#include <16F877.h>            // register definition header file for
#fuses HS,NOWDT                // a Microchip PIC16F877 and fuses

unsigned char z;       // declare z

void main(void)
{
     port_b_pullups(TRUE); // enable internal pull-ups for
                         // use on FlashPIC development board

     set_tris_d(0x00); // set all bits of port D for output
     set_tris_b(0xFF); // set port B for all input
     while(1)
     {
       z = input_b();    // read the binary value on the
                         // port B pins (i.e., input from port B
       output_d(z + 1); // write the binary value read from B
                         // plus 1 to port D)
     }
}
```

The example above shows the methods to both read and write a parallel port. *z* is declared as an unsigned character size variable because the port is an 8-bit port and an unsigned character variable will hold 8-bit data.

The TRISx registers of the PIC processor is used to determine which bits of port x (A, B, and so on depending on the processor) are to be used for output. Upon reset, all of the I/O ports default to input by the microcontroller writing a 1 into all the bits of the TRISx registers. The programmer then sets the TRISx bits depending on which bits are to be used for output. For example,

```
set_tris_b(0xc3);        // set the upper 2 and lower 2 bits of
                         // port B for input, the rest for output
```

This example configures the upper two bits and the lower two bits to be used as input bits.

1.6 OPERATORS AND EXPRESSIONS

An expression is a statement in which an operator links identifiers such that when evaluated, the result may be true, false, or numeric in value. Operators are symbols that indicate to the compiler which type of operation is to be performed using its surrounding identifiers. There are rules that apply to the precedence or order in which operations are performed. When you combine operators in a single expression, those rules of precedence must be applied to obtain the desired result.

1.6.1 ASSIGNMENT AND ARITHMETIC OPERATORS

Once variables have been declared, operations can be performed on them using the assignment operator, an equal sign (=). A value assigned to a variable can be a constant, a variable, or an expression. An expression in the C language is a combination of operands (identifiers) and operators. A typical assignment may appear as follows:

```
dog = 35;
val  = dog;
dog = dog + 0x35;
val  = (2 * (dog + 0172)) + 6;
y = (m * x) + b;
```

The compiled program arithmetically processes expressions just as you would process them by hand. Start from inside the parentheses and work outward from left to right: In the above expression $y = (m * x) + b$, the m and x would be multiplied together first and then added to b; finally, y would be assigned the resulting value. In addition to the parentheses, there is an inherent precedence to operators themselves. Multiplication and division are performed first, followed by addition and subtraction. Therefore, the statement

```
y = m * x + b;
```

is evaluated the same as

```
y = (m * x) + b;
```

It should be noted that the use of the parentheses improves the readability of the code.

Table 1.4 shows the typical arithmetic operators in order of precedence.

Multiply	*
Divide	/
Modulo	%
Addition	+
Subtraction or Negation	–

Table 1.4 *Arithmetic Operators*

There are other types of operators besides arithmetic and assignment operators. These include bitwise, logical, relational, increment, decrement, compound assignment, and conditional.

Bitwise Operators

Bitwise operators perform functions that will affect the operand at the bit level.

These operators work on non–floating point operands: **char, int,** and **long.** Table 1.5 lists the bitwise operators in order of precedence.

<table>
<tr><td>Ones Complement</td><td>~</td></tr>
<tr><td>Left Shift</td><td><<</td></tr>
<tr><td>Right Shift</td><td>>></td></tr>
<tr><td>AND</td><td>&</td></tr>
<tr><td>Exclusive OR</td><td>^</td></tr>
<tr><td>OR (Inclusive OR)</td><td>|</td></tr>
</table>

Table 1.5 *Bitwise Operators*

The following is a description of each bitwise operator:

- The ones complement operator converts the bits within an operand to 1 if they were 0, and to 0 if they were 1.
- The left shift operator will shift the left operand to the left, in a binary fashion, the number of times specified by the right operand. In a left shift operation, zero is always shifted in replacing the lower bit positions that would have been "empty." Each shift left effectively multiplies the operand by 2.
- The right shift operator will shift the left operand to the right, in a binary fashion, the number of times specified by the right operand. Each right shift effectively performs a division by 2. When a right shift is performed, signed and unsigned variables are treated differently. The sign bit (the left or most significant bit) in a signed integer will be replicated. This sign extension allows a positive or negative number to be shifted right while maintaining the sign. When an unsigned variable is shifted right, zeros will always be shifted in from the left.
- The AND operator will result in a 1 at each bit position where both operands were 1.
- The exclusive OR operator will result in a 1 at each bit position where the operands differ (a 0 in one operand and a 1 in the other).
- The OR (inclusive OR) operator will result in a 1 at each bit position where either of the operands contained a 1.

Table 1.6 gives some examples.

Assume that an unsigned character y = 0xC9.

Operation	Result
x = ~y;	x = 0x36
x = y << 3;	x = 0x48
x = y >> 4;	x = 0x0C
x = y & 0x3F;	x = 0x09
x = y ^ 1;	x = 0xC8
x = y | 0x10;	x = 0xD9

Table 1.6 *Examples of Bitwise Operations*

The AND and OR bitwise operators are useful in dealing with parallel ports. Observe the following example:

```
#include <16F877.h>     // register definition header file for
#fuses HS,NOWDT         // a Microchip PIC16F877 and fuses

unsigned char z;        // declare z

void main(void)
{
   port_b_pullups(TRUE);
   set_tris_d(0x00); // set all bits of port D for output
   set_tris_b(0xFF); // set port B for all input

   while(1)
   {
      z = input_b() & 0x6;
                           // read the binary value on the
                           // port B pins ANDed with 00000110.
      output_d(input_d() | 0x60);
                           // write the binary value from port D
                           // ORed with 01100000 back to port D
      output_d(input_d() & 0xfe);
                           // write the binary value from port D
                           // ANDed with 0xfe to port D.
   }
}
```

The example above demonstrates *masking* and *bitwise port control.* Masking is the technique used to determine the value of certain bits of a binary value. In this case the masking is accomplished by ANDing the unwanted bits with 0 ('x' AND 0 always results in 0) and the bits of interest with 1 ('x' AND 1 always results in 'x'). In this way all the bits except for the center two bits of the lower nibble are eliminated (set to 0).

Bitwise port control is a method changing one or more bits of a parallel port without affecting the other bits of the port. The first *output_d* line demonstrates using the OR operator to force two bits of the port high without affecting the other bits of the port. The second *output_d* line shows how to force a bit of the port low by ANDing the bit with a 0.

1.6.2 LOGICAL AND RELATIONAL OPERATORS

Logical and relational operators are all binary operators but yield a result that is either TRUE or FALSE. TRUE is represented by a non-zero value, and FALSE by a zero value. These operations are usually used in control statements to guide the flow of program execution.

Logical Operators

Table 1.7 shows the logical operators in order of precedence.

AND	&&
OR	||

Table 1.7 *Logical Operators*

These differ greatly from the bitwise operators in that they deal with the operands in a TRUE and FALSE sense. The AND logical operator yields a TRUE if both operands are TRUE, otherwise a FALSE is the result. The OR logical operator yields a TRUE if either of the operators is TRUE. In the case of an OR, both operators must be FALSE in order for the result to be FALSE. To illustrate the difference:

Assume x = 5 and y = 2;

(x && y) is TRUE, because both are non-zero.

(x & y) is FALSE, because the pattern 101b and 010b ANDed bitwise are zero.

(x || y) is TRUE, because either value is non-zero.

(x | y) is TRUE, because the pattern 101b and 010b Inclusive-ORed bitwise is 111b (non-zero).

Relational Operators

Relational operators use comparison operations. As in the logical operators, the operands are evaluated left to right and a TRUE or FALSE result is generated. They effectively "ask" about the relationship of two expressions in order to gain a TRUE or FALSE reply.

"Is the left greater than the right?"

"Is the left less than or equal to the right?"

Table 1.8 shows the relational operators.

Is Equal to	==
Is Not Equal to	!=
Less Than	<
Less Than or Equal to	<=
Greater Than	>
Greater Than or Equal to	>=

Table 1.8 *Relational Operators*

Examples are presented in Table 1.9.

Assume that x =3 and y = 5.

Operation	Result
(x == y)	FALSE
(x != y)	TRUE
(x < y)	TRUE
(x <= y)	TRUE
(x > y)	FALSE
(x >= y)	FALSE

Table 1.9 *Examples of Relational Operations*

1.6.3 INCREMENT, DECREMENT, AND COMPOUND ASSIGNMENT

When the C language was developed, a great effort was made to keep things concise but clear. Some "shorthand" operators were built into the language to simplify the generation of statements and shorten the keystroke count during program development. These operations include the increment and decrement operators, as well as compound assignment operators.

Increment Operators

Increment operators allow for an identifier to be modified, in place, in a pre-increment or post-increment manner. For example,

```
x = x + 1;
```

is the same as

```
++x;            // pre-increment operation
```

and as

```
x++;            // post-increment operation
```

In this example the value is incremented by 1. *++x* is a pre-increment operation, whereas *x++* is a post-increment operation. This means that during the evaluation of the expression by the compiled code, the value is changed pre-evaluation or post-evaluation.

For example,

```
i = 1;
k = 2 * i++;     // at completion, k = 2  and  i = 2

i = 1;
k = 2 * ++i;     // at completion, k = 4  and  i = 2
```

In the first case, *I* is incremented after the expression has been resolved. In the second case, *I* was incremented before the expression was resolved.

Decrement Operators

Decrement operators function in a similar manner, causing a subtraction-of-one operation to be performed in a pre-decrement or post-decrement fashion:

```
j--;             // j = j-1
--j;             // j = j-1
```

Compound Assignment Operators

Compound assignment operators are another method of reducing the amount of syntax required during the construction of a program. A compound assignment is really just the combining of an assignment operator (=) with an arithmetic or logical operator. The expression is processed right to left, and syntactically it is constructed somewhat like the increment and decrement operators. Here are some examples:

```
a += 3;   // a = a + 3
b -= 2;   // b = b - 2
c *= 5;   // c = c * 5
d /= a;   // d = d / a
```

This combining of an assignment with another operator works with modulo and bitwise operators (%, >>, <<, &, |, and ^) as well as the arithmetic operators (+, –, *, and /), as shown below:

```
a |= 3;        // a = a OR 3
b &= 2;        // b = b AND 2
c ^= 5;        // c = c exclusively OR-ed with 5

PORTC &= 3; // write the current value on PORTC
            // ANDed with 3 back to port C. Forcing
            // all of the bits except the lower 2 to 0
            // and leaving the lower 2 bits unaffected.
```

1.6.4 THE CONDITIONAL EXPRESSION

The conditional expression is probably the most cryptic and infrequently used of the operators. It was definitely invented to save typing time, but in general, it is not easy for beginning programmers to follow. The conditional expression is covered here in the "Operators and

Expressions" section due to its physical construction, but it is really more closely associated with control statements covered later in this chapter in Section 1.7, "Control Statements." If the control sequence

```
if(expression_A)
        expression_B;
else
        expression_C;
```

is reduced to a conditional expression, it reads as follows:

```
expression_A ? expression_B : expression_C;
```

In either of the forms shown above, the logical expression *expression_A* is evaluated: If the result is TRUE, then *expression_B* is executed; otherwise, *expression_C* is executed.

In a program, a conditional expression might be written as follows:

```
(x < 5) ? y = 1 : y = 0;      // if x is less than 5, then
                              // y = 1, else y = 0
```

1.6.5 OPERATOR PRECEDENCE

When multiple expressions are in a single statement, the operator precedence establishes the order in which expressions are evaluated by the compiler. In all cases of assignments and expressions, the precedence, or order of priority, must be remembered. When in doubt, you should either nest the expressions with parentheses, to guarantee the order of process, or look up the precedence for the operator in question. Some of the operators listed previously actually share an equal level of precedence. Table 1.10 shows the operators, their precedence, and the order in which they are processed (left to right or right to left). This processing order is called grouping or association.

Some operators in the table are covered later in this chapter in the "Pointers and Arrays" and "Structures and Unions" sections. These would include the primary operators like dot (.), bracket ([]), and indirection (->), which are used to indicate the specifics of an identifier, such as an array or structure element, as well as pointer and indirection unary operators, such as contents-of (*) and address-of (&). The following examples help to make this clear:

```
y = 3 + 2 * 4;    //would yield y = 11 because the * is higher
                  //in precedence and would be done before the +.

y = (3 + 2) * 4;  //would yield y = 20 because the () are the
                  //highest priority and force the + to be done
                  //first.

y = 4 >> 2 * 3;   //would yield y = 4 >> 6 because the * is
                  //higher in priority than the shift right
                  //operation.

y = 2 * 3 >> 4;   //would yield y = 6 >> 4 due to the precedence of
                  //the * operator.
```

Name	Level	Operators	Grouping
Primary	1 (High)	() . [] ->	Left to Right
Unary	2	! ~ - (type) * & ++ — sizeof	Right to Left
Binary	3	* / %	Left to Right
Arithmetic	4	+ -	Left to Right
Shift	5	<< >>	Left to Right
Relational	6	< <= > >=	Left to Right
Equality	7	== !=	Left to Right
Bitwise	8	&	Left to Right
Bitwise	9	^	Left to Right
Bitwise	10	|	Left to Right
Logical	11	&&	Left to Right
Logical	12	||	Left to Right
Conditional	13	? :	Right to Left
Assignment	14 (low)	= += -= /= *= %= <<= >>= &= ^= |=	Right to Left

Table 1.10 *Operator Precedence*

The precedence of operators forces the use of a rule similar to the one for casting: If in doubt, use many parentheses to ensure that the math will be accomplished in the desired order. Extra parentheses do not increase the code size and do increase its readability and the likelihood that it will correctly accomplish its mission.

1.7 CONTROL STATEMENTS

Control statements are used to control the flow of execution of a program. **if/else** statements are used to steer or branch the operation in one of two directions. **while, do/while,** and **for** statements are used to control the repetition of a block of instructions. **switch/case** statements are used to allow a single decision to direct the flow of the program to one of many possible blocks of instructions, in a clean and concise fashion.

1.7.1 WHILE LOOP

The **while** loop appears early in the descriptions of C language programming. It is one of the most basic control elements. The format of the **while** statement is as follows:

```
while (expression)      or      while(expression)
{                                          statement;
```

```
        statement1;
        statement2;
        ...
    }
```

When the execution of the program enters the top of the **while** loop, the expression is evaluated. If the result of the expression is TRUE (non-zero), then the statements within the **while** loop are executed. The 'loop' associated with the **while** statement are those lines of code contained within the braces {} following the **while** statement. Or, in the case of a single statement **while** loop, the statement following the **while** statement. When execution reaches the bottom of the loop, the program flow is returned to the top of the **while** loop, where the expression is tested again. Whenever the expression is TRUE, the loop is executed. Whenever the expression is FALSE, the loop is completely bypassed and execution continues at the first statement following the **while** loop. Consider the following:

```
#include <16F877.h>       // register definition file for
#fuses HS,NOWDT           // a Microchip PIC16F877
#use delay(clock=10000000) //Setup the RS232 port for standard output
#use rs232(baud=9600,parity=N,xmit=PIN_C6,rcv=PIN_C7,stream=RS232,bits=8)

#include <stdio.h>

void main(void)
{
      char c;

      c = 0;
      printf(" Start of program \n");
      while(c < 100)              // if c less than 100 then ..
      {
            printf("c = %d\n",(int)c);      // print c's value each
                                            // time through the loop
            c++;                      // increment c
      }
      printf(" End of program \n");  // indicate that the
                                     //program is finished
      while(1)    // because 1 is always TRUE, then just sit here..
        ;
}
```

In this example, *c* is initialized to 0 and the text string "Start of program" is printed. The **while** loop will then be executed, printing the value of *c* each pass as *c* is incremented from 0 to 100. When *c* reaches 100, it is no longer less than 100 and the **while** loop is bypassed. The "End of program" text string is printed and the program then sits forever in the *while(1)* statement.

The functioning of the *while(1)* statement should now be apparent. Because the '1' is the expression to be evaluated and is a constant (1 is always non-zero and, therefore, considered to be TRUE), the **while** loop, even a loop with no instructions as in the example above, is entered and is never left because the 1 always evaluates to TRUE. In this case the *while(1)* is used to stop execution by infinitely executing the loop, so that the processor does not keep executing non-existent code beyond the program.

Also note the cast of *c* to an integer inside the *printf()* function, inside the **while** loop. This is necessary because the *printf()* function in most embedded C compilers will handle only integer size variables correctly.

The **while** loop can also be used to wait for an event to occur on a parallel port.

```
void main(void)
{
        while (input_b() & 0x02) //hangs here waiting while the
           ;                     //bit 1 of port B is high.
        while(1) // because 1 is always TRUE, then just sit here..
           ;
}
```

In this example the **while** loop is used to await a bit going low. The expression being evaluated is *'input_b() & 0x02'*, which will mask all but the second bit (bit 1) of the data read from Port B. While the bit is at logic 1, the result will be non-zero (TRUE), and so the program will remain in the **while** loop until the value on the second bit drops to zero. At a later point it should become clear that in real-time programming, this construction is not appropriate, but it is a correct use of a **while** statement.

1.7.2 DO/WHILE LOOP

The **do/while** loop is very much like the **while** loop, except that the expression is tested after the loop has been executed one time. This means that the instructions in a **do/while** loop are always executed *once* before the test is made to determine whether or not to remain in the loop. In the **while** construct, the test is made *before* the instructions in the loop are executed even once. The format of the **do/while** statement is as follows:

```
do                            or            do
{                                                   statement;
           statement1;                      while (expression);
           statement2;
           ...
} while (expression);
```

When execution reaches the bottom of the **do/while** construct, the expression is evaluated. If the result of the expression is TRUE (non-zero), then the program flow is returned to the top of the **do/while** loop. Each time execution reaches the bottom of the construct, the expression is tested again. Whenever the expression is TRUE, the loop is executed, but if the expression is FALSE, the program continues on with the instructions that follow the construct. The previous example, coded using the **do/while** construct, would appear as follows:

```
#include <16F877.h>       // register definition file for
#fuses HS,NOWDT           // a Microchip PIC16F877
#use delay(clock=10000000) //Setup the RS232 port for standard output
#use rs232(baud=9600,parity=N,xmit=PIN_C6,rcv=PIN_C7,stream=RS232,bits=8)

#include <stdio.h>

void main(void)
{
      char c;

      c = 0;
      printf(" Start of program \n");

      do
      {
           printf("c = %d\n",(int)c);      // print c's value each
                                            // time through the loop
           c++;                        // increment c
      } while(c < 100);                // if c less than 100 then
                                       //repeat the operation

      printf(" End of program \n"); // indicate that the
                                       //program is finished

      while(1)    // because 1 is always TRUE, then just sit here..
        ;
}
```

In this example, *c* is initialized to 0 and the text string "Start of program" is printed. The **do/while** loop will then be executed, printing the value of *c* each pass as *c* is incremented from 0 to 100. When *c* reaches 100, it is no longer less than 100 and the **do/while** loop is bypassed. The "End of program" text string is printed and the program then sits forever in the *while(1)* statement.

1.7.3 FOR LOOP

A **for** loop construct is typically used to execute a statement or a statement block a specific number of times. A **for** loop can be described as an initialization, a test, and an action that leads to the satisfaction of that test. The format of the **for** loop statement is as follows:

```
for (expr1; expr2; expr3)     or     for(expr1; expr2; expr3)
{                                           statement;
      statement1;
      statement2;
      ...
}
```

expr1 will be executed only one time at the entry of the **for** loop. *expr1* is typically an assignment statement that can be used to initialize the conditions for *expr2*.

expr2 is a conditional control statement that is used to determine when to remain in the **for** loop. *expr3* is another assignment that can be used to satisfy the *expr2* condition.

When the execution of the program enters the top of the **for** loop, *expr1* is executed. *expr2* is evaluated and if the result of *expr2* is TRUE (non-zero), then the statements within the **for** loop are executed—the program stays in the loop. When execution reaches the bottom of the construct, *expr3* is executed, and the program flow is returned to the top of the **for** loop, where the *expr2* expression is tested again. Whenever *expr2* is TRUE, the loop is executed. Whenever *expr2* is FALSE, the loop is completely bypassed. The **for** loop structure could be represented with a **while** loop in this fashion:

```
expr1;
while(expr2)
{
        statement1;
        statement2;
        ...
        expr3;
}
```

Here is an example:

```
#include <16F877.h>       // register definition file for
#fuses HS,NOWDT           // a Microchip PIC16F877
#use delay(clock=10000000) //Setup the RS232 port for standard output
#use rs232(baud=9600,parity=N,xmit=PIN_C6,rcv=PIN_C7,stream=RS232,bits=8)

#include <stdio.h>

void main(void)
{
      char c;

      printf(" Start of program \n");
      for(c = 0; c < 100; c++)          // if c less than 100 then ..
      {
           printf("c = %d\n",(int)c); // print c's value each time
                                        // through the loop
      }                                 // c++ is executed before the
                                        //loop returns to the top

      printf(" End of program \n"); // indicate that the program is
                                     // finished
```

```
        while(1)      // because 1 is always TRUE, then just sit here..
            ;
}
```

In this example, the text string “Start of program” is printed. *c* is then initialized to 0 within the **for** loop construct. The **for** loop will then be executed, printing the value of *c* each pass as *c* is incremented from 0 to 100, also within the **for** loop construct. When *c* reaches 100, it is no longer less than 100 and the **for** loop is bypassed. The “End of program” text string is printed, and the program then sits forever in the *while(1)* statement.

1.7.4 IF/ELSE

if/else statements are used to steer or branch the operation of the program based on the evaluation of a conditional statement.

If Statement

An **if** statement has the following form:

```
if (expression)          or     if(expression)
{                                       statement;
        statement1;
        statement2;
        ...
}
```

If the expression result is TRUE (non-zero), then the statement or block of statements is executed. Otherwise, if the result of the expression is FALSE, then the statement or block of statements is skipped.

If/Else Statement

An **if/else** statement has the following form:

```
if(expression)           or     if(expression)
{                                       statement1;
        statement1;             else
        statement2;                     statement2;
        ...
}
else
{
        statement3;
        statement4;
        ...
}
```

The **else** statement adds the specific feature to the program flow that the statement or block of statements associated with the **else** will be executed only if the expression result is FALSE. The block of statements will be skipped if the expression result is TRUE.

An **else** statement must always follow the **if** statement it is associated with.

A common programming technique is to cascade **if/else** statements to create a selection tree:

```
if(expr1)
        statement1;
else if (expr2)
        statement2;
else if(expr3)
        statement3;
else
        statement4;
```

This sequence of **if/else** statements will select and execute only one statement. If the first expression, *expr1*, is TRUE, then *statement1* will be executed and the remainder of the statements will be bypassed. If *expr1* is FALSE, then the next statement, *if (expr2)*, will be executed. If *expr2* is TRUE, then *statement2* will be executed and the remainder bypassed, and so on. If *expr1, expr2*, and *expr3* are all FALSE, then *statement4* will be executed.

Here is an example of **if/else** operation:

```
#include <16F877.h>       // register definition file for
#fuses HS,NOWDT           // a Microchip PIC16F877
#use delay(clock=10000000) //Setup the RS232 port for standard output
#use rs232(baud=9600,parity=N,xmit=PIN_C6,rcv=PIN_C7,stream=RS232,bits=8)

#include <stdio.h>

void main(void)
{
        char c;

        printf(" Start of program \n");

        for(c = 0; c < 100; c++)    // while c is less than 100 then ..
        {
                if(c < 33)
                        printf("0<c<33   "); // use if/else to show the
                        range of
                else if((c >32) && (c < 66)) // numbers that c is in.
                        printf("32<c<66   ");
                else
                        printf("66<c<100 ");
                        printf("c= %d\n",(int)c); // print c's value
                        each time
                                                  //through the loop
        }
```

```
        printf(" End of program \n"); // indicate that the program
        is finished

        while(1)    // because 1 is always TRUE, then just sit here..
            ;
}
```

In this example, the text string “Start of program” is printed. *c* is then initialized to 0 within the **for** loop construct. The **for** loop will then be executed, printing the value of *c* each pass as *c* is incremented from 0 to 100, also within the **for** loop construct.

If the value of *c* is less than 33, then the text string “0<c<33” is printed. If the value of *c* is between 33 and 66, the text string “32<c<66” is printed. If the value of *c* is not within either of the preceding cases, the text string “66<c<100” is printed.

When *c* reaches 100, it is no longer less than 100, and the **for** loop is bypassed. The “End of program” text string is printed and the program then sits forever in the *while(1)* statement.

Using the constructs and other techniques covered so far, it is possible to create a program that efficiently tests each bit of an input port and prints a message to tell the state of the bit.

```
#include <16F877.h>      // register definition file for
#fuses HS,NOWDT    // a Microchip PIC16F877
#use delay(clock=10000000)
#use rs232(baud=9600,parity=N,xmit=PIN_C6,rcv=PIN_C7,stream=RS232,bits=8)

#include <stdio.h>

#define test_port input_b()

void main(void)
{
      unsigned char cnt, bit_mask;   //variables

      port_b_pullups(TRUE);
      set_tris_b(0xFF); // set port B for all input

      bit_mask = 1;        //start with lowest bit

      for (cnt=0;cnt<8;cnt++)    //for loop to test 8 bits
      {
              // the instructions below test port bits
              // and print result
          if (test_port & bit_mask)
              printf("Bit  %d is high.\n",(int)cnt);
          else
              printf("Bit  %d is low.\n",(int)cnt);
```

```
        bit_mask <<= 1;         //shift bit to be tested
      }

      while(1) // because 1 is always TRUE, then just sit here..
        ;
}
```

The example above uses a **for** loop that is set to loop eight times, once for each bit to be tested. The variable *bit_mask* starts at a value of 1 and is used to mask all of the bits except the one being tested. After being used as a mask, it is shifted one bit to the left, using the compound notation, to test the next bit during the next pass through the **for** loop. During each loop the **if/else** construct is used to print the correct statement for each bit. The conditional statement in the **if** construct is a bitwise AND using *bit_mask* to mask unwanted bits during the test.

Conditional Expression

Another version of the **if/else** is the conditional expression, designed to save the programmer time and steps through a simplified syntax. The **if/else** sequence

```
if(expression_A)
        statement1;
else
        statement2;
```

can be reduced to a conditional expression that would read as follows:

```
expression_A ? statement1 : statement2;
```

In both of the forms shown above, the logical expression *expression_A* is evaluated and if the result is TRUE, then *statement1* is executed; otherwise *statement2* is executed.

In a program, a conditional expression might be written as follows:

```
(x < 5) ? y = 1 : y = 0;     // if x is less than 5, then
                             // y = 1, else y = 0
```

1.7.5 SWITCH/CASE

The **switch/case** statement is used to execute a statement, or a group of statements, selected by the value of an expression. The form of this statement is as follows:

```
switch (expression)
{
       case const1:
              statement1;
              statement2;
       case const2:
              statement3;
              ...
              statement4;
```

```
        case constx:
                statement5;
                statement6;
        default:
                statement7;
                statement8;
    }
```

The *expression* is evaluated and its value is then compared against the constants (*const1*, *const2*,... *constx*). Execution begins at the statement following the constant that matches the value of the expression. The constants must be integer or character values. All of the statements following the matching constant will be executed, to the end of the **switch** construct. Because this is not normally the desired operation, **break** statements can be used at the end of each block of statements to stop the "fall-though" of execution and allow the program flow to resume after the **switch** construct.

The default case is optional, but it allows for statements that need to be executed when there are no matching constants.

```
#include <16F877.h>       // register definition file for
#fuses HS,NOWDT           // a Microchip PIC16F877
#use delay(clock=10000000)
#use rs232(baud=9600,parity=N,xmit=PIN_C6,rcv=PIN_C7,stream=RS232,bits=8)

#include <stdio.h>

void main(void)
{
        unsigned char c;

        port_b_pullups(TRUE);
        set_tris_b(0xFF); // set port B for all input

        while(1)
        {
                c = input_b() & 0xf; // read the lower nibble of port B
                switch(c )
                {
                        case '0':
                        case '1':    // you can have multiple cases
                        case '2':    // for a set of statements..
                        case '3':
                                printf(" c is a number less than 4 \n);
                                break; // break to skip out of the loop
                        case '5':    // or just one is ok..
                                printf(" c is a 5 \n");
```

```
                        break;
                default:
                        printf(" c is 4 or is > 5 \n");
                }
            }
        }
    }
```

This program reads the value on Port B and masks the upper four bits. It compares the value of the lower nibble from Port B against the constants in the **case** statement. If the character is a 0, 1, 2, or 3, the text string "c is a number less than 4" will be printed to the standard output. If the character is a 5, the text string "c is a 5" will be printed to the standard output. If the character is none of these (a 4, or a number greater than 5), the **default** statements will be executed and the text string "c is 4 or is > 5" will be printed. Once the appropriate **case** statements have been executed, the program will return to the top of the **while** loop and repeat.

1.7.6 BREAK, CONTINUE, AND GOTO

The **break**, **continue**, and **goto** statements are used to modify the execution of the **for, while, do/while,** and **switch** statements.

Break

The **break** statement is used to exit from a **for, while, do/while,** or **switch** statement. If the statements are nested one inside the other, the **break** statement will exit only from the immediate block of statements.

The following program will print the value of c to the standard output, as it is continuously incremented from 0 to 100, then reinitialized to 0. In the inner **while** loop, *c* is incremented until it reaches 100, and then the **break** statement is executed. The **break** statement causes the program execution to exit the inner **while** loop and continue execution of the outer **while** loop. In the outer **while** loop, *c* is set to 0, and control is returned to the inner **while** loop. This process continues to repeat itself forever.

```
#include <16F877.h>      // register definition file for
#fuses HS,NOWDT    // a Microchip PIC16F877
#use delay(clock=10000000)
#use rs232(baud=9600,parity=N,xmit=PIN_C6,rcv=PIN_C7,stream=RS232,bits=8)

#include <stdio.h>

void main(void)
{
        int c;

        while(1)
        {
                while(1)
```

```
            {
                    if (c > 100)
                          break; // this will take us out of this while
                    ++c;          // block, clearing c..
                    printf("c = %d\n",c);
            }
            c = 0;      // clear c and then things will begin again.
        }               // printing the values 0-100, 0-100, etc..
}
```

Continue

The **continue** statement will allow the program to start the next iteration of a **while**, **do/while**, or **for** loop. The **continue** statement is like the **break** statement in that both stop the execution of the loop statements at that point. The difference is that the **continue** statement starts the loop again, from the top, where **break** exits the loop entirely.

```
#include <16F877.h>      // register definition file for
#fuses HS,NOWDT   // a Microchip PIC16F877
#use delay(clock=10000000)
#use rs232(baud=9600,parity=N,xmit=PIN_C6,rcv=PIN_C7,stream=RS232,bits=8)

#include <stdio.h>

void main(void)  .
{
        int c;

        while(1)
        {
                c = 0;
                while(1)
        {
                        if (c > 100)
                                continue; // this will cause the printing to
                                          // stop when c>100, because the
                                          // continue will cause the rest of
                                          // this loop to be skipped!!
                        ++c;
                        printf("c = %d\n",c);// no code after the continue
                                                  //will be executed
                }
        }
}
```

In this example, the value of *c* will be displayed until it reaches 100. At that point the program will appear as if it has stopped, when in fact, it is still running. It is simply skipping the increment and *printf()* statements.

Goto

The **goto** statement is used to literally "jump" execution of the program to a label marking the next statement to be executed. This statement is very unstructured and is aggravating to the "purist," but in an embedded system it can be a very good way to save some coding and the memory usage that goes with it. The label can be before or after the **goto** statement in a function. The form of the **goto** statement is as follows:

```
goto identifier;             or          identifier:
    ...                                       statement;
                                              ...
identifier:
    statement;                           goto identifier;
```

The label is a valid C name or identifier, followed by a colon (:).

```
#include <16F877.h>       // register definition file for
#fuses HS,NOWDT   // a Microchip PIC16F877
#use delay(clock=10000000)
#use rs232(baud=9600,parity=N,xmit=PIN_C6,rcv=PIN_C7,stream=RS232,bits=8)

#include <stdio.h>

void main(void)
{
        int c,d;

        while(1)
        {

        start_again:

                c = 0;
                d = -1;
                while(1)
                {
                        if (d == c)
                                goto start_again; // (stuck? bail out!)
                        d = c;      // d will remember where we were
                        if (c > 100)
                                continue;    // this will reinitiate this
                                             // while loop
                        ++c; // d will be checked to see if c is stuck
                        printf("c = %d\n",c);
                                    // because c won't change ( > 100!!)
                }
        }
}
```

In this example, *c* and *d* are initialized to different values. The *if (d == c)* statement checks the values and, as long as they are not equal, execution falls through to the assignment *d = c*. If *c* is less than or equal to 100, execution falls through to the increment *c* statement. The value of *c* is printed and the **while** loop begins again. The values of *c* and *d* will continue to differ by 1 until *c* becomes greater than 100. With *c* at a value of 101, the **continue** statement will cause the increment *c* and *printf()* statements to be skipped. The *if (d==c)* will become TRUE, because they are now equal, and the **goto** will cause the program execution to jump to the *start_again* label. The result would be the value of *c* printing from 0 to 100, over and over again.

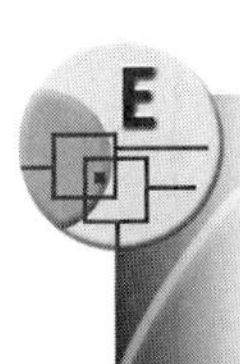

Chapter 1 Example Project: Part A

This is the first portion of a project that will be used to demonstrate some of the concepts presented here in Chapter 1. This example project will be based on a simple game according to the following situation:

You have been asked to build a Pocket Slot Machine as a mid-semester project by your favorite instructor, who gave you some basic specifications that include the following:

1. The press and release of a button will be the "pulling the arm of the One Armed Bandit."
2. The duration of the press and the release can be used to generate a pseudo random number to create three columns of figures.
3. A flashing light is used to indicate the machine is "moving."
4. There are four kinds of figures that can appear in a column:
   a. Bar
   b. Bell
   c. Lemon
   d. Cherry
5. The payout of the machine has the following rules:
   a. Three of a kind pays a Nickel.
   b. A Cherry anywhere pays a Dime.
   c. Three Cherries are a Jackpot.
   d. ...Everything else loses.

In this first step we will develop a brief program that touches on several of the concepts. We will develop a *main()* that contains various types of control statements to create the desired operation. This first pass will get the press and release, work up a pseudo random number, and print the results in a numeric form using standard I/O functions. **while**, **do/while**, and **for** loops will be demonstrated as well as **if/else** control statements.

EXAMPLE

```
//
// "Slot Machine" - The Mini-Exercise
//
//   Phase 1..

#include <16F877.h>
#use delay(clock=10000000)
#fuses HS,WDT
#use rs232(baud=9600,parity=N,xmit=PIN_C6,rcv=PIN_C7,stream=RS232,bits=8)

#include <stdio.h>      /* this gets the printf() definition */

// Global Variable Declarations..
char    first,second,third;    // Columns in the slot machine.
char    seed;                  // Number to form a seed for random #s.
char    count;                 // General purpose counter.
long    delay;                 // A variable for delay time.

void main(void)
{

   set_tris_b(0x00); // port B is all input
   port_b_pullups(TRUE);
   set_tris_c(0xFB); // port C.2 is an LED

   while(1)      // do forever..
   {
      while(input_b() & 1)    // Wait for a button and
         seed++;      // let the counter roll over and
                              // over to form a seed.

      first = second = third = seed; // preload the columns

      do                      // mix up the numbers
      {                       // while waiting for button release.
         first ^= seed>>1; // Exclusive ORing in the moving seed
         second^= seed>>2; // can really stir up the numbers.
         third ^= seed>>3;
         seed++;              // keep rolling over the seed pattern
      } while((input_b() & 1) == 0); // while the button is pressed

      for(count = 0; count < 5; count++) // flash light when done..
      {
         for(delay = 0; delay < 10000; delay++)
            ;                       // just count up and wait
```

```
            output_bit(PIN_C2,0); // turn the LED on..
            for(delay = 0; delay < 10000; delay++)
                ;
            output_bit(PIN_C2,1); // turn the LED off..
        }

        first  &= 3;        // limit number to values from 0 to 3
        second &= 3;
        third  &= 3;
                            // show the values..
        printf("-> %d, %d ,%d <-\n", first, second, third);

                            // determine a payout..
        if((first == 3) && (second == 3) && (third == 3))
            printf("Paid out: JACKPOT!!\n\r"); // Three "Cherries"
        else if((first == 3) || (second == 3) || (third == 3))
            printf("Paid out: One Dime\n\r");   // One "Cherry"
        else if((first == second) && (second == third))
            printf("Paid out: One Nickel\n\r"); // Three of a kind
        else
            printf("Paid out: ZERO\n\r");       // Loser..
    }
}
```

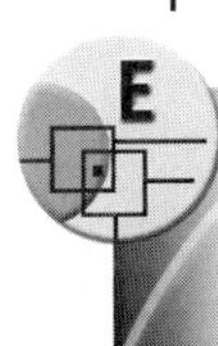

EXAMPLE

1.8 FUNCTIONS

A function is an encapsulation of a block of statements that can be used more than once in a program. Some languages refer to functions as subroutines or procedures. Functions are generally written to break up the operational elements of a program into its fundamental parts. This allows a programmer to debug each element and then use that element again and again.

One of the prime advantages in using functions is the development of a "library." As functions are developed that perform a certain task, they can be saved and used later by a different application, or even a different programmer. This saves time and maintains stability, because the functions developed get used and reused, tested and retested.

A function may perform an isolated task requiring no parameters whatsoever. A function may accept parameters in order to have guidance in performing its designed task. A function may not only accept parameters but return a value as well. Even though a function may accept multiple parameters, it can only return one. Some examples are as follows:

```
sleep();      // this function performs a 1 second delay,
              // with no parameters, in or out

printf("this is a parameter %d",x);
              // printf will print its parameters
```

```
c = getchar();
            // getchar will return a value from the
            // standard input
```

The standard form of a function is

```
type function_name ( type param1, type param2, … )
{
      statement1;
      statement2;
            …
      statementx;
}
```

which is a type, followed by a name, followed by the primary operator *( )*. The parentheses, or function operator, indicate to the compiler that the name is to be executed as a function.

The type of a function or its parameters can be any valid variable type such as **int**, **char**, or **float**. The type can also be empty or void. The type **void** is a valid type and is used indicate a parameter or returned value of *zero size*. The default type of a function is **int**.

So a typical function declaration might be as follows:

```
unsigned char getchar(void)
{
     while((PIR1 & 0x20) == 0)
          ;      // wait for a character to arrive
     return RCREG;
               // return its unsigned char value to the caller
}
```

In this example, *getchar()* is a function that requires no parameters and returns an unsigned character value when it has competed execution. The *getchar()* function is one of the many library functions provided in the C compiler. These functions are available for the programmer's use and will be discussed in more depth later.

1.8.1 PROTOTYPING AND FUNCTION ORGANIZATION

Just as in the use of variables and constants, the function type and the types of its parameters must be declared before it is called. This can be accomplished in a couple of ways: one is the order of the declarations of the functions, and the other is the use of function prototypes.

Ordering the functions is always a good idea. It allows the compiler to have all the information about the function in place before it is used. This would also mean that all programs would have the following format:

```
// declaration of global variables would be first
int var1, var2, var3;
```

```
// declarations of functions would come next
int function1(int x, int y)
{
}

void function2(void)
{
}

// main would be built last
void main(void)
{

        var3 = function1(var1, var2);
        function2();
}
```

This is all nice and orderly but sometimes impossible. There are many occasions when functions use one another to accomplish a task and you simply cannot declare them in such a way that they are all in a top-down order.

Function prototypes are used to allow the compiler to know the requirements and the type of the function before it is actually declared, reducing the need for a top-down order.

The previous example can be organized differently:

```
int var1, var2, var3; // declaration of global variables

void main(void)        // main
{
        // these functions are not yet known to the
        // compiler
        var3 = function1(var1, var2);

        function2();
}

// declarations of functions here now generate a
// "Function Redefined Error"
int function1(int x, int y)
{
}

void function2(void)
{
}
```

This organization would typically lead to an error message from the compiler. The compiler simply would not have enough information about the functions that are called in *main()*, or their format, and would be unable to generate the proper code. The prototypes of the functions can be added to the top of the code like this:

```
// the prototype of a function tells the compiler what to expect
int function1(int, int);

void function2(void);

int var1, var2, var3; // declaration of global variables

void main(void)          // main
{
        var3 = function1(var1, var2);
        function2();
}
// the declaration of the functions here is
// perfectly OK, because
// the format of the functions is presented in the prototypes

int function1(int x, int y)
{
}

void function2(void)
{
}
```

The compiler now has all the required information about each function as it processes the function name. As long as the actual declared function matches the prototype, everything is fine.

The C compiler provides many library functions. They are made available to the program by the ***#include*** *<filename.h>*" statement, where *filename.h* will be the file containing the function prototypes for the library functions. In previous examples we used the function *printf()*. The *printf()* function itself is declared elsewhere but its prototype exists in the stdio.h header file. This allows the programmer to simply include the header file at the top of the program and start writing code. The library functions are each defined in Appendix A, Library Functions Reference.

1.8.2 FUNCTIONS THAT RETURN VALUES

In many cases a function is designed to perform a task and return a value or status from the task performed. The control word **return** is used to indicate an exit point in a function or, in the case of a non-void type function, to select the value that is to be returned to the caller.

In a function of type **void**

```
int z;    // global variable z

void sum(int x, int y)
{
        z = x + y;          // z is modified by the function, sum()
}
```

the **return** is implied and is at the end of the function. If a **return** were placed in the **void** function like this:

```
void sum(int x, int y)
{
        return;             // this would return nothing..
        z = x + y;          // and this would be skipped
}
```

the return statement would send the program execution back to the caller, and the statements following the return would not be executed.

In a function whose type is not **void**, the **return** control word will also send the execution back to the caller. In addition, the **return** will place the value of the expression, to the right of the **return**, on the stack in the form of the type specified in the declaration of the function. The following function is of type **float**. So a **float** will be placed on the stack when the return is executed:

```
float cube(float v)
{
        return (v*v*v); // this returns a type float value
}
```

The ability to return a value allows a function to be used as part of an expression, such as an assignment. For example, in the program below,

```
void main(void)
{
        float a;

        b = 3.14159;        // put PI into b

        a = cube(b);        // pass b to the cubed function, and
                            // assign its return value to a

        printf("a = %f, b = %f \n",a ,b);  // print the result

        while(1)            // done
            ;
}
```

a would be assigned the value of the function *cube(b)*. The result would appear as

a = 31.006198, b = 3.14159

1.8.3 RECURSION

A recursive function is one that calls itself. The ability to generate recursive code is one of the more powerful aspects of the C language. When a function is called, its local variables are placed on a stack or heap-space, along with the address of the caller, so that it knows how to get back. Each time the function is called, these allocations are made once again. This makes the function "reentrant" because the values from the previous call would be left within the previous allocations, untouched.

The most common example of a recursive operation is calculating factorials. A factorial of a number is the product of that number and all of the numbers that lead up to it. For example,

$$5! = 5 * 4 * 3 * 2 * 1 = 120$$

This can be stated algebraically as

$$n! = n * (n-1) * (n-2) * \ldots * 2 * 1 \quad \textbf{\textit{or}} \quad n! = n * (n-1)\,!$$

So a program to demonstrate this operation may look like this:

```
#include <stdio.h>

int fact(int n)
{
        if(n == 0)
        return 1;           // if n is zero, return a one, by
                            // definition of factorial

        else
            return (n * fact(n-1));    // otherwise, call myself with
                                       // n - 1 until n = 0, then
                                       // return n * the result the
                                       // call to myself
}
void main(void)
{
         int n;

         n = fact(5);     // compute 5!, recursively

         printf(" 5! = %d \n", n);     // print the result

         while(1)                      // done.
               ;
}
```

When the function *fact()* is called with argument *n*, it calls itself *n*-1 times. Each time it calls itself it reduces *n* by 1. When *n* equals 0, the function returns instead of calling itself again. This causes a "chain reaction" or "domino effect" of returns, which leads back to the call made in *main()* where the result is printed.

As powerful as recursion can be to perform factorials, quick sorts, and linked-list searches, it is a memory-consuming operation. Each time the function calls itself it allocates memory for the local variables of the function, the return value, the return address for the function, and any parameters that are passed during the call. In the previous example this would include the following:

```
int n     (passed parameter)  2 bytes
Return value                  2 bytes
Return address                2 bytes
          Total               6 bytes, per recursion, minimum.
```

In the case of "5!" at ***least*** a total of 30 bytes of memory would be allocated during the factorial operation and de-allocated upon its return. This makes this type of operation dangerous and impractical for a small microcontroller. If the depth of the allocation due to recursion becomes too great, the stack or heap-space will overflow, and the program's operation will become unpredictable.

Chapter 1 Example Project: Part B

In this phase of the development of the Pocket Slot Machine, we are going to create a couple of functions to break up the main loop into smaller chunks of code and make the code slightly more readable. The parts of code that flash the lights and compute payout will be moved into functions, shortening the *main()* function and making it a bit easier to follow. Also, a **switch/case** statement will be used in the indication of payout.

```
//
// "Slot Machine" – The Mini-Exercise
//
//    Phase 2..

#include <16F877.h>
#use delay(clock=10000000)
#fuses HS,WDT
#use rs232(baud=9600,parity=N,xmit=PIN_C6,rcv=PIN_C7,stream=RS232,bits=8)

#include <stdio.h>      /* this gets the printf() definition */

// Global Variable Declarations..
char    first,second,third; // Columns in the slot machine.
char    seed;                 // Number to form a seed for random #s.
char    count;                // General purpose counter.
long    delay;                // A variable for delay time.

#define JACKPOT   3     /* this defines give different payouts */
#define DIME      2     /* names in order to make the code more */
#define NICKEL    1     /* readable to humans */
#define ZERO      0
```

```
void flash_lights(char n)    // flash the lights a number of times
{
        for(count = 0; count < n; count++) // flash light when done..
        {
           for(delay = 0; delay < 10000; delay++)
               ;                    // just count up and wait
           output_bit(PIN_C2,0); // turn the LED on..
           for(delay = 0; delay < 10000; delay++)
               ;
           output_bit(PIN_C2,1); // turn the LED off..
        }
}

int get_payout(void)
{
     if((first == 3) && (second == 3) && (third == 3))
          return JACKPOT;    // if all "cherries"..
     else if((first == 3) || (second == 3) || (third == 3))
          return DIME;  // if any are "cherries"..
     else if((first == second) && (second == third))
          return NICKEL;  // if three are alike, of any kind..
     else
          return ZERO;  // otherwise - you lose..
}

void main(void)
{
    int i;                 // declare a local variable for
                                  // temporary use..

    set_tris_b(0x00); // port B is all input
    port_b_pullups(TRUE);
    set_tris_c(0xFB); // port C.2 is an LED

    while(1)               // do forever..
    {
          while(input_b() & 1)   // Wait for a button and
               seed++;     // let the counter roll over and
                                   // over to form a seed.

          first = second = third = seed; // preload the columns

          do                         // mix up the numbers
          {                          // while waiting for button release.
```

```
        first ^= seed>>1;   // Exclusive ORing in the moving seed
        second^= seed>>2;   // can really stir up the numbers.
        third ^= seed>>3;
        seed++;             // keep rolling over the seed pattern
    } while((input_b() & 1) == 0); // while the button is pressed


    flash_lights(5);        // flash the lights 5 times..

    first  &= 3;        // limit number to values from 0 to 3
    second &= 3;
    third  &= 3;
                        // show the values..
    printf("-> %d, %d ,%d <-\n\r", first, second, third);

                        // determine a payout..
    i = get_payout();

    switch(i)         // now print the payout results
    {
            case    ZERO:
                printf("Paid out: ZERO\n\r");
                break;

            case    NICKEL:
                printf("Paid out: One Nickel\n\r");
                break;

            case    DIME:
                printf("Paid out: One Dime\n\r");
                break;

            case    JACKPOT:
                printf("Paid out: JACKPOT!!\n\r");
                break;
    }
  }
}
```

EXAMPLE

1.9 POINTERS AND ARRAYS

Pointers and arrays are widely used in the C language because they allow programs to perform more generalized and more efficient operations. Operations that require gathering data may use these methods to easily access and manipulate the data without moving the data around in memory. They also allow for the grouping of associated variables such as communications buffers and character strings.

1.9.1 POINTERS

Pointers are variables that contain the address or location of a variable, constant, function, or data object. A variable is declared to be a pointer with the indirection or de-referencing operator (*):

```
char *p;      // p is a pointer to a character
int *fp;      // fp is a pointer to an integer
```

The pointer data type allocates an area in memory large enough to hold the machine address of the variable. For example, in a typical microcontroller the address of a memory location will be described in 16 bits. So in a typical microcontroller, a pointer to a character will be a 16-bit value, even though the character itself is only an 8-bit value.

Once a pointer is declared, you are now dealing with the address of the variable it is pointing to, not the value of the variable itself. You must think in terms of locations and contents-of locations. The address operator (&) is used to gain access to the address of a variable. This address may be assigned to the pointer and is the pointer's value. The indirection or de-referencing operator (*) is used to gain access to the data located at the address contained in the pointer. For example,

```
char *p;          // p is a pointer to a character
char a, b;        // a and b are characters

p = &a;           // p is now pointing to a
```

In this example, *p* is assigned the address of *a*, so *p* is "pointing to" *a*.

To get to the value of the variable that is pointed to by *p*, the indirection operator (*) is used. When executed, the indirection operator causes the value of *p*, an address, to be used to look up a location in memory. The value at this location is then read from or written to according to the expression containing the indirection operator. So, in the following code, **p* would cause the value located at the address contained in p to be read and assigned to *b*.

```
b = *p;   // b equals the contents pointed to by p
```

Therefore, the combined code of the previous two examples would produce the same result as

```
b = a;
```

The indirection operator can also appear on the right side of an assignment.

In this example,

```
char *p;          // p is a pointer to a character
char a, b;        // a and b are characters

p = &a;

*p = b;           // the location pointed to by p, is
                  // assigned the value of b
```

the memory location, at the address stored in *p* is assigned the value of *b*. This would produce the same result as

```
a = b;
```

Whenever you read these operations, try to read them as "*b* is assigned the value pointed to by *p*" and "*p* is assigned the address of *a*". This helps to avoid making the most common mistake with pointers:

```
b = p;    // b will be assigned a value of p, an address,
          // not what p points to.

p = a;    // p will be assigned the value of a, not its address.
```

These two assignments are allowed because they are syntactically correct. Semantically speaking, they are most likely not what was intended.

With power and simplicity comes the opportunity to make simple and powerful mistakes. Pointer manipulation is one of the leading causes of programming misfortune. But exercising a little care, and reading the syntax aloud, can greatly reduce the risk of changing memory in an unintended fashion.

Pointers are also an excellent method of accessing a peripheral in a system, such as an I/O port. For instance, if we had an 8-bit parallel output port located at 0x1010 in memory, that port could be accessed through indirection:

```
unsigned char *my_port; // declare my_port as a pointer

my_port = 0x1010;        // assign my_port with the address value

*my_port = 0xaa;         // now assign my_port's address a value
```

In this code, the location pointed to by *my_port* would be assigned the value 0xAA. It can also be described as any value assigned to **my_port* will be written to memory address 0x1010.

By the structure of the C language, it is possible to have pointers that point to pointers. In fact, there really is no limit to the depth of this type of indirection, except for the confusion that it may cause. For example,

```
int *p1;          // p1 is a pointer to an integer
int **p2;         // p2 is a pointer, to a pointer to an integer
int ***p3;        // p3 is a pointer, to a pointer, to a pointer
                  // to an integer
int i, j;

p1 = &i;          // p1 is assigned that address of i
p2 = &p1;         // p2 is now pointing to the pointer to i
p3 = &p2;         // p3 is pointing to the pointer that is pointing
                  // to i
```

```
                    // Therefore,
j = ***p3;          // j is assigned the value pointed to by the value
                    // pointed to by the value pointed to by p3
```

yields the same result as

```
j = i;                  // any questions??
```

Because pointers are effectively addresses, they offer the ability to move, copy, and change memory in all sorts of ways with very little instruction. When it comes to performing address arithmetic, the C compiler makes sure that the proper address is computed based on the type of the variable or constant being addressed. For example,

```
int *ptr;
long *lptr;

ptr = ptr + 1;          // moves the pointer to the next integer
                        //     (2 bytes away)

lptr = lptr + 1;        // moves the pointer to the next long integer
                        //     (4 bytes away)
```

ptr and *lptr* are incremented by one location, which in reality is 2 bytes for *ptr* and 4 bytes for *lptr*, because of their subsequent types. This is also true when using increment and decrement operations:

```
ptr++;    // moves the pointer to the next integer location
          //     (2 bytes)

—lptr;    // moves the pointer back to the preceding long
          // integer location
          //     (-4 bytes)
```

Because the indirection (*) and address (&) operators are unary operators and are at the highest precedence, they will always have priority over the other operations in an expression. Because increment and decrement are also unary operators and share the same priority, expressions containing these operators will be evaluated left to right. For example, listed below are pre-increment and post-increment operations that are part of an assignment. Take note of the comments in each line to see how the same precedence level affects the outcome of the operation:

```
char c;
char *p;

c = *p++;.        // assign c the value pointed to by p, and then
                  // increment the address p

c = *++p;         //increment the address p, then assign c the
                  //value pointed to by p
```

```
c = ++*p;       // increment the value pointed to by p, then
                // assign it to c, leaving the value of p untouched

c = (*p)++;     // assign c the value pointed to by p, and then
                // increment the value pointed to by p, leaving
                // the value of p untouched
```

Pointers can be used to extend the amount of information returned from a function. Because a function inherently can return only one item using the **return** control, passing pointers as parameters to a function allows the function an avenue for returning additional values. Consider the following function *swap2()*:

```
void swap2 (int *a, int *b)
{
        int temp;

        temp = *b;  // place value pointed to by b into temp
        *b = *a;    // move the value pointed to by a into location
                    // pointed to by b
        *a = temp;  // now set the value of location a to the
                    // value of temp
}
```

This sample function swaps the values of a and b. The caller provides the pointers to the variables it wishes to transpose like this:

```
int v1,v2;

swap2( &v1, &v2);      // pass the addresses of v1 and v2
        ...
```

Because the *swap2()* function is using the addresses that were passed to it, it swaps the values in variables *v1* and *v2* directly. This process of passing pointers is frequently used and can be found in standard library functions like *scanf()*. The *scanf()* function (defined in stdio.h) allows multiple parameters to be gathered from the standard input in a formatted manner and stores them in the specified locations of memory. A typical call to s*canf()* looks like this:

```
int x,y,z;

scanf("%d %d %d", &x, &y, &z);
```

This *scanf()* call will retrieve three decimal integer values from the standard input and place these values in *x*, *y*, and *z*.

1.9.2 ARRAYS

An array is another system of indirection. An array is a data set of a declared type, arranged in order. An array is declared like any other variable or constant, except for the number of required array elements:

```
int digits[10]; // this declares an array of 10 integers
char str[20];   // this declares an array of 20 characters
```

The referencing of an array element is handled by an index or subscript. The index can range from 0 to the length of the declared array less 1.

```
str[0],   str[1], str[2], . . . . . str[19]
```

Array declarations can contain initializers. In a variable array the initialization values will be placed in the program memory area and copied into the actual array before *main()* is executed. A constant array differs in that the values will be allocated in program memory, saving RAM memory, which is usually in short supply on a microcontroller. A typical initializer would appear as

```
int array[5] = {12, 15, 27 56, 94 };
```

In this case, *array[0]* = 12, *array[1]* = 15, *array[2]* = 27, *array[3]* = 56,

and *array[4]* = 94.

The C language has no provision for checking the boundaries of an array. If the index were to be assigned a value that exceeds the boundaries of an array, memory could be altered in an unexpected way leading to unpredictable results. For example,

```
char digits[10]={0,1,2,3,4,5,6,7,8,9};// an array of characters

numb = digits[12]; // this reads outside the array
```

Arrays are stored in sequential locations in memory. Reading *digits[5]* in the example above will cause the processor to go to the location of the first index of the array and then read the data 5 locations above that. Therefore, the second line of code above will cause the processor to read the data 12 spaces above the starting point for the array. Whatever data exists at that location will be assigned to *numb* and may cause many strange results. So, as in many other programming areas, some caution and forethought should be exercised.

A primary difference in the use of an array versus that of a pointer is that, in an array, an actual memory area has been allocated for the data. With a pointer, only an address reference location is allocated, and it is up to the programmer to declare and define the actual memory areas (variables) to be accessed.

The most common array type is the character array. It is typically referred to as a string or character string. A string variable is defined as an array of characters, whereas a constant string is typically declared by placing a line of text in quotes. In the case of a constant string, the C compiler will null-terminate, or add a zero to the end of the string. When you declare character strings, constant or variable, the declared array size should be one more than what is needed for the contents in order to allow for the null terminator:

```
char str[12];               // variable string

printf("Hello World!");  // constant string in program memory
```

```
const cstr[16] = "Constant String";
                        // constant string in program memory
```

cstr in the example above is set to contain 16 values because the string itself contains fifteen characters and one more space must be allowed for the terminator (the character, not the movie).

An array name followed by an index may reference the individual elements of an array of any type. It is also possible to reference the first element of any array by its name alone. When no index is specified, the array name is treated as the address of the first element in the array. Given the following declarations,

```
char stng[20];
char *p;
```

the assignment

```
p = stng;       // p is pointing to stng[0]
```

is the same as

```
p = &stng[0];   // p is pointing to stng[0]
```

Character strings often need to be handled on a character-by-character basis. Sending a message to a serial device or a liquid crystal display (LCD) are examples of this requirement. The example below shows how array indexing and pointer indirection function nearly interchangeably. This example uses the library function *putchar()* to send one character at a time to the standard output device, most likely, a serial port:

```
#include <16F877.h>       // register definition file for
#fuses HS,NOWDT           // a Microchip PIC16F877
#use delay(clock=10000000)
#use rs232(baud=9600,parity=N,xmit=PIN_C6,rcv=PIN_C7,stream=RS232,bits=8)
#include <stdio.h>

const char s[15] = {"This is a test"};

char i;
char *p;

void main(void)
{
   for(i=0; i<15; i++)   // print each character of the array
          putchar(s[i]);        // by using i as an index
   p = s;                // point to string as a whole
   for(i=0; i<15; i++)   // print each character of the array
       putchar(*p++);    // and move to the next element
                              // by incrementing the pointer p

   while(1)
       ;
}
```

The first portion on this program uses a **for** loop to output each character of the array individually. The **for** loop counter is used as the index to retrieve each character of the array so that it can be passed to the *putchar()* function. The second portion uses a pointer to access each element of the array. The line '*p = s;*' sets the pointer to the address of the first character in the array. The **for** loop then uses the pointer (post-incrementing it each time) to retrieve the array elements and pass them to *putchar()*.

1.9.3 MULTIDIMENSIONAL ARRAYS

The C language supports multidimensional arrays. When a multidimensional array is declared, it should be thought of as arrays of arrays. Multidimensional arrays can be constructed to have two, three, four, or more dimensions. The adjacent memory locations are always referenced by the right-most index of the declaration.

A typical two-dimensional array of integers would be declared as

```
int two_d[5][10];
```

In memory, the elements of the array would be stored in sequential rows like this:

```
two_d[0][0], two_d[0][1], two_d[0][2], ... two_d[0][9],
two_d[1][0], two_d[1][1], two_d[1][2], ... two_d[1][9],
two_d[2][0], two_d[2][1], two_d[2][2], ... two_d[2][9],
two_d[3][0], two_d[3][1], two_d[3][2], ... two_d[3][9],
two_d[4][0], two_d[4][1], two_d[4][2], ... two_d[4][9]
```

When a two-dimensional array is initialized, the layout is the same, sequential rows:

```
int matrix[3][4] = {  0, 1,  2,  3,
                      4, 5,  6,  7,
                      8, 9, 10, 11 };
```

Multidimensional arrays are useful in operations such as matrix arithmetic, filters, and look-up-tables (LUTs).

For instance, assume we have a telephone keypad that generates a row and column indication, or scan-code, whenever a key is pressed. A two-dimensional array could be used as a look-up-table to convert the scan-code to the actual ASCII character for the key. We will assume for this example that the routine *getkeycode()* scans the keys and sets the *row* and *col* values to indicate the position of the key that is pressed. The code to perform the conversion operation may look something like this:

```
#include <16F877.h>        // register definition file for
#fuses HS,NOWDT            // a Microchip PIC16F877
#use delay(clock=10000000)
#use rs232(baud=9600,parity=N,xmit=PIN_C6,rcv=PIN_C7,stream=RS232,bits=8)
#include <stdio.h>

#define TRUE 1
#define FALSE 0
```

```
char getkeycode(char *row, char *col);
                    // the getkeycode routine gets the key press and
                    // returns TRUE if a key is pressed, FALSE if not

/* look up table for ASCII values */
const char keys[4][3] = {      '1','2','3',
                               '4','5','6',
                               '7','8','9',
                               '*','0','#' };

void main(void)
{
        char row, col;
        char i;

        while(1)    // while forever....
        {
                i = getkeycode(&row, &col);
                                // TRUE, if there was a key pressed
                                // and row and col contain which key
                if(i == TRUE)
                                // only print the key value that is pressed
                   putchar(keys[row][col]);
        }
}
```

Another, more common form of a two-dimensional array is an array of strings. This example declares an array of strings initialized to the days of the week:

```
char day_of_the_week[7][10] = {
            "Sunday",
            "Monday",
            "Tuesday",
            "Wednesday",
            "Thursday",
            "Friday",
            "Saturday" };
```

In the array shown above, the strings are of different lengths. The compiler places the null terminator after the last character in the string no matter how long the string may be. Any wasted locations in memory are left uninitialized. Functions such as *printf()* print the string until they encounter the null-terminator character, thus being able to print strings of various length correctly.

To access the fourth day of the week and print the associated string, we will use *printf()*. *printf()* requires the address of the first character of a string:

```
printf("%s", &day_of_the_week[3][0]); // prints Wednesday
```

The name of a string is considered to be the address of the first character of that string. Stating only the first dimension is effectively referencing the entire string as the second dimension. So, the same string within the array that was just shown can be accessed as

```
printf("%s", day_of_the_week[3]);   // prints Wednesday
```

1.9.4 POINTERS TO FUNCTIONS

One of the more esoteric and powerful aspects of pointers and indirection is that they can also be applied to functions. Using a pointer to a function allows for functions to be called from the result of look-up-table operation. Pointers to functions also allow functions to be passed by reference to other functions. This can be used to create a dynamic flow of execution, which is sometimes called "self-modifying" code.

Consider an example that calls a function from a table of pointers to functions. In the following example, the *scanf()* function gets a value from the standard input. The value is checked to make sure it is in range (1–6). If it is an appropriate value, func_number is used as an index into an array of function pointers. The array value at the *func_number* index is assigned to *fp*, which is a pointer to a function of type **void**.

Remember from Section 1.8, "Functions," that the standard form of a function is

```
type function_name ( type param1, type param2, … )
{
        statement1;
        statement2;
                …
}
```

which is a type, followed by a name, followed by the primary operator *( )*. The parentheses, or function operator, indicate to the compiler that the identifier is to be treated as a function. In this case, *fp* contains the address of a function. The indirection operator is added to obtain the address of the function from the pointer *fp*. Now the function can be called by simply adding the function operator *( )* like this:

```
(*fp)();     // execute the function pointed to by fp
```

Here is the entire program:

```
#include <stdio.h>

void do_start_task(void)
{
    printf("start selected\n");
}

void do_stop_task(void)
{
    printf("stop selected\n");
}
```

```
void do_up_task(void)
{
    printf("up selected\n");
}

void do_down_task(void)
{
    printf("down selected\n");
}

void do_left_task(void)
{
    printf("left selected\n");
}

void do_right_task(void)
{
    printf("right selected\n");
}

void (*task_list[6])(void) = {
   do_start_task,
   do_stop_task,          /* an array of pointers to functions */
   do_up_task,
   do_down_task,
   do_left_task,
   do_right_task,
};

void main(void)
{
       int func_number;
       void (*fp) (void);      /* fp is a pointer to a function */
       while(1)
       {
              printf("\nSelect a function 1-6 :");
              scanf("%d",&func_number);

              if((func_number > 0) && (func_number < 7))
              {
                     fp = task_list[func_number-1];
                     (*fp)();
                     /* assign the function address to fp */
              }      /* and call the selected function */
       }
}
```

Some PIC compilers permit pointers to functions, but the CCS-PICC compiler does not. Because the PIC has a limited stack space, function parameters are passed in a special way that requires knowledge at compile time of the function that is being called. Sometimes you will want to use function pointers to create a state machine or control flow like the one shown earlier. The following example shows how to perform a similar operation without the availability of function pointers:


```
enum tasks {taskA, taskB, taskC};

run_task(tasks task_to_run) {

      switch(task_to_run) {
            case taskA :
                  taskA_main();
                  break;

            case taskB :
                  taskB_main();
                  break;

            case taskC :
                  taskC_main();
                  break;
      }

}
void main()
{
       run_task (taskB);      // run the second task (taskB)

      while(1)
          ;
}
```

Chapter 1 Example Project: Part C

In this phase of the development of the Pocket Slot Machine, we are going to use enumerations and multidimensional arrays to "polish" the project's appearance. Instead of printing numbers for the column values and referencing numbers in our payout calculations, we will give these values names.

```
//
// "Slot Machine" – The Mini-Exercise
//
//     Phase 3..

#include <16F877.h>
```

```
#use delay(clock=10000000)
#fuses HS,WDT
#use rs232(baud=9600,parity=N,xmit=PIN_C6,rcv=PIN_C7,stream=RS232,bits=8)

#include <stdio.h>      /* this gets the printf() definition */

// Global Variable Declarations..
char     first,second,third; // Columns in the slot machine.
char     seed;                 // Number to form a seed for random #s.
char     count;                // General purpose counter.
long     delay;                // A variable for delay time.

#define JACKPOT   3      /* this defines give different payouts */
#define DIME      2      /* names in order to make the code more */
#define NICKEL    1      /* readable to humans */
#define ZERO      0

enum { BAR, BELL, LEMON, CHERRY };   // the values for each kind


char kind[4][8] = {              // for kinds of names..
     "BAR",
     "BELL",
     "LEMON",
     "CHERRY"
};


void flash_lights(char n)    // flash the lights a number of times
{
        for(count = 0; count < n; count++) // flash light when done..
        {
           for(delay = 0; delay < 10000; delay++)
               ;                     // just count up and wait
           output_bit(PIN_C2,0); // turn the LED on..
           for(delay = 0; delay < 10000; delay++)
               ;
           output_bit(PIN_C2,1); // turn the LED off..
        }
}

int get_payout(void)
{
    if((first == 3) && (second == 3) && (third == 3))
        return JACKPOT;    // if all "cherries"..
```

```
    else if((first == 3) || (second == 3) || (third == 3))
        return DIME;  // if any are "cherries"..
    else if((first == second) && (second == third))
        return NICKEL;  // if three are alike, of any kind..
    else
        return ZERO;  // otherwise - you lose..
}

void main(void)
{
      int i;                    // declare a local variable for
                                // temporary use..

   set_tris_b(0x00); // port B is all input
   port_b_pullups(TRUE);
   set_tris_c(0xFB); // port C.2 is an LED

   while(1)       // do forever..
      {
        while(input_b() & 1)  // Wait for a button and
            seed++;     // let the counter roll over and
                                // over to form a seed.

        first = second = third = seed; // preload the columns

        do                        // mix up the numbers
        {                         // while waiting for button release.
           first ^= seed>>1; // Exclusive ORing in the moving seed
           second^= seed>>2; // can really stir up the numbers.
           third ^= seed>>3;
           seed++;              // keep rolling over the seed pattern
        } while((input_b() & 1) == 0); // while the button is pressed


        flash_lights(5);       // flash the lights 5 times..

        first  &= 3;    // limit number to values from 0 to 3
        second &= 3;
        third  &= 3;
                      // show the values.. BY NAME!!
                      // simply change the %d to %s
                      // and pass the pointer to the string from
                      // the 2D array kind[] to printf()
        printf("-> %s, %s ,%s <-\n\r",
                kind[first], kind[second], kind[third]);
```

```
                    // determine a payout..
        i = get_payout();

        switch(i)       // now print the payout results
        {
                case    ZERO:
                   printf("Paid out: ZERO\n\r");
                   break;

                case    NICKEL:
                   printf("Paid out: One Nickel\n\r");
                   break;

                case    DIME:
                   printf("Paid out: One Dime\n\r");
                   break;

                case    JACKPOT:
                   printf("Paid out: JACKPOT!!\n\r");
                   break;
        }
    }
}
```

EXAMPLE

1.10 STRUCTURES AND UNION

Structures and unions are used to group variables under one heading or name. Because the word "object" in C programming generally refers to a group or association of data members, structures and unions are the fundamental elements in *object-oriented programming*. Object-oriented programming refers to the method in which a program deals with data on a relational basis. A structure or union can be thought of as an object. The members of the structure or union are the properties (variables) of that object. The object name then provides a means to identify the association of the properties to the object throughout the program.

1.10.1 STRUCTURES

A structure is a method of creating a single data object from one or more variables. The variables within a structure are called members. This method allows for a collection of members to be referenced from a single name. Some high-level languages refer to this type of object as a record or based-variable. Unlike an array, the variables contained within a structure do not need to be of the same type.

A structure declaration has the form:

```
Struct structure_tag_name {
       type member_1;
```

```
struct structure_tag_name {
       type member_1;
```

```
        type member_2;              or              type member_2;
             ...                                         ...
        type member_x;                              type member_x;
    };                                          } structure_var_name;
```

Once a structure template has been defined, the *structure_tag_name* serves as a common descriptor and can be used to declare structures of that type throughout the program. Declared below are two structures, *var1* and *var2*, and an array of structures *var3*:

```
struct structure_tag_name var1, var2, var3[5];
```

Structure templates can contain all sorts of variable types, including other structures, pointers to functions, and pointers to structures. It should be noted that when a template is defined, no memory space is allocated. Memory is allocated when the actual structure variable is declared.

Members within a structure are accessed using the member operator (.). The member operator connects the member name to the structure that it is associated with:

```
structure_var_name.member_1  structure_var_name.member_x
```

Like arrays, structures can be initialized by following the structure name with a list of initializers in braces:

```
struct DATE {
        int month;          // declare a template for a
        int day;            // date structure
        int year;
}

// declare a structure variable and initialize it..

struct DATE date_of_birth = { 2, 21, 1961 };
```

which yields the same result as these assignments:

```
date_of_birth.month = 2;
date_of_birth.day = 21;
date_of_birth.year = 1961;
```

Because structures themselves are a valid type, there is no limit to the nesting of members within a structure. For example, if a set of structures is declared as

```
struct LOCATION {
        int x;
        int y;    // this is the location coordinates x and y
};

struct PARTS {
        char part_name[20];     // a string for the part name
        long int sku;           // a SKU number for the part
        stuct LOCATION bin;     // its location in the warehouse
```

```
    } widget;
```

to access the location of a "widget," you would provide a reference like this:

```
x_location = widget.bin.x;
                    // the x coordinate of the bin of the widget
y_location = widget.bin.y;
                    // the y coordinate of the bin of the widget
```

To assign the location of the "widget," the same rules apply:

```
widget.bin.x = 10;      // the x coordinate of the bin of the widget
widget.bin.y = 23;      // the y coordinate of the bin of the widget
```

A structure can be passed to a function as a parameter as well as returned from a function. For example, the function

```
struct PART new_location( int x, int y)
{
     struct PART temp;

     temp.part_name = "";   // initialized the name to NULL
     temp.sku = 0;          // and zero the sku number
     temp.bin.x = x;        // set the location to the passed x and y
     temp.bin.y = y;


     return temp;        // and then returns the structure to the
caller
}
```

would return a PART structure with the *bin.x* and *bin.y* location members assigned to the parameters passed to the function. The *sku* and *part_name* members would be also cleared before the structure was returned. The function above could then be used in an assignment like

```
widget = new_location(10,10);
```

The result would be that the *part_name* and *sku* would be cleared and the *widget.bin.x* and *widget.bin.y* values would be set to 10.

1.10.2 ARRAYS OF STRUCTURES

As with any other variable type, arrays of structures can also be declared. The declaration of an array of structures appears as follows:

```
struct PART {
     char part_name[20];    // a string for the part name
     long int sku;          // a SKU number for the part
     stuct LOCATION bin;    // its location in the warehouse
} widget[100];
```

The access of a member is still the same. The only difference is in the indexing of the structure variable itself. So, to access a "particular widget's location," a reference like this may be used:

```
x_location = widget[12].bin.x;
                // the x coordinate of the bin of the widget 12
y_location = widget[12].bin.y;
                // the y coordinate of the bin of the widget 12
```

In this example there is a character string, *part_name*, which can be accessed as strings normally are:

```
widget[12].part_name;          // the name of widget 12

        or

widget[12].part_name[0];
            // the first character in the name of widget 12
```

Arrays of structures can be initialized by following the structure name with a list of initializers in braces; there simply needs to be an initializer for each structure element within each array element:

```
struct DATE {
        int month;
        int day;
        int year;
}

struct DATE birthdates[3] = {      2, 21, 1961,
                                   8,  8, 1974,
                                   7, 11, 1997 };
```

1.10.3 POINTERS TO STRUCTURES

Sometimes it is desirable to manipulate the members of a structure in a generalized fashion. One method of doing this is to use a pointer to reference the structure, for example, passing a pointer to a structure to a function instead of passing the entire structure.

A pointer to a structure is declared:

```
struct structure_tag_name *structure_var_name;
```

The pointer operator (*) states that *structure_var_name* is a pointer to a structure of type *structure_tag_name*. Just as with any other type, a pointer must be assigned a value that points to something tangible, like a variable that has already been declared. A variable declaration guarantees that memory has been allocated for a purpose.

The following code would be used to declare the structure variable, widget, and a pointer to a structure variable, *this_widget*. The final line in the example assigns the address of *widget* to the pointer *this_widget*.

```
struct LOCATION {
      int x;
      int y;    // this is the location coordinates x and y
};

struct PART {
      char part_name[20];     // a string for the part name
      long int sku;           // a SKU number for the part
      stuct LOCATION bin;     // its location in the warehouse
};

struct PART widget, *this_widget;
          // declare a structure and a pointer to a structure

  . . .

this_widget = &widget;
          // assign the pointer the address of a structure
```

When a pointer is used to reference a structure, the structure pointer operator ->, minus-greater-than, is used to access the members of the structure through indirection:

```
this_widget->sku = 1234;
```

This could also be stated using the indirection operator to first locate the structure and then using the member operator to access the *sku* member:

```
(*this_widget).sku = 1234;
```

Because *this_widget* is a pointer to *widget*, both methods of assignment, shown above, are valid. The parentheses around *this_widget* are required because the member operator has a higher precedence than the indirection (*) operator. If the parentheses were omitted, the expression would be misinterpreted as

```
*(this_widget.partname[0])
```

which is the actual address of *widget* (&widget).

Structures can contain pointers to other structures but can also contain pointers to structures of the same type. A structure cannot contain itself as a member because that would be a recursive declaration, and the compiler would lack the information required to resolve the declaration. Pointers are always the same size regardless of what they point to. Therefore, by pointing to a structure of the same type, a structure can be made "Self Referential." A very basic example would be as follows:

```
struct LIST_ITEM {
      char *string;    // a text string
      int position;    // its position in a list
      struct LIST_ITEM *next_item; // a pointer to another
                                   // structure of the same type
```

```
} item, item2;

item.next_item = &item2;
        // assign the pointer with the address of the item2
```

Now

```
item.next_item->position
```

is the *position* member of the structure pointed to by *next_item*. This would be equivalent to

```
item2.position
```

Self-referential structures are typically used for data manipulations like linked-lists and quick sorts.

1.10.4 UNIONS

A union is declared and accessed much like a structure. A union declaration has the form:

```
Union union _tag_name {                union union _tag_name {
        type member_1;                         type member_1;
        type member_2;      or                 type member_2;
             ...                               ...
        type member_x;                         type member_x;
};                                     } union _var_name;
```

The primary difference between a union and a structure is in the way the memory is allocated. The members of a union actually share a common memory allocated to the largest member of that union:

```
union SOME_TYPES {
        char character;
        int integer;

        long int long_one;
} my_space;
```

In this example, the total amount of memory allocated to *my_space* is equivalent to the size of the long int *long_one* (4 bytes). If a value is assigned to the long int:

```
my_space.long_one = 0x12345678L;
```

then the value of *my_space.character* and *my_space. integer* are also modified. In this case their values would now be

```
my_space.character = 0x12;
my_space.integer = 0x1234;
```

Unions are sometimes used as a method of preserving valuable memory space. If there are variables that are used on a temporary basis, and there is never a chance that they will be used at the same time, a union is a method for defining a "scratch pad" area of memory.

More often, a union is used as a method of extracting smaller parts of data from a larger data object. This is shown in the previous example. The position of the actual data depends on the data types used and how a particular compiler handles numbers larger than type char (8 bits). The example above assumes big-endian (most significant byte first) storage. Compilers vary in how they store data. The data could be byte order swapped, word order swapped, or both! This example could be used as a test for a compiler in order to find out how the data is organized in memory.

Union declarations can save steps in coding to convert the format of data from one organization to another. Shown below are two examples where two 8-bit input ports are combined in one 16-bit value. The first method uses shifting and combining; the second method uses a union:

```
#include <16F877.h>                // register definition file for
#fuses HS,NOWDT                    // a Microchip PIC16F877
#use delay(clock=10000000)
#use rs232(baud=9600,parity=N,xmit=PIN_C6,rcv=PIN_C7,stream=RS232,bits=8)

#include <stdio.h>

unsigned long port_w;   // CCS-PICC default 'long' is 16 bits!!

void main(void)
{

      set_tris_a(0xFF); // set ports A and B to input..
      set_tris_b(0xFF);

      while(1)
      {
            port_w = input_a();  // get port A into the 16 bit value
            port_w <<= 8;        // shift it up..
            port_w |= input_b(); // now combine in port B

            printf("16-Bits = %04LX\n\r",port_w);
      }                          // put the combined value out to
}                                // the standard output
```

Now the same results using a **union** declaration:

```
#include <16F877.h>                      // register definition file for
#fuses HS,NOWDT                    // a Microchip PIC16F877
#use delay(clock=10000000)
#use rs232(baud=9600,parity=N,xmit=PIN_C6,rcv=PIN_C7,stream=RS232,bits=8)

#include <stdio.h>

   // declare the two types of data in
   // a union to occupy the same space..
union
```

```
{
     unsigned char port_b[2];
     unsigned long port_w;     // CCS-PICC default 'long' is 16 bits!!
} value;

void main(void)
{
     set_tris_a(0xFF);  // set ports A and B to input..
     set_tris_b(0xFF);

     while(1)
     {
          value.port_b[0] = input_a();     // get port A
          value.port_b[1] = input_b();     // get port B
                                  // the union eliminates the
                                  // need to combine the data
                                  // manually

          printf("16-Bits = %04LX\n\r",value.port_w);
     }                            // put the combined value out to
}                                 // the standard output
```

1.10.5 TYPEDEF OPERATOR

The C language supports an operation that allows for creating new type names. The **typedef** operator allows for a name to be declared synonymous with an existing type. For example:

```
typedef  unsigned char byte;
typedef  unsigned int word;
```

Now the aliases "byte" and "word" can be used to declare other variables that are actually of type **unsigned char** and **unsigned int.**

```
byte var1;      // this is the same as an unsigned char
word var2;      // this is the same as an unsigned int
```

This method of alias works for structures and unions as well:

```
typedef struct
{

     char name[20];
     char age;
     int home_room_number;
} student;

student Bob;    // these allocate memory in the form of
student Sally;          // a structure of type student
```

The **#define** statement is sometimes used to perform this alias operation through a text substitution in the compiler's preprocessor. **typedef** is evaluated by the compiler directly and can work with declarations, castings, and usages that would exceed the capabilities of the preprocessor.

1.10.6 BITS AND BITFIELDS

Bits and bitfields are often used when memory space is at a premium. Some compilers support a type **bit**, or **int1**, which is automatically allocated by the compiler and is referenced as a variable of its own. For example,

```
// the typdef sets up 'bit' as type 'int1' for the CCS PIC-C
typedef bit int1;

bit running;      // the compiler allocates a single bit of
                  // storage for this flag..

running = 1;      // the only two possibilities for value are 1 and 0.
running = 0;

if(running)
    ;             // the value can be tested as well as assigned
```

Unlike bits, bitfields are more common for larger, more generalized systems and are not always supported by embedded compilers. Bitfields are associated with structures because of the form of their declaration:

```
struct structure_tag_name {              struct structure_tag_name {
     unsigned int bit_1:1;                    unsigned int bit_1:1;
     unsigned int bit_2:1;       or           unsigned int bit_2:1;
        ...                                      ...
     unsigned int bit_15:1;                   unsigned int bit_15:1;
};                                       } struct_var_name;
```

A bitfield is specified by a member name (of type **unsigned int** only), followed by a colon (:) and the number of bits required for the value. The width of the member can be from 1 to the size of type **unsigned int** (16 bits). This allows several bitfields within a structure to be represented in a single unsigned integer memory location.

```
struct {
     unsigned int    running : 1;
     unsigned int    stopped : 1;
     unsigned int    counter : 4;
} machine;
```

These bitfields can be accessed by member name, just as you would for a structure:

```
machine.stopped = 1;  // these are single bits.. so only
                      // 1 and 0 are allowed..

machine.running = 0;
```

```
machine.counter++;      // this is 4-bit value, so
                        // 0-15 are allowed..
```

Sometimes in embedded systems bitfields are used to describe I/O port pins:

```
typedef struct
{
        unsigned bit_0: 1;
        unsigned bit_1: 1;
        unsigned bit_2: 1;
        unsigned bit_3: 1;
        unsigned bit_4: 1;
        unsigned bit_5: 1;
        unsigned bit_6: 1;
        unsigned bit_7: 1;
} bits;
```

The following **#define** labels the contents of memory at location 0x08. (Remember, **#define** is treated as a textual substitution directive by the preprocessor.) This allows the programmer to access memory location 0x08 by the name *PORTD*, as if it were a variable, throughout the program:

```
// PORTB is the contents of address 0x08, a bitfield "bits"
#define PORTD (*(bits *) 0x08)
```

The *bits* bitfield allows the individual bits of I/O port *PORTD* to be accessed independently (sometimes called "bit banged"):

```
// bitfields can be used in assignments
PORTD.bit_3 = 1;
PORTD.bit_5 = 0;

// bitfields can also be used in conditional expressions
if(PORTD.bit_2 || PORTD.bit_1)
        PORTD.bit_4 = 1;
```

1.10.7 SIZEOF OPERATOR

The C language supports a unary operator called **sizeof**. This operator is a compile-time feature that creates a constant value related to the size of a data object or its type. The forms of the **sizeof** operation are as follows:

```
sizeof( type_name )     // type_name could be keyword int,
                        // char, long, etc.

sizeof( object )        // object could be a variable, array,
                        // structure, or union variable name
```

These operations produce an integer that reveals the size (in bytes) of the type or data object located in the parentheses. For example, assuming these declarations,

```
int value,x;
long int array[2][3];

struct record
{
      char  name[24];
      int      id_number;
} students[100];
```

here are some of the resulting possibilities:

```
x = sizeof(int);    // this would set x=2, since an int is 2 bytes
x = sizeof(value); // this would set x=2, since value is an int
x = sizeof(long);  // this would set x=4, since a long is 4 bytes

x = sizeof(array);
            // x = sizeof(long)*array_width*array_length= 4*2*3 = 24

x = sizeof(record);
            // x = 24+2 for the character string plus the integer

x = sizeof(students);
            // x = 100 Elements *  (24 characters + sizeof(int))
            // x = 100*(24+2) = 2600 !!
```

1.11 MEMORY TYPES

The architecture of a microprocessor may require that variables and constants be stored in different types of memory. Data that will not change should be stored in one type of memory, whereas data that must be read from and written to repetitively in a program should be stored in another type of memory. A third type of memory can be used to store variable data that must be retained even when power is removed from the system. When special memory types such as pointers and register variables are accessed, additional factors must be considered.

1.11.1 CONSTANTS AND VARIABLES

The PIC microcontroller was designed using Harvard architecture, with separate address spaces for data (SRAM), program (FLASH or EPROM), and EEPROM memory. The default or automatic allocation of variables, where no memory descriptor keyword is used, is in SRAM. Constants can be placed in FLASH memory (program space) with the const keyword. For variables to be placed in EEPROM, sets of functions are provided in a library to handle the complex procedure required to get data to and from this area.

The following declarations place physical data directly in program memory (FLASH). These data values are all constant and cannot be changed in any way by program execution:

```
const long int integer_constant = 1234 + 5;
```

```
const char char_constant = 'a';
const int32 long_int_constant1 = 99L;
const int32 long_int_constant2 = 0x10000000;
const int  integer_array1[] = {1,2,3};

// The first two elements will be 1 and 2, the rest will be 0.
const int  integer_array2[10] = {1,2};

const int  multidim_array[2][3] = {{1,2,3},{4,5,6}};
const char string_constant1[] = "This is a string constant";

const struct {
               long  a;
               char b[3], c[3];
} sf = {0x000a,{0xb1,0xb2,0xb3},{0xb1,0xb2,0xb3}};
```

PIC processors, with few exceptions, ***do not*** allow for direct access to their program memory space. Some of the new FLASH-based controllers allow direct reading and writing of the memory space, but the technique is a bit complex and is performed using methods similar to EEPROM access discussed later in this section.

The PIC RISC instruction set includes an instruction, RETLW (Return Literally in W). This instruction, when called by the program, immediately returns with a constant value loaded in the W (Working) register from the program memory location where the RETLW is located. The actual constant data is part of the instruction itself. So, for example, when a constant is defined as:

```
int  const  table[5]  = { 1,3,5,7,9 };
```

The table will be put into FLASH program memory space, and you can access it as:

```
x = table[i];
```

However, the compiler, in reality, implements it something like this:

```
int table(int index) {
   switch(i) {
     case  0  : return 1;
     case  1  : return 3;
     case  2  : return 5;
     case  3  : return 7;
     case  4  : return 9;
   end;
}

x = table(i);   // x is returned a value from the function!
```

The resulting assembly code for the constant definition in memory looks something like this:

```
.................... int  const  table[5]  = { 1,3,5,7,9 };
0004:  BCF    0A.0    ; set up code space page selects
```

```
0005:   BCF     0A.1
0006:   BCF     0A.2
0007:   ADDWF   02,F     ; add the table index to the PC
0008:   RETLW   01
0009:   RETLW   03
000A:   RETLW   05       ; return the value when the PC
000B:   RETLW   07       ; jumps based on the index
000C:   RETLW   09
```

EEPROM space is a nonvolatile yet variable area of memory. Variables are allocated and accessed manually by the programmer. The use of structures, enumerations, and **#define** statements can greatly improve the readability of your code, particularly if many parameters need to be maintained in this special memory area. The CCS-PICC compiler library provides for access to the EEPROM with the following functions:

```
unsigned char read_eeprom (unsigned char address);
void write_eeprom (unsigned char address, unsigned char value);
```

In both the *read_eeprom()* and *write_eeprom()* functions, the address is a value from 0 to 255, describing which EEPROM location is to be accessed. The range of the address varies with the selection of processor. For example, the PIC16F873 has 128 bytes of onboard EEPROM, whereas the PIC16F877 has 256 bytes. Also, the data passed to *write_eeprom( )*, or returned from the *read_eeprom()* function is 8 bits in width. This means that long integers and floating point numbers require multiple fetches in order to obtain the entire value, and care must be taken that these values not be altered before they are moved in their entirety to their final location.

In a simple system you may wish to load some parameters into working variables at startup like this:

```
char current_set,prop_gain,int_gain,dif_gain; // working vars

// use an enumeration to create addresses for the values
// that have meaningful names.. it gets tough to remember
// what was in location 5.. 6 months later..
enum {t_setpoint, alarm_high, alarm_low, pid_p, pid_i, pid_d};

void load_values(void) // load values from EEPROM to SRAM
{
   current_set = read_eeprom(t_setpoint);
   prop_gain = read_eeprom(pid_p);
   int_gain = read_eeprom(pid_i);
   dif_gain = read_eeprom(pid_d);
}

void save_values(void) // save values from SRAM to EEPROM
{
```

```
        write_eeprom(t_setpoint, current_set);
        write_eeprom(pid_p, prop_gain);
        write_eeprom(pid_I, int_gain);
        write_eeprom(pid_d, dif_gain);
    }
```

If the application requires variant data of sizes like **long int, int16, float,** or **int32,** a more intelligent procedure may be necessary to move the data to and from the EEPROM. A function like the example *save_it()* below, can be called with a base address, data, and a number of bytes to move:

```
void save_it(unsigned char add, unsigned char dat,
        unsigned char size)
  {
        switch(size)
        {
           case  sizeof(float):        // 4 byte move
             write_eeprom(add+3, dat);
             write_eeprom(add+2, dat);
           case  sizeof(long):         // 2 byte move
             write_eeprom(add+1, dat);
           case  sizeof(char):         // 1 byte move
             write_eeprom(add, dat);
        }
}
```

In application, the *save_it()* function may be used like this:

```
float fancy_fraction;
char something_small;
long sixteen_bitter;

unsigned char ee_index = 0;

. . .
save_it(ee_index, fancy_fraction, sizeof(fancy_fraction));
ee_index += sizeof(fancy_fraction);
                    // save a 4 byte variable and move to the
                    // next available address

save_it(ee_index, something_small, sizeof(something_small));
ee_index += sizeof(something_small);
                    // save a 1 byte variable and move to the
                    // next available address

save_it(ee_index, sixteen_bitter, sizeof(sixteen_bitter));
ee_index += sizeof(sixteen_bitter);
                    // save a 2 byte variable and move to the
```

```
                    // next available address
    . . .
```

An expansion on this concept could lead to some very flexible, readable, and generalized access to the EEPROM area.

These permanent (FLASH) and semipermanent (EEPROM) memory areas have many system-specific uses in the embedded world. FLASH space is an excellent area for non-changing data. The program code itself resides in this region. Declaring items such as text strings and arithmetic look-up tables in this region directly frees up valuable SRAM space.

If a string is declared with an initializer like

```
char mystring[30] = "This string is placed in SRAM";
```

30 bytes of SRAM will be allocated, and the text, "This string is placed in SRAM" is physically placed in FLASH memory with the program. On startup this FLASH-resident data is copied to SRAM and the program works from SRAM whenever accessing *mystring*. This is a waste of 30 bytes of SRAM unless the string is intended for alteration by the program during run time. To prevent this loss of SRAM space, the string could be stored in FLASH memory directly by the declaration:

```
const char mystring[30] = "This string is NOT in SRAM";
```

The EEPROM area is called nonvolatile, meaning that when power is removed from the microprocessor the data will remain intact, but it is semipermanent in that the program can alter the data located in this region. EEPROM also has a life—it has a maximum number of write cycles that can be performed before it will electrically fail. This is due to the way that EEPROM memory itself is constructed, a function of electrochemistry. In many cases, this memory area will have a maximum rating of 10,000 write operations. Newer technologies are being developed all the time, increasing the number of write operation to the hundreds of thousands, even millions. There are no limitations on the number of times the data can be read.

It is important to understand this physical constraint when designing software that uses the EEPROM. Data that needs to be kept and does not change frequently can be stored in this area. This region is great for low-speed data logging, calibration tables, runtime hour meters, and software setup and configuration values.

1.11.2 REGISTER VARIABLES

The SRAM area of the PIC microcontroller includes a region called the Register File. This region contains I/O ports, timers, and other peripherals as well as some "working" or "scratch pad" area. To instruct the compiler to allocate a variable to a register or registers, the storage class modifier **register** must be used.

```
register int abc;
```

The compiler may choose to automatically allocate a variable to a register or registers, even if this modifier is not used. In order to prevent a variable from being allocated to a register or registers, the **volatile** modifier must be used. This warns the compiler that the variable may be subject to outside change during evaluation.

```
volatile int abc;
```

The **volatile** modifier is frequently used in applications where variables are stored in SRAM while the microcontroller sleeps. (Sleep is a stopped, low-power mode typically used in battery applications.) This allows the value in SRAM to be valid each time the microcontroller is awakened and returned to normal operation.

Global variables that have not been allocated to registers are stored in the General or Global Variable area of SRAM. Local variables that have not been allocated to registers are stored in dynamically allocated space in the Data Stack or Heap Space area of SRAM.

#bit and #byte

The I/O ports and peripherals are located in the register file. Therefore, special instructions are used to indicate to the compiler the difference between an SRAM location used as a variable and that of an I/O port, or another peripheral in the register file. The **#bit** and **#byte** keywords indicate to the compiler that specific addresses and access instructions are to be used to access the PIC microcontroller's I/O registers.

```
                          // define port names and bit names
#byte PORTD = 0x08
#bit   PORTD0 =   PORTD.0
#bit   PORTD1 =   PORTD.1
#bit   PORTD2 =   PORTD.2
#bit   PORTD3 =   PORTD.3
#bit   PORTD4 =   PORTD.4
#bit   PORTD5 =   PORTD.5
#bit   PORTD6 =   PORTD.6
#bit   PORTD7 =   PORTD.7

#byte PORTC = 0x07
#bit   PORTC2 = PORTC.2

void main(void)
{
      set_tris_d(0x00);        // set all of port D to output
      set_tris_c(0xFB);        // set port C bit 2 to output

      while(1)
      {
            PORTD++;          // lights!!

            if(PORTD4)
               PORTC2 = 1;
            else
```

```
                PORTC2 = 0;
        }

        delay_ms(10);
    }
```

The bit-level access to the I/O registers is allowed by using bit selectors appended after the name of the I/O register. Because bit-level access to I/O registers is done using the **BSF**, **BCF**, **BTFSC**, and **BTFSS** assembly language instructions, the definition must be performed using the #bit. For example,

```
#include <16F877.h>              // register definition file for
#fuses HS,NOWDT                  // a Microchip PIC16F877

#byte PORTA=0x05
#byte DDRA=0x85

#bit     DDRA0  = DDRA.0
#bit     DDRA1  = DDRA.1
#bit     PORTA0 = PORTA.0
#bit     PORTA1 = PORTA.1

void main(void)
{
     DDRA0    = 0;      // set bit 0 of Port A as output
     DDRA1    = 1;      // set bit 1 of Port A as input
     PORTA0   = 1;      // set bit 0 of Port A to a 1

     while(1)
     {
        if (PORTA1)     // test bit 1 input of Port A
           PORTA0 = 0;
        else
           PORTA0 = 1;
        }
}
```

Usually the **#bit** and **#byte** keywords are found in a header files included at the top of the program with the **#include** preprocessor directive. These header files are typically related to a particular processor. The header files provide predetermined names for the I/O and other useful registers in the microcontroller being used in a particular application.

Listed below is a section of an include file that might be provided with a compiler (this portion is from the CCS-PICC, 16F877.h file):

```
// CCP Functions: SETUP_CCPx, SET_PWMx_DUTY
// CCP Variables: CCP_x, CCP_x_LOW, CCP_x_HIGH
// Constants used for SETUP_CCPx() are:
```

```
#define CCP_OFF                              0
#define CCP_CAPTURE_FE                       4
#define CCP_CAPTURE_RE                       5
#define CCP_CAPTURE_DIV_4                    6
#define CCP_CAPTURE_DIV_16                   7
#define CCP_COMPARE_SET_ON_MATCH             8
#define CCP_COMPARE_CLR_ON_MATCH             9
#define CCP_COMPARE_INT                      0xA
#define CCP_COMPARE_RESET_TIMER              0xB
#define CCP_PWM                              0xC
#define CCP_PWM_PLUS_1                       0x1c
#define CCP_PWM_PLUS_2                       0x2c
#define CCP_PWM_PLUS_3                       0x3c

long CCP_1;
#byte    CCP_1     =                         0x15
#byte    CCP_1_LOW=                          0x15
#byte    CCP_1_HIGH=                         0x16
long CCP_2;
#byte    CCP_2     =                         0x1B
#byte    CCP_2_LOW=                          0x1B
#byte    CCP_2_HIGH=                         0x1C
```

Note that in this section of definitions that CCP_1 is declared long (16 bits in ICC-PICC) and that its address is defined at 0x15. Additionally, the upper and lower halves of the CCP_1 are declared independently. This allows the peripheral to be accessed either as two 8-bit values or a single 16-bit value.

1.12 REAL-TIME METHODS

Real-time programming is sometimes misconstrued as some sort of complex and magical process that can only be performed on large machines with operating systems like Linux, QNX, or Unix. Not so! Embedded systems can, in many cases, perform on a more real-time basis than a large system.

A simple program may run its course over and over. It may be able to respond to changes to the hardware environment it operates in, but it will do so in its own time. The term real-time is used to indicate that a program function is capable of performing all of its functions in a regimented way within a certain allotment of time. The term may also indicate that a program has the ability to respond immediately to outside (hardware input) stimulus.

The PIC, with its rich peripheral set, has the ability to not only respond to hardware timers, but to input changes as well. The ability for a program to respond to these real-world changes is called interrupt or exception processing.

1.12.1 USING INTERRUPTS

An interrupt is just that, an exception, change of flow, or interruption in the program operation caused by an external or internal hardware source. An interrupt is, in effect, a hardware-generated function call. The result is that the interrupt will cause the flow of execution to pause while the interrupt function, called the interrupt service routine (ISR), is executed. Upon completion of the ISR, the program flow will resume, continuing from where it was interrupted.

In a PIC, an interrupt will cause the status register and program counter to be placed on the stack, and based on the source of the interrupt, the program counter will be assigned a value from a table of addresses. These addresses are referred to as vectors.

Once a program has been redirected by interrupt vectoring, it can be returned to normal operation through the machine instruction **RETFIE** (Return from Interrupt). The **RETFIE** instruction restores the status register to its pre-interrupt value and sets the program counter to the next machine instruction following the one that was interrupted.

There are many sources of interrupts available on the PIC microcontroller. The larger the PIC, the more sources that are available. Listed below are some of the possibilities. These definitions are usually found in a header file, at the top of program, and are specific to a given PIC microprocessor. These would be found in a header for a PIC16F877, 16F877.h:

```
//////////////////////////////////////////////////////////////// INT
// Interrupt Functions: ENABLE_INTERRUPTS(), DISABLE_INTERRUPTS(),
//                      EXT_INT_EDGE()
//
// Constants used in EXT_INT_EDGE() are:
#define L_TO_H              0x40
#define H_TO_L                 0
// Constants used in ENABLE/DISABLE_INTERRUPTS() are:
#define GLOBAL                     0x0BC0
#define INT_RTCC                   0x0B20
#define INT_RB                     0x0B08
#define INT_EXT                    0x0B10
#define INT_AD                     0x8C40
#define INT_TBE                    0x8C10
#define INT_RDA                    0x8C20
#define INT_TIMER1                 0x8C01
#define INT_TIMER2                 0x8C02
#define INT_CCP1                   0x8C04
#define INT_CCP2                   0x8D01
#define INT_SSP                    0x8C08
#define INT_PSP                    0x8C80
#define INT_BUSCO                  0x8D08
#define INT_EEPROM                 0x8D10
#define INT_TIMER0                 0x0B20
```

This list contains a series of masks associated with a name describing the interrupt source. To create an ISR, the function that is called by the interrupt system, the ISR, is declared using the reserved word **#int_xxx** as a function-type modifier.

```
#int_ad
void adc_handler(void)
{
      adc_active=FALSE;
}

      or

#int_rtcc  noclear
void tmr0_isr(void)
{
      rtcc++;
}
```

The **#int_xxx** keyword is, in part, an identifier that is used to tie the service routine that follows it to the interrupt source. With a PIC processor, there is really only one interrupt service routine (two in the case of some of the higher level parts like the PIC18) physically located in program memory. The compiler organizes the defined service routines into a single package that resides within the single interrupt routine in program memory. The masks in the interrupt definitions are then used to determine the source of the interrupt in order to sort out which service routine code is to execute. If multiple sources are present, then multiple service routine functions will be executed during the single interrupt service request.

There are many possible names made available to the programmer by the compiler to tie the service routines back to the main ISR. For the CCS-PICC compiler the list of names are as follows:

```
#INT_AD          Analog to digital conversion complete
#INT_ADOF        Analog to digital conversion timeout
#INT_BUSCOL      Bus collision
#INT_BUTTON      Pushbutton
#INT_CCP1        Capture or Compare on unit 1
#INT_CCP2        Capture or Compare on unit 2
#INT_COMP        Comparator detect
#INT_EEPROM      write complete
#INT_EXT         External interrupt
#INT_EXT1        External interrupt #1
#INT_EXT2        External interrupt #2
#INT_I2C         I2C interrupt (only on 14000)
#INT_LCD         LCD activity
#INT_LOWVOLT     Low voltage detected
#INT_PSP         Parallel Slave Port data in
#INT_RB          Port B any change on B4-B7
```

```
#INT_RC          Port C any change on C4-C7
#INT_RDA         RS232 receiver data available
#INT_RTCC        Timer 0 (RTCC) overflow
#INT_SSP         SPI or I2C activity
#INT_TBE         RS232 transmit buffer empty
#INT_TIMER0      Timer 0 (RTCC) overflow
#INT_TIMER1      Timer 1 overflow
#INT_TIMER2      Timer 2 overflow
#INT_TIMER3      Timer 3 overflow
```

ISRs can be executed at any time, once the interrupt sources are initialized and the global interrupts are enabled. The ISR cannot return any values, because technically there is no "caller," and it is always declared as type **void**. It is for this same reason that nothing can be passed to the ISR.

Interrupts in an embedded environment can genuinely create a real-time execution. It is not uncommon in a peripheral-rich system to see an empty *while(1)* loop in the *main()* function. In these cases, *main()* simply initializes the hardware, and the interrupts perform all the necessary tasks when they need to happen.

```
#include <16F877.h>    // include appropriate header file
#use delay(clock=10000000)
#fuses HS,WDT          // and set the fuses

int var0,var1,var2;

#int_TIMER0
void TIMER0_isr(void)
{
   if(var0++ & 0x08)
      output_bit(PIN_D0,1);    // flash LED 0 with timer 0
   else
      output_bit(PIN_D0,0);
}

#int_TIMER1
void TIMER1_isr(void)
{
   if(var1++ & 0x08)
      output_bit(PIN_D1,1);    // flash LED 1 with timer 1
   else
      output_bit(PIN_D1,0);
}

#int_TIMER2
void TIMER2_isr(void)
```

```
{
   if(var2++ & 0x08)
      output_bit(PIN_D2,1);    // flash LED 2 with timer 2
   else
      output_bit(PIN_D2,0);
}

void main()
{
   setup_counters(RTCC_INTERNAL,RTCC_DIV_256);     // setup timer 0
   setup_timer_1(T1_INTERNAL|T1_DIV_BY_4);         // setup timer 1
   setup_timer_2(T2_DIV_BY_16,255,1);              // setup timer 2

   enable_interrupts(INT_TIMER0); // enable each timer interrupt
   enable_interrupts(INT_TIMER1);
   enable_interrupts(INT_TIMER2);
   enable_interrupts(global); // allow the PIC to be interrupted

   while(1)     // just sit here.. and let the ISRs do the work
      ;
}
```

In this example, *main()* is used to initialize the system and enable the TIMER1 overflow interrupt. The interrupt function *TIMER0_isr* blinks LED0, interrupt function *TIMER1_isr* blinks LED1, and interrupt function *TIMER2_isr* blinks LED2. Each LED blinks at a different rate due to the different physical timer settings. The *var0*, *var1*, and *var2* variables are used to scale the output to LEDs to make them slower and easier to see. Note that *main()* sits in a *while(1)* loop while all the blinking is going on.

1.12.2 STATE MACHINES

State machines are a common method of structuring a program such that it never sits idle waiting for input. State machines are generally coded in the form of a **switch/case** construct, and flags are used to indicate when the process is to move from its current state to its next state. State machines also offer a better opportunity to change the function and flow of a program without a rewrite, simply because states can be added, changed, and moved without impacting the other states that surround it.

State machines allow for the primary logical operation of a program to happen somewhat in background. Because typically very little time is spent actually processing each state, more free CPU time is left available for time-critical tasks like gathering analog information, processing serial communications, and performing complex mathematics. The additional CPU time is often devoted to communicating with humans: user interfaces, displays, keyboard services, data entry, alarms, and parameter editing.

Figure 1.1 and Figure 1.2 show an example state machine used to control an "imaginary" traffic light:

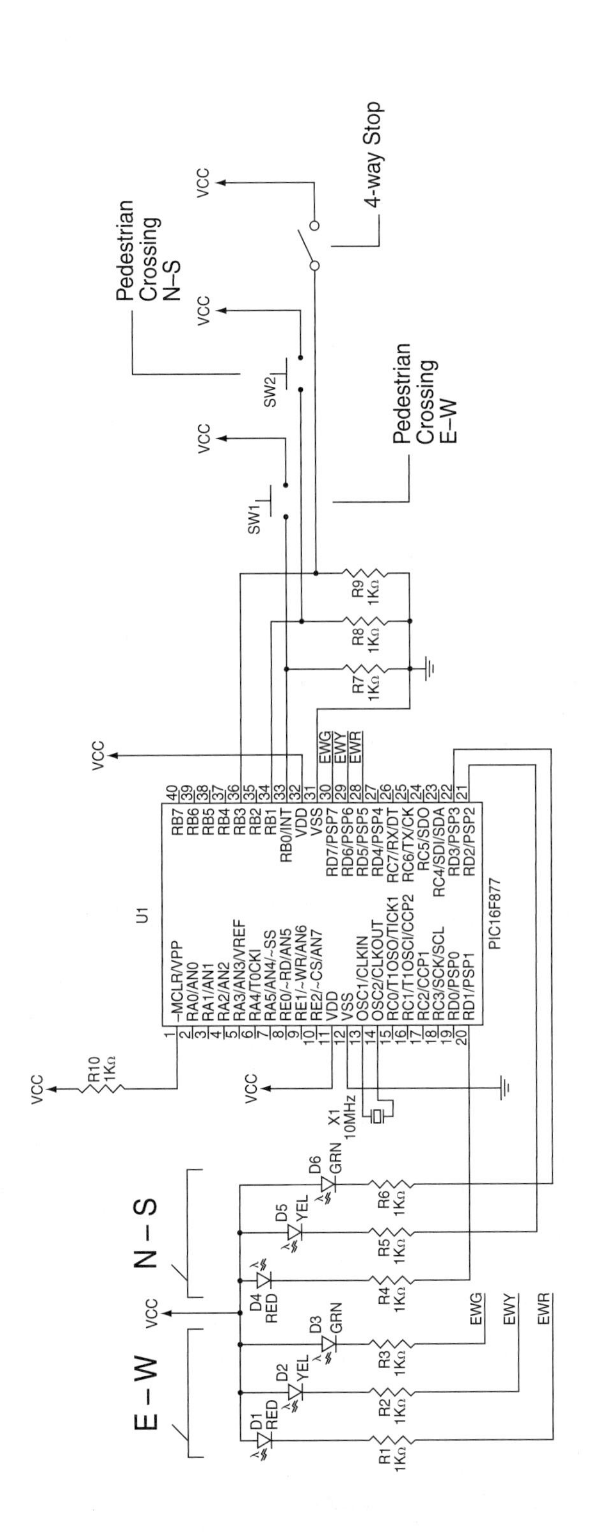

Figure 1.1 *Imaginary Traffic Light Schematic*

```
#include <16F877.h>
#use delay(clock=10000000)
#fuses HS,WDT

#byte    PORTD = 0x08                    /* output port D definition */
#bit     EW_RED_LITE     = PORTD.5    /* definitions to actual outputs */
#bit     EW_YEL_LITE     = PORTD.6    /* used to control the lights */
#bit     EW_GRN_LITE     = PORTD.7
#bit     NS_RED_LITE     = PORTD.1
#bit     NS_YEL_LITE     = PORTD.2
#bit     NS_GRN_LITE     = PORTD.3

#byte    PINB  = 0x06                  /* input port B definition */
#bit     PED_XING_EW     = PINB.0   /* pedestrian crossing push button */
#bit     PED_XING_NS     = PINB.1   /* pedestrian crossing push button */
#bit     FOUR_WAY_STOP = PINB.3     /* switch input for 4-Way Stop */

char time_left;          // time in seconds spent in each state
int  current_state;      // current state of the lights
char flash_toggle;       // toggle used for FLASHER state

// This enumeration creates a simple way to add states to the machine
// by name. Enumerations generate an integer value for each name
// automatically, making the code easier to maintain.

enum { EW_MOVING , EW_WARNING , NS_MOVING , NS_WARNING , FLASHER };

// The actual state machine is here..
void Do_States(void)
{
       switch(current_state)
       {
              case  EW_MOVING:          // east-west has the green!!
                    EW_GRN_LITE = 0;
                    NS_GRN_LITE = 1;
                    NS_RED_LITE = 0;   // north-south has the red!!
                    EW_RED_LITE = 1;
                    EW_YEL_LITE = 1;
                    NS_YEL_LITE = 1;

                    if(PED_XING_EW || FOUR_WAY_STOP)
                    {      // pedestrian wishes to cross, or
                           // a 4-way stop is required
                           if(time_left > 10)
```

Figure 1.2 *"Imaginary Traffic Light" Software (Continues)*

```
                time_left = 10;     // shorten the time
        }
        if(time_left != 0)              // count down the time
        {
                --time_left;
                return;                 // return to main
        }                             // time expired, so..
        time_left = 5;        // give 5 seconds to WARNING
        current_state = EW_WARNING;
                                    // time expired, move
        break;                      // to the next state

case    EW_WARNING:
        EW_GRN_LITE = 1;
        NS_GRN_LITE = 1;
        NS_RED_LITE = 0;   // north-south has the red..
        EW_RED_LITE = 1;
        EW_YEL_LITE = 0;   // and east-west has the yellow
        NS_YEL_LITE = 1;

        if(time_left != 0)              // count down the time
        {
                --time_left;
                return;                 // return to main
        }                  // time expired, so..
        if(FOUR_WAY_STOP) // if 4-way requested then start
                current_state = FLASHER; // the flasher
        else
        {                           // otherwise..
                time_left = 30;     // give 30 seconds to MOVING
                current_state = NS_MOVING;
        }                           // time expired, move
        break;                      // to the next state

case    NS_MOVING:
        EW_GRN_LITE = 1;
        NS_GRN_LITE = 0;
        NS_RED_LITE = 1;   // north-south has the green!!
        EW_RED_LITE = 0;   // east-west has the red!!
        EW_YEL_LITE = 1;
        NS_YEL_LITE = 1;

        if(PED_XING_NS || FOUR_WAY_STOP)
        {       // if a pedestrian wishes to cross, or
```

Figure 1.2 *"Imaginary Traffic Light" Software (Continues)*

```
            // a 4-way stop is required..
            if(time_left > 10)
                  time_left = 10;    // shorten the time
      }
      if(time_left != 0)             // count down the time
      {
            --time_left;
            return;                  // return to main
      }                  // time expired, so..
      time_left = 5;     // give 5 seconds to WARNING
      current_state = NS_WARNING;    // time expired, move
      break;                           // to the next state

case  NS_WARNING:
      EW_GRN_LITE = 1;
      NS_GRN_LITE = 1;
      NS_RED_LITE = 1;   // north-south has the yellow..
      EW_RED_LITE = 0;
      EW_YEL_LITE = 1;   // and east-west has the red..
      NS_YEL_LITE = 0;

      if(time_left != 0)                // count down the time
      {
            --time_left;
            return;                  // return to main
      }                  // time expired, so..
      if(FOUR_WAY_STOP) // if 4-way requested then start
      current_state = FLASHER; // the flasher
      else
      {                          // otherwise..
            time_left = 30;      // give 30 seconds to MOVING
            current_state = EW_MOVING;
                                 // time expired, move
      break;           // to the next state

case  FLASHER:
      EW_GRN_LITE = 1;   // all yellow and
      NS_GRN_LITE = 1;   // green lites off
      EW_YEL_LITE = 1;
      NS_YEL_LITE = 1;

      flash_toggle ^= 1;         // toggle LSB..
      if(flash_toggle & 1)
      {
```

Figure 1.2 *"Imaginary Traffic Light" Software (Continues)*

```
                        NS_RED_LITE = 0;  // blink red lights
                    EW_RED_LITE = 1;
                    }
                    else
                    {
                        NS_RED_LITE = 1;  // alternately
                    EW_RED_LITE = 0;
                    }
                    if(!FOUR_WAY_STOP)      // if no longer a 4-way stop
                        current_state = EW_WARNING;
                    break;                  // then return to normal

                default:
                    current_state = NS_WARNING;
                    break;      // set any unknown state to a good one!!
        }
}

void main(void)
{
   port_b_pullups(TRUE);
   set_tris_b(0xFF); // port B is all input
   set_tris_d(0x00); // port D is all output

   current_state = NS_WARNING;        // initialize to a good starting
                                      // state (as safe as possible)
   while(1)
   {
        delay_ms(250);  // 1 second delay.. this time could
        delay_ms(250);  // be used for other needed processes
        delay_ms(250);
        delay_ms(250);

        Do_States();    // call the state machine, it knows
                        // where it is and what to do next
   }
}
```

Figure 1.2 *"Imaginary Traffic Light" Software (Continued)*

The state machine in this example uses PORTD to drive the red, green, and yellow lights in the North-South and East-West directions in a current-sink mode (0=ON, 1=OFF). Note that in *main()* the *delay_ms()* function is used to control the time. This keeps the example simple. In real life, this time could be used for a myriad of other tasks and functions, and the timing of the lights could be controlled by an interrupt from a hardware timer.

The state machine *Do_States()* is called every second, but it executes only a few instructions before returning to *main()*. The global variable *current_state* controls the flow of the execution through *Do_States()*. The *PED_XING_EW*, *PED_XING_NS*, and *FOUR_WAY_STOP* inputs create exceptions to the normal flow of the machine by altering either the timing, the path of the machine, or both.

The example system functions as a normal stoplight. The North-South light is green when the East-West light is red. The traffic is allowed to flow for thirty-second periods in each direction. There is a five-second yellow warning light during the change from a green light to a red light condition. If a pedestrian wishes to cross the flowing traffic, pressing the *PED_XING_EW* or *PED_XING_NS* button will cause the time remaining for traffic flow to be shortened to no longer than ten seconds before changing directions. If the *FOUR_WAY_STOP* switch is turned on, the lights will convert over to all four flashing red. This conversion will happen only during the warning (yellow) light transition to make it safe for traffic.

User interfaces, displays, keyboard services, data entry, alarms, and parameter editing can also be performed using state machines. The more "thinly sliced" a program becomes through the use of flags and states, the more real time it is. More things are being dealt with continuously without becoming stuck waiting for a condition to change.

CHAPTER SUMMARY

This chapter has provided a foundation for you to begin writing C language programs.

The beginning concepts demonstrated the basic structure of a C program. Variables, constants, enumerations, their scope and construction, both simple, as well as in arrays, structures, and unions, have been shown to be useful in defining how memory is to be allocated and the data within that memory is to be interpreted by a C program.

Expressions and their operators, including I/O operations, were discussed to provide a basis for performing arithmetic operations and determining logical conditions. These operations and expressions were also used with control constructs such as **while** and **do/while** loops, **for** loops, **switch/case**, and **if/else** statements to form functions as well as guide the flow of execution in a program.

The advanced concepts of real-time programming using interrupts and state machines were explored to demonstrate how to streamline the execution of programs, improve their readability, and provide a timely control of the processes within a C language project.

1.14 EXERCISES

1. Define the terms variable and constant (Section 1.4).

*2. Create an appropriate declaration for the following for both Standard Embedded C sizes and the CCS-PICC compiler (Section 1.4):

   A. A constant called 'x' that will be set to 789.

   B. A variable called 'fred' that will hold numbers from 3 to 456.

   C. A variable called 'sensor_out' that will contain numbers from –10 to +45.

   D A variable array that will have 10 elements each holding numbers from –23 to 345.

   E. A character string constant that will contain the string 'Press here to end'.

   F. A pointer called 'array_ptr' that will point to an array of numbers ranging from 3 to 567.

   G Use an enumeration to set 'uno', 'dos', 'tres' to 21, 22, 23, respectively.

3. Evaluate the following for both Standard Embedded C sizes and the CCS-PICC compiler (Section 1.6):

   A. unsigned char t; t = 0x23 * 2; // t = ?

   B. unsigned int t; t = 0x78 / 34; // t = ?

   C. unsigned char x; x = 678; // x = ?

   D. char d; d = 456; // d = ?

   E. enum {start = 11, off, on , gone}; // gone = ?

   F. x = 0xc2; y = 0x2; z= x ^ y; // z = ?

   G. e = 0xffed; e = e >> 4; // e = ?

   H. e = 27; e += 3; // e = ?

   I. f = 0x90 | 7; // f = ?

   J. x = 12; y = x + 2; // y = ? x = ?

*4. Evaluate as true or false as if used in a conditional statement (Section 1.6):

   For all problems: x = 0x45; y = 0xc6

   A. (x == 0x45)

   B. (x | y)

   C. (x > y)

   D. (y – 0x06 == 0xc){end LL}

5. Evaluate the value of the variables after the fragment of code runs for both Standard Embedded C sizes and the CCS-PICC compiler (Section 1.7):

```
unsigned char loop_count;
unsigned int value = 0;
for (loop_count = 123; loop_count< 133;loop_count++)
        value += 10;
                 //value = ??
```

*6. Evaluate the value of the variables after the fragment of code runs (Section 1.7):

```
unsigned char cntr = 10;
unsigned int value = 10;

do
{
     value++;
} while  (cntr < 10);
             // value = ??       cntr = ??
```

7. Evaluate the value of the variables after the fragment of code runs (Section 1.7):

```
unsigned char cntr = 10;
unsigned int value = 10;

while  (cntr < 10)
{
    value++;
}
       // value = ??       cntr = ??
```

8. Given: unsigned char num_array[] = {1,2,3,4,5,6,7,8,9,10,11,12}; (Section 1.9):

```
//  num_array[3] = ??
```

9. Given: unsigned int *ptr_array; (Section 1.9):

```
unsigned  int num_array[] = {10,20,30,40,50,60,70};
unsigned int x,y;
ptr_array = num_array;
x = *ptr_array++;
y = *ptr_array++;
    // x= ??      y = ??       ptr_array = ??{end UL}
```

*10. Write a fragment of C code to declare an appropriate array and then fill the array with the powers of 2 from 21 to 26 (Section 1.7).

"*" preceding an exercise number indicates that the question is answered or partially answered in the appendix.

1.15 LABORATORY ACTIVITIES

1.

A. Create, compile, and test a program to print 'C Rules and Assembler Drools!!' on the standard output device.

B. Modify the program to use a **for** loop to print the same phrase four times.

C. Repeat using a **while** loop to print the phrase three times.

D. And yet again using a **do/while** to print the phrase five times.

2. Create a program to calculate the capacity of a septic tank. The three dimensions height, width, and length should be read from three parallel Ports A, B, and C (in feet) and the result printed to the standard output device. The maximums the program should handle are 50 feet of width, 50 feet of length, and 25 feet of depth (it's a big tank!). Use appropriate variable sizes and use casting where required to correctly calculate the capacity and print the result as follows (x, y, z, and qq should be the actual numbers):

```
The capacity of a tank x feet long by y feet high by z
feet wide is qq cubic feet of 'stuff'.
```

The program should run once, print the results, and stop.

3. Modify the program in Activity 2 so that if any of the three inputs exceeds the allowed range, the program says

```
The input exceeds the program range.
```

4. Create a program to print the powers of 2 from 2^0 to 2^{12}.

5. Use a **switch/case** construct to create a program that continuously reads Port B and provides output to Port D according to Table 1.11:

Input	*Output*
0x00	0b00000000
0x01	0b00001111
0x02	0b11110000
0x03	0b01010101
0x04	0b10101010
0x05	0b11000011
0x06	0b00111100
0x07	0b10000001
other	0b11111111

Table 1.11 *Switch/Case Exercise*

6. Create a program to determine the order of storage of long variables according to the suggestion in Section 1.10, "Structures and Unions," relative to unions (near the middle of the section).
7. Create a program that can search an array of structures for specific values. Declare an array of structures containing the members 'month,' 'day,' and 'year.' Initialize the array to contain one and only one instance of your own birthday. However, the month, day, and year of your birthday must show up in other places in the array, just not all at the same point. The program must search the array to find your birthday and then print a message to indicate the index in the array where the birthday resides.
8. Modify the program in Activity 7 to search the array of structures using pointers.
9. Create a program to use the standard input function *getchar()* to retrieve one character and compare its value to the value placed on Port B. If they match, print a message to that effect on the standard output device using *printf()*. Otherwise, output a 0x01 to Port D if Port B is higher in value than the character received, or output a 0x80 if Port B is lower.
10. Use an external hardware interrupt to print a message to the standard output device each time an interrupt occurs. Also print the number of times the interrupt has occurred. You ***may not*** use a global variable to count the interrupt occurrences.

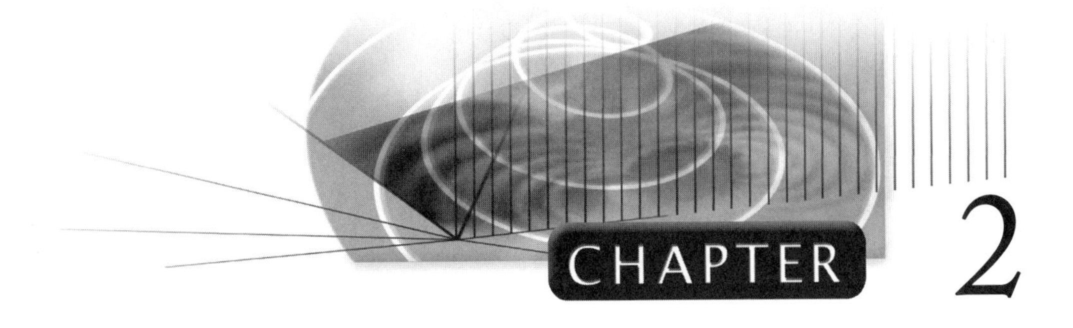

PIC Microcontroller Hardware

2.1 OBJECTIVES

At the conclusion of this chapter, you should be able to:

- Define, describe the uses of, and identify the memory spaces and registers of the PIC microcontrollers.
- Allocate the microcontroller resources to a project.
- Apply interrupts to C programs.
- Write C programs to complete simple I/O tasks.
- Apply the major peripherals of the PIC microcontrollers to solve interface problems to electronic devices.
- Write C programs to accomplish serial communication.
- Read and determine the results of PIC assembly language programs.

2.2 INTRODUCTION

This chapter provides the hardware background needed by an engineer to successfully apply the Microchip PIC microcontrollers to a problem. It is based on the premise that in order to be successful, a designer must be as intimately familiar with the hardware as she or he is with the programming language.

The chapter begins with the basic overview of the processor, its architecture, and its functions. Next, the memory spaces are discussed in detail so that you can identify and understand where your variables are stored and why. The uses of interrupts are discussed along with the most common built-in peripherals. These topics are then combined into the process of allocating processor resources for a project.

A variety of the peripherals common to many of the PIC microcontrollers are discussed. Many of the PIC parts have further peripherals and /or more functions in their peripherals. Consult the specification for your particular part to discover the full range of peripheral functionality for your particular microcontroller. Each of the peripherals is shown as they apply to projects. Lastly, the assembly language is described as it pertains to understanding the microcontroller's functions.

2.3 ARCHITECTURAL OVERVIEW

Although the PIC series of microcomputers may be used as either general purpose microcomputers or as microcontrollers, they are most usually used as microcontrollers.

A microcontroller is a complete computer system optimized for hardware control that encapsulates the entire processor, memory, and all of the I/O peripherals on a single piece of silicon. Being on the same piece of silicon means that speed is enhanced because internal I/O peripherals take less time to read or write than do external devices.

The following figure shows the internal architecture of the PIC16F877. The architecture is typical of microcontrollers in that it has an internal bus structure to communicate with the many built-in peripherals.

Being optimized for hardware control means that the machine level instruction set provides convenient instructions to allow easy control of I/O devices; that is, there will be instructions to allow setting, clearing, or interrogating a single bit of a parallel I/O port or a register. Being able to set or clear a bit is typically used to turn a hardware device off or on. In a more usual processor not optimized for hardware control, the setting, clearing, or reading the individual bits of an I/O port (or any other register) would require additional AND, OR, XOR, or other instructions to affect a single bit of the port.

As an example, consider the following code fragment that sets the third bit of Port A high:

```
output_high(PIN_A2);        //set bit 2 high
```

Although this line of code is correct for use with either a standard microprocessor or microcontroller, the actual results are different for the two devices.

The following snippet shows some assembly code generated for a standard microcomputer and for a microcontroller to execute the line shown above:

```
Standard microcomputer (from a non-PIC microcomputer)
   IN    R30,0x1B      ;read the current contents of the Port
   MOV   R26,R30       ;get the data into a useful register
   LDI   R30,LOW(4)    ;set bit 4 in the output register
   OR    R30,R26       ;Logically OR the previous contents of the port
   OUT   0x1B,R30      ;output the new data to the port register
Microcontroller:
   BSF   08.0          ;set the desired bit high
```

As shown above, the standard microprocessor takes five instructions to set the bit high, whereas the microcontroller takes only one. This equates to a 5 times speed increase for port oper-

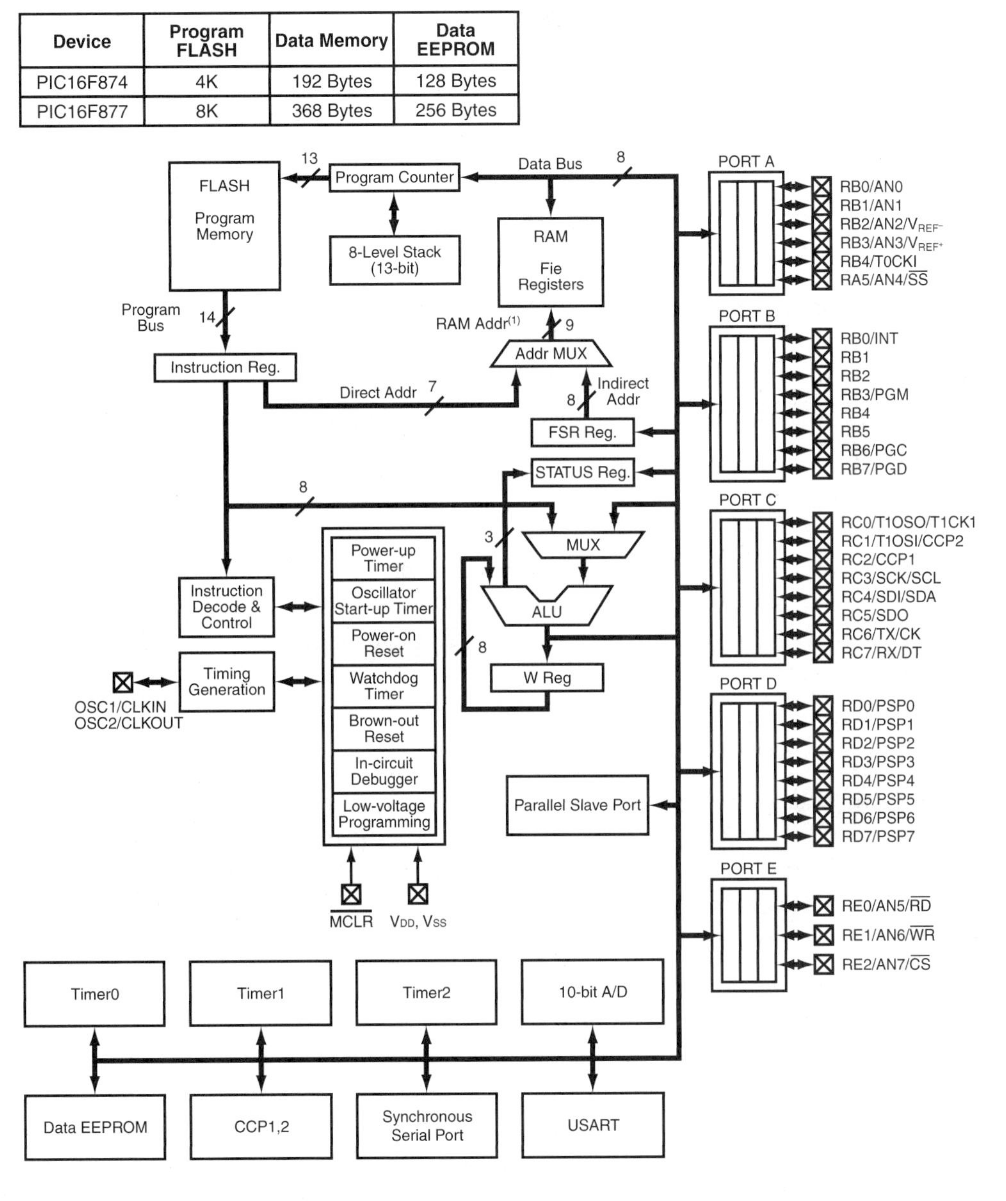

Device	Program FLASH	Data Memory	Data EEPROM
PIC16F874	4K	192 Bytes	128 Bytes
PIC16F877	8K	368 Bytes	256 Bytes

PIC16F877 Internal Architecture. (Reprinted with permission from MicroChip)

ations in the microcontroller and approximately a 5 times decrease in code size for these operations. Microcontrollers allow for reduced code size and increased execution speed by providing instructions that are directly applicable to hardware control.

2.4 MEMORY ORGANIZATION

PIC microcontrollers employ Harvard-style memory architecture. Each of the memory spaces is interfaced on a separate bus structure. The advantage to this arrangement is that access to the memory is faster. In the PIC microcontrollers, the data and code memory can actually be accessed simultaneously to further increase processing speed.

The memory system being discussed is one that is typically found in a PIC processor. The exact memory details for any particular processor will be found in the specification for that processor.

2.4.1 DATA MEMORY

The data memory of PIC processors is composed of two distinct areas: *general purpose registers* and *special function registers*. The general purpose registers are used to store intermediate data and variables, whereas the special function registers are used to control the operation of the microcontroller.

Data memory in the PIC coprocessors is typically broken up into a number of banks. Each bank contains some general purpose registers and some special function registers. The exact manner in which the data memory is broken up and the method by which the general purpose registers and the special function registers are distributed will vary widely among the PIC processors. Fortunately, the C language compiler will take care of any problems caused by the manner in which the memory is broken up. Such information will only be required when doing assembly language programming.

2.4.2 FLASH MEMORY

The FLASH code memory section is a contiguous block of FLASH memory that starts at location 0x0000 and is of a size that is dependent on the particular PIC microcontroller in use. The FLASH code memory is *nonvolatile* (it retains its data even when power is removed) memory, and it is used to store the executable code and constants because they must remain in the memory even when power is removed from the device.

In the 'PIC14' and 'PIC16' series devices, the code memory is *paged*. That is, the code memory is divided into pages of a smaller size than the entire code memory. The *program counter*, the register that determines what instruction is executed next, is limited to, say, 12 bits. Such a register can only address 2^{12} address locations and so the processor would have pages of $2^{12} = 4096$ bytes. The processor cannot execute code that lies across page boundaries, making it difficult to locate and fit the code into the code memory space and still keep each code segment within page boundaries. The '18' series of the PICC devices have a single contiguous memory space and a program counter sized to address the entire space (i.e., a 21-bit program counter that can address a 2 MB space). In any event, one of the many advantages to the CCS-PICC compiler is that it takes care of the code paging issues in those processors with paged memory.

The PIC microcontroller FLASH memory contains all of the executable instructions and the vectors for RESET and for interrupt(s). The block of FLASH typically looks like the one shown in Figure 2.1.

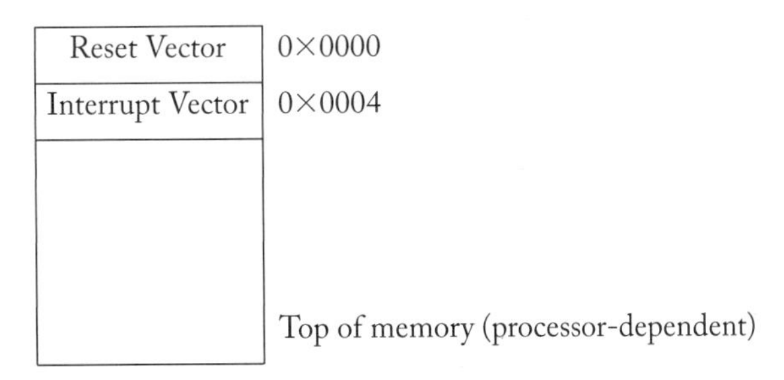

Figure 2.1 *FLASH Memory Map*

Although the FLASH memory can be programmed and reprogrammed with executable code, there is only limited provision to write to the FLASH via an executable program; it is usually programmed by external means. Consequently, it is viewed as read-only memory from the programmer's perspective and is, therefore, only used to store constant-type variables along with the executable code.

2.4.3 RETURN ADDRESS STACK

The return address stack is a set of registers used to store return addresses for function calls and for interrupts. In the PIC processors, a special memory segment contains the stack. This tiny area of memory is limited to, for instance, thirty-two cells in the PIC18 series of devices, and only eight cells in the PIC16 series. The return address stack cells have the same width (number of bits as the program counter in the processor). This feature is necessary to store whatever size address the processor can handle.

Chapter 2 Example Project – Part A

This is the first part of a project example that will be used to further demonstrate the concepts presented in this chapter. The example project will center on a data collection system for a stock car according to the following scenario:

Assume that your boss hands you a data collection system based on a PIC18F458 microcontroller. He explains that this system was designed to collect data for a drag racer over a 15-second run, but that he would like you to modify the system to collect data once per second over an entire 2-minute lap for his stock car. The collected data is then uploaded to a PC for analysis. The system was originally designed to record engine rpm based on a pulse from an engine rpm sensor, drive shaft rpm from a magnetic sensor reading a magnet attached to the drive shaft, and engine temperature from an RTD thermocouple attached at the cylinder head. Your boss explains that he tried to have the programming modified once before, but, unfortunately, the inept programmer only succeeded in erasing the entire program (he didn't have access to this text!). Happily for you, this means that all of the conditioning circuitry for the engine signals exists and that you only need to provide a program to collect the data and send it to the PC.

The first task for you to do is to verify that the processor has sufficient resources to do the job.

First, we verify that the memory is big enough to hold the data and processor variables. The data is to be saved every second for 2 minutes, so the number of sets of data is:

2 minutes * 60 seconds/minute *1 data set / second = 120 sets of data

So what you now need to know is the number of bytes in each set of data. Engine rpm is in the range of 2000–10,000 (according to your boss), shaft rpm is in the range of 1000–2000 rpm, and engine temperature is in the range of 100–250°F. Your boss specifies that he wants the rpm accurate to 100 rpm and the temperature accurate to 1°F.

Therefore, you will be storing two integer-sized numbers (2 bytes each for a total of 4 bytes) for the two rpm readings and one byte-size number for the temperature in each set of data. The total storage needed is:

120 sets of data * 5 bytes /set = 600 bytes

In addition to the data storage, you will need about 100 bytes for miscellaneous data storage and the like. Checking the PIC18F458 specification, you discover that the microcontroller has 11,536 bytes of SRAM memory, so the memory resources are sufficient for the task (whew!). Normally, at this point, you would also consider whether the peripheral resources are sufficient as well. In this case, however, you don't know about the peripherals yet and, besides, the system worked for drag racers, so the peripheral resources must be there.

EXAMPLE

2.5 INTERRUPTS AND RESET

Interrupts, as discussed in Chapter 1, Embedded C Language Tutorial, are essentially hardware-generated function calls. The purpose of this section is to describe how the interrupts function and how *RESET*, which is a special interrupt, functions.

Interrupts, as their name suggests, *interrupt* the flow of the processor program and cause it to branch to an *interrupt service routine* (ISR) that does whatever is supposed to happen when the interrupt occurs. Interrupts are useful for those situations in which the processor must immediately respond to the interrupt or in those cases where it is very wasteful for the processor to poll for an event to occur. Examples of the need for immediate response include using interrupt to keep track of time (the interrupt may provide a tick for the clock's time base) or an emergency off button that *immediately* stops a machine when an emergency occurs.

Unlike some microcontrollers, the PIC microcontrollers do not treat the sources of interrupts as separate calls. There is typically only one interrupt vector (the address to which all interrupts are directed) and one reset vector. (The PIC18 family has two interrupt vectors for high and low priority interrupts.)

It is the job of the programmer to provide an ISR that sorts out which interrupt is demanding service and to direct the program flow to the code appropriate to the interrupt that occurred. Fortunately, the CCS-PICC compiler provides the basic ISR for you based on which functions of your program are defined as the service routines for the various interrupts; that is, you do not need to provide the "sorting out" code because you are using the CCS-PICC compiler.

When the interrupt occurs, the *return address* (the address of the next instruction to be executed when the interrupt is finished) is stored on the return address stack. The last instruction in the interrupt service routine is an RETFIE assembly language instruction, which is a *return from interrupt*. This instruction causes the return address to be popped off the stack and program execution to continue from the point at which it was interrupted.

The interrupts, as with all of the PIC peripherals, are actually controlled by the bits of one or more *special function control registers*. The bits may be manually set or cleared by the programmer to cause the peripherals (the interrupts, in this case) to behave as desired. However, in most cases, programmers find that the built-in functions provided by the CCS-PICC compiler are a more convenient way to control the peripherals. In the case of the interrupts in a PIC16F877, the registers of interest include the OPTION_REG register; the INTCON register; and the PIE1, PIE2, PIR1 and PIR2 registers. CCS-PICC handles the masking and unmasking for the interrupts as necessary for each peripheral device.

The example hardware and software shown in Figure 2.2 and Figure 2.3, respectively, create a rudimentary stopwatch. The program uses Timer1 to create interrupts at a 1-millisecond rate to keep track of time. A low-going signal attached to the external interrupt is used to start and to display the time (in seconds in Port D) and to restart the timing for the next period.

The program in Figure 2.3 implements the example stopwatch. Note particularly the *#int_TIMER1* and *#int_EXT* lines. These tell the compiler that the following functions are

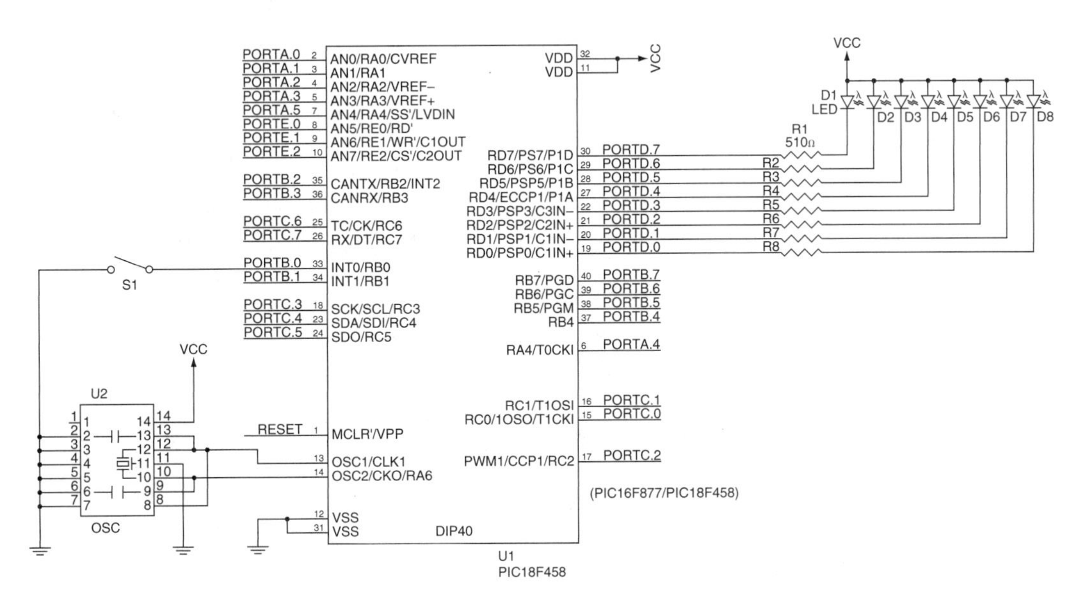

Figure 2.2 *Interrupt Example Hardware*

```
#include <16F877.h>          // include appropriate header file
#use delay(clock=10000000)
#fuses HS,NOWDT                         // and set the fuses

int16 msecs;    //milliseconds counter
int secs;       //elapsed seconds counter

#int_TIMER1     //denote the following function as the ISR for
                //Timer1 overflow
void TIMER1_isr(void)
{
   set_timer1(63036);
   if (++msecs == 1000)
   {
      if (++secs == 60) secs = 0;
      msecs = 0;
   }
}

#int_EXT    //the following function is the ISR for a falling edge on
INT0
void ext_isr(void)
{
   output_D(~secs);
   set_timer1(63036);  //reset to start timing
   msecs = 0;     //clear milliseconds counter
   secs = 0;     //clear seconds counter
   delay_ms(100);    //wait for switch bounce to stop
}

void main()
{
   port_b_pullups(TRUE);  //activate puulups on port B for ext interrupt
   ext_int_edge(0,H_TO_L);  //set external interrupt to falling edge
   setup_timer_1(T1_INTERNAL|T1_DIV_BY_1);       // setup timer 1
   enable_interrupts(INT_EXT); // enable each interrupt
   enable_interrupts(INT_TIMER1);
   enable_interrupts(global); // allow the PIC to be interrupted

   while(1)      // just sit here.. and let the ISRs do the work
      ;
}
```

Figure 2.3 *Interrupt Example Software*

the ISRs for the Timer0 overflow interrupt and for the external interrupt. Also note that each interrupt must be enabled by a statement, such as *enable_interrupts(INT_EXT)*. The labels for the individual interrupts may be found by clicking the "valid interrupt" button on the CCS-PICC IDE and then selecting your processor.

The *enable_interrupts(global)* line is used to enable all the interrupts at one time. To be active, any individual interrupt must be enabled by its own *enable_interrupt* line. All the interrupts also must be enabled globally. *The global enable only turns on those interrupts that have been enabled by their own enable interrupt line.* The global enable/disable is provided to allow the programmer an easy way to turn the enabled interrupts on or off as needed by a single statement.

The PIC18 devices have two interrupts that function in a manner similar to those described above. The major difference is that one of the interrupts is a high-priority interrupt and the other one is a low-priority interrupt. This arrangement gives the programmer more control over critical operations. The high-priority interrupt can interrupt either the regular program or a low-priority interrupt, but the low-priority interrupt can only interrupt the regular program code. This means the high-priority interrupt should be used for the most critical operations.

The balance of the example program will become clear as the other peripherals are discussed in succeeding sections.

2.5.1 RESET

RESET is a special form of interrupt in that it causes the processor to stop what it is doing and to restart at location zero. This, of course, completely restarts the program.

2.6 I/O PORTS

Of the I/O devices, the parallel I/O ports are the most commonly used. Each of the ports has two associated I/O control registers: PORTx and TRISx. The *PORTx* (where 'x' is A, B, C, D, etc., depending on the specific device) register is used to read or to write via the I/O ports and the associated *TRISx* register. The *TRISx* register is used to control whether each bit of the port is used for input or for output. Writing a logic '1' to bit 2 of *TRISA*, for example, will make bit 2 of *PORTA* act as an input. Writing a '0' into a bit of the TRISx register will cause the corresponding bit of PORTx to act as an output.

Although it is possible for the programmer to handle all the writes to the TRIS registers (something that might be done to increase processing speed), CCS-PICC does most of the work for you.

For example, refer to the following line from the example program in Figure 2.2:

```
output_D(~secs);
```

This causes CCS-PICC to include instructions to set all of TRISD to logic 0 so all of the PORTD bits are set to output data and then to output the data passed to the output function. In this case the data is being used to light LEDs on PORTD that are turned on by a logic 0, and the '~' (1s complement) is used to complement the data so a lighted LED represents a logic 1.

Similarly, CCS-PICC provides functions to control individual port bits. For example:

```
output_bit(PIND_0,1);
```

would cause PORTD bit 0 to set for output (via TRISD) and a logic 1 to be written to the PORTD bit 0 data latch and hence to the port pin.

PORTB is slightly different in that it is possible to engage weak pull-ups on the PORTB pins when they are used for input. This is accomplished by writing a logic 0 to the control bit RBPU. CCS-PICC provides a function to handle the pull-ups as shown below:

```
port_b_pullups(TRUE); //note 'true' will activate
                      // pull-ups, 'false' will deactivate them.
```

Figure 2.4 and Figure 2.5 show the hardware and software, respectively, of a program to illustrate the use of the I/O ports. The scenario is that you are asked to create a simple matching quiz game. In this game the contestant puts a probe on a port bit representing the question being answered and another probe on an input bit representing the answer to the question. If the question and answer match, a green LED is lit; if not, a red LED is lit. It is desirable that the LEDs only light when both a question and an answer bit are touched. The quiz board is shown below (Note: P1 connects to the contact points and D4 and D5 are the red and green LEDs):

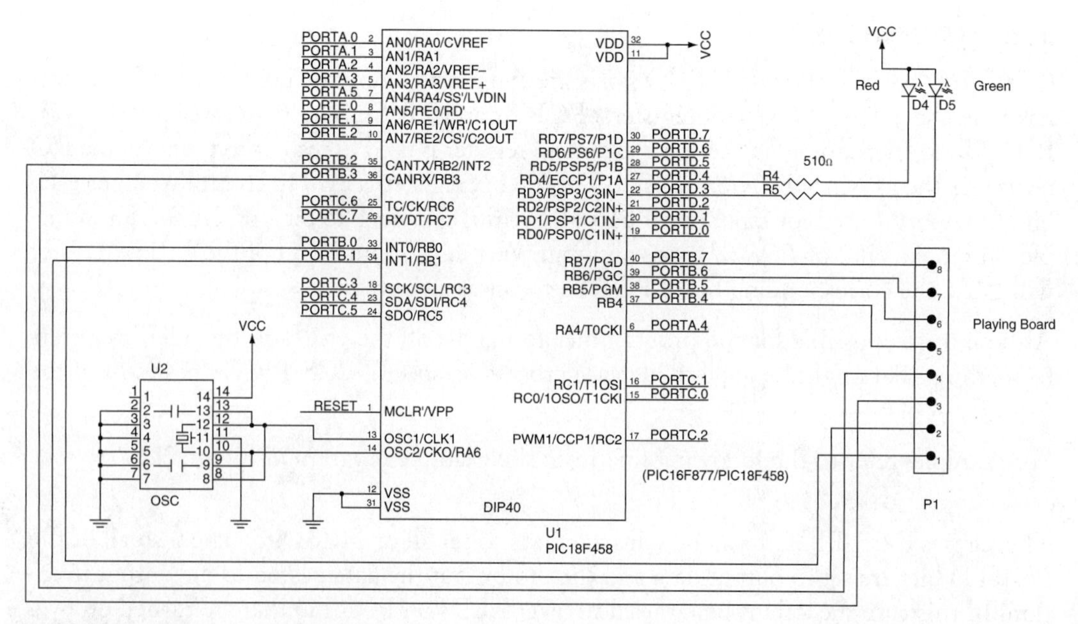

Figure 2.4 *Quiz Game Hardware*

```
#16F877.h>// include appropriate header file
#use delay(clock=10000000)
#fuses HS,NOWDT                    // and set the fuses

int correct;

void main()
{
   port_b_pullups(TRUE);  //activate puulups on port B
   //Note 'question' and 'answer' probes tie a ground to the PORTB pins

   while(1)     //In this program all the work happens in main
   {
      //check for both a question and an answer being pressed.
      if (((input_B() & 0xF) != 0xF) && ((input_B() & 0xF0) != 0xF0))
      {
         if (!input(PIN_B0) && !input(PIN_B5)) correct = 1;
         else if (!input(PIN_B1) && !input(PIN_B7)) correct = 1;
         else if (!input(PIN_B2) && !input(PIN_B4)) correct = 1;
         else if (!input(PIN_B3) && !input(PIN_B6)) correct = 1;
         else correct = 0;          //no correct answer

         if (correct)   output_bit(PIN_D3,0);  //its right!
         else output_low(PIN_D4);      //not right!
      }
      else output_D(0xff);    //extinguish both LED's
   }
      ;
}
```

Figure 2.5 *Quiz Game Software*

(PORTD Bit 3) ? Correct !!	Ooops ? (PORTD bit 4)
(PORTB Bit 0) ? 2 + 2 = ??	6 ? (PORTB Bit 4)
(PORTB Bit 1) ? 2 + 3 = ??	4 ? (PORTB Bit 5)
(PORTB Bit 2) ? 2 + 4 = ??	3 ? (PORTB Bit 6)
(PORTB Bit 3) ? 2 + 1 = ??	5 ? (PORTB Bit 7)

The software in Figure 2.5 shows a variety of the input and output functions for use with the parallel port I/O. The *input_B()* function reads all of PORTB and then is masked by the '& 0xF'. Another *input_B()* function is used to get the PORTB data and so that the upper nibble can be separated out in a like manner. Separating the nibbles allows each half to be independently checked for a probe touch. The probes are connected to ground so a probe touch provides a logic 0 input to one port pin.

The *input(PIN_xx)* function is used to interrogate an individual port pin. These are used in the *if* statements to see if a correct pair is being touched simultaneously. Any correct pair sets the *correct* variable to 1. The final *else* clears the *correct* variable if no correct answer is found.

The bottom *if-else* is used to interrogate the *correct* variable and light the appropriate LED. Two forms of output functions are shown. The *output_bit(PIN_xx, y)* is used to output a 0 to PORTD bit 3 to light the green LED in the case of a correct result. A slightly different form, the *output_low(PIN_xx)* is used to light the red LED in the case of an incorrect result. The *output_bit* function would be more appropriate where a variable is included in the function as the level to be output, and the *output_low* function would be more appropriate where a 0 is always to be output.

Notice that the *port_b_pullups(true)* function is used so that external pull-ups are not required on the inputs. Also it is interesting to note that if the answer bits are all connected together, the answer will always show as correct. Why?

Many of the port bits have alternate functions. That is, they can function either as simple I/O or as a special bit. PORTB bit 0, for instance, is the input for interrupt 0 (see the preceding section). The alternate functions will be discussed as they are used with each peripheral device.

2.6.1 PARALLEL SLAVE PORT MODE

Parallel slave port (PSP) is a special mode in which Port D is made to act like a register that can be attached to another microcontroller's bus. In this way, the other microcontroller and the PIC device can exchange data and/or commands as appropriate to the needs of the system.

PSP mode implements chip select (CS), Read (RD), and Write (WR) signals exactly as they would be implemented on a memory or peripheral device that is designed to be attached to a microcontroller bus. The external system writes to the PSP port by asserting WR and CS low or reads from the PSP port by asserting RD and CS low. Figure 2.6 shows the signals that control data transfer via a PSP port.

Figure 2.7 shows the block diagram of a system to illustrate PSP usage. One PIC is configured to operate in PSP mode, and another is configured to provide the control signals to communicate with the PSP-mode PIC.

Figure 2.7 shows a pair of PICs configured to demonstrate PSP operation. The PSP PIC reads the digital input from its Port B and transfers the data to the PSP port. The test PIC continuously reads the PSP port and then echoes the value read back to the PSP port. The PSP PIC takes whatever data it receives via the PSP port, inverts it (to drive sink-mode LEDs correctly), and drives it out Port C. That is, the PSP device reads the data on Port B, allows it to read via the PSP port, and then takes whatever data it gets via the PSP port and outputs the data through Port C.

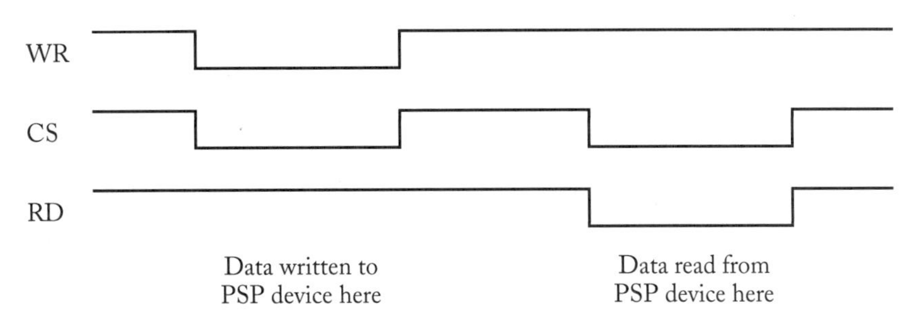

Figure 2.6 *PSP Control Waveforms*

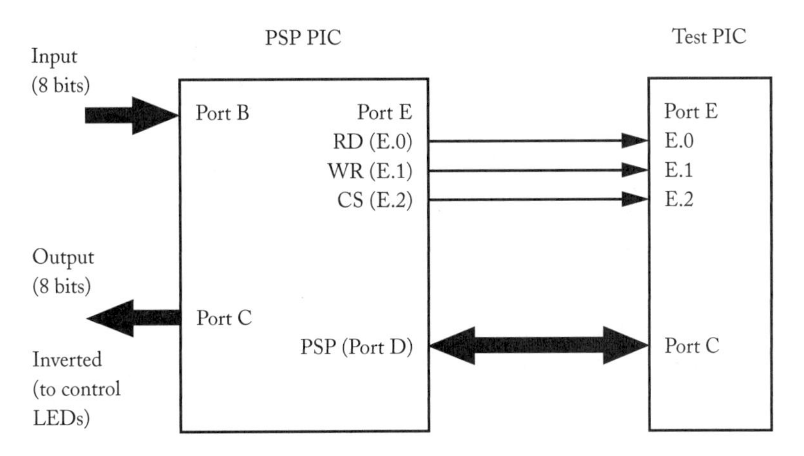

Figure 2.7 *PSP Illustration Diagram*

The software for the PSP demonstration is shown in Figure 2.8 and Figure 2.9.

Figure 2.8 shows the software for the PSP mode PIC. The PSP transfers are handled on an interrupt basis as is appropriate for real-time processing. The PIC does not know when or if a PSP transfer may occur and so the interrupt is used to prevent tying up the processor polling for a PSP event.

The *if* statement in the PSP ISR is used to detect whether the PSP event is a read event or a write event so that the data is handled appropriately. The *if* uses the input buffer full (IBF) bit of TRISE to determine if the PSP event was a write event to the PSP device. If it was, the IBF bit will be set to indicate that data has been received via the PSP. If the IBF is not set, then it must have been a read event and new data should be transferred to the PSP port. A programmer could also read the output buffer full (OBF) bit to determine when a PSP read event has occurred.

```
#include <16F877.h>
#use delay(clock=10000000)
#fuses HS,NOWDT
#BYTE TRISE = 0x89     //identify TRISE

#int_PSP
PSP_isr()
{
   output_high(PIN_A0);//*****************
   if (TRISE & 0x80)          //check for write to PSP
      output_C(~input_D());   //write data from PSP to LED's on C
   else                       //must be a read from PSP so get new data
      output_D(input_B());    //from port B
   output_low(PIN_A0); //*********************************
}

void main()
{
   port_b_pullups(TRUE);          //port B pullups on
   setup_adc_ports(NO_ANALOGS); //make sure port E bits are not set for
analog
   setup_psp(PSP_ENABLED);        //enable PSP mode
   enable_interrupts(INT_PSP);    //unmask PSP interrupt
   enable_interrupts(global);     //enable all unmasked interrupts
   while(1);                      //Do nothing
}
```

Figure 2.8 *PSP PIC Software*

Figure 2.9 shows the code for the Test PIC. It continuously reads from the PSP device and then echoes the result back in a write to the PSP device. The task is accomplished by manually creating the waveforms from Figure 2.6.

2.7 TIMERS

Timer/counters are probably the most commonly used complex peripheral in a microcontroller. They are highly versatile, being able to measure time periods, determine pulse width, measure speed, measure frequency, or provide output signals. Example applications might include measuring the rpm of a car's engine; timing an exact period of time, such as necessary to time the speed of a bullet; producing tones to create music or to drive the spark ignition system of a car; or providing a pulse-width or variable-frequency drive to control a motor's speed. In this section, timer/counters are discussed in the generic sense, and then typical **timer /counters** in the PIC microcontrollers will be discussed.

```
#include <16F877.h>
#use delay(clock=10000000)
#fuses HS,WDT
int ch;    //variable to hold data

void main() {
   setup_adc_ports(NO_ANALOGS);  //make sure port E does digital - not
                                 //analog
   setup_counters(RTCC_INTERNAL,WDT_72MS);
   output_E(0x7);     //set PSP control bits high
   while(1)
   {
      restart_wdt();     //reset WDT
      //first read from the remote PSP
      output_E(0x2);    //CS and RD low on PSP device for read
      ch = input_c();   //get data from PSP device
      output_E(0x7);    //all control bits high
      delay_us(50);     //allow time twixt PSP operations
      //now output to PSP device
      output_C(ch);     //data out to PSP device
      output_E(0x1);    //CS and WR low to cause write to PSP
      output_E(0x7);    //control signals high
      delay_us(50);     //allow time twixt PSP operations
   }
}
```

Figure 2.9 *Test PIC Software*

Although used in two distinctly different modes, timing and counting, timer/counters are simply binary up-counters. When used in timing mode, the binary counters are counting time periods applied to their input; in counter mode, they are counting the events or pulses or something of this nature. For instance, if the binary counters had 1-millisecond pulses as their input, a time period could be measured by starting the counter (from zero) at the beginning of an event and stopping the counter at the end of the event. The ending count in the counter would be the number of milliseconds that had elapsed during the event.

As a counter, the events to be counted are applied to the input of the binary counter, and the number of events occurring is counted. For instance, the counter could be used to count the number of cans of peas coming down an assembly line by applying one pulse to the input of the counter for each can of peas. At any time the counter could be read to determine how many cans of peas had gone down an assembly line.

PIC microcontrollers have both 8-bit and 16-bit /counters. In either case, an important issue for the program is to know when the counter reaches its maximum count and *rolls over*. In the case of an 8-bit counter, this occurs when the count reaches 255, in which case the next pulse will cause the counter to *roll over* to 0. In the case of a 16-bit counter, the same thing occurs at 65,535. The rollover events are extremely important for the program to be able to accurately read the results from a timer/counter. In fact, rollovers are so important that an interrupt, which is provided, will occur when a timer/counter rolls over.

The PIC16F877, as an example, has two 8-bit timers (Timer0 and Timer2) and one16-bit timer (Timer1). Other PIC microcontrollers will have varying numbers and types of timer/counter peripherals. The following sections will discuss the most common uses for each timer/counter even though most timers have more functions than those discussed next. For any specific PIC microcontroller, check the specifications to determine all of the various functions possible with the timer/counters.

2.7.1 GENERAL

This section describes the common applications of timer/counter peripherals. These are described in a general fashion and are followed by specific examples in the sections that follow. The applications being described include a timer used as a "tick" to provide a time base or delay function, a timer used to measure pulse width or frequency of an incoming signal, and a timer/counter used to provide output in the form of periodic signals or pulse-width modulation (PWM) to control motor speed or the like. The descriptions provided are the basic schemes being used to achieve the desired results. Most of these may also be created through special hardware included with the timers on the PIC microcontrollers. Specific examples and applications follow the descriptions.

2.7.1.1 Timers as "Ticks"

A timer may be used to provide a signal on a regular, repeating basis. This signal may be counted to determine how much time has passed, or as the basis for some time-critical function such as providing an output signal on a regular basis. An example might be firing the spark plugs of an engine at the correct instant.

The essential idea behind timer ticks is to allow the timer to count up to its maximum count and then overflow. The overflow provides an interrupt signal so that the timer can be reset for the next tick period and any time-critical functions can be carried out. Because the timer is a hardware device (not dependent on software), it will accurately keep time in the background while other functions are occurring.

Essentially the scheme is as diagrammed in Figure 2.10.

As shown in Figure 2.10, the timer is clocked by a signal derived from the system clock. The signal may be at the same frequency as the system clock, or it may be *prescaled* to a lower frequency. The timer is preloaded with a specific value (e.g., 56) and begins counting as clock pulses occur. Eventually, the timer counts up to its maximum value then rolls over, causing an interrupt to occur. At this point, the ISR reloads the timer with 56 and the whole process starts over. The ISR is the tick that can be used to control time-critical events.

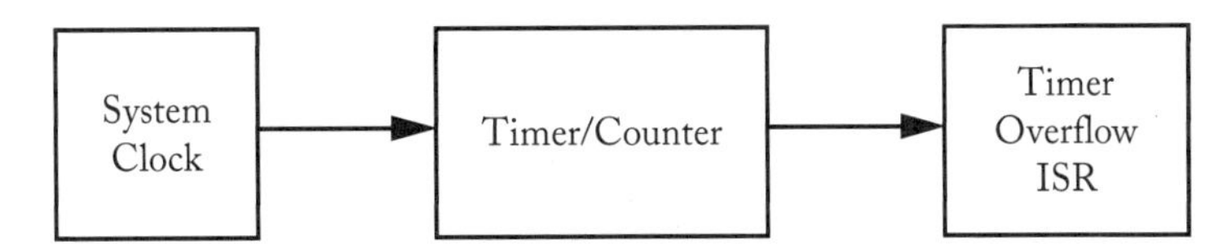

Figure 2.10 *Timer "Tick" Diagram*

An important bit of knowledge in this process is that of computing the time period of the ticks so they can be used. For the example given above, if we assume the system clock to be 4 MHz, the tick time would be computed as follows:

1. The time for each clock pulse is computed:

   $Time_{clockpulse} = 1 / F_{clock} = 1 / 4$ MHz = 0.25 microsecond

2. The number of clock pulses that must occur to cause rollover is computed:

   Maximum timer count – starting timer value = 256 – 56 = 200 counts

3. The tick time can then be computed:

   #counts * time/count = 200 counts * 0.25 microsecond/count = 50 microseconds

Therefore, for this example, the tick occurs every 50 microseconds. The ticks could be used as a time delay by counting them and waiting until the desired delay time had passed.

It is often useful to calculate the number of timer counts needed for a certain tick period. In this instance, reverse the math to find the number of clock pulses required for a certain period, then calculate the starting number for the timer that will provide the desired tick rate.

2.7.1.2 Timers Measuring Pulse Widths or Frequencies

Timers that are used to measure pulse widths (or frequency by inverting the period of the pulses) operate in a radically different manner than timers used for creating ticks and time delays. In the case of pulse measurement, the timers are typically running and are never stopped, started, or reset as they are when creating ticks. This difference in philosophy makes a large difference in their use.

To measure a pulse width, the program usually records the timer contents at the start of the pulse and at the end of the pulse and then subtracts the starting count from the ending count to see how long the pulse was. This is exactly what you would do if you wanted to know how long it took to walk from your residence to the bookstore: you would record the time you left home, the time you arrived at the bookstore, and then use subtraction to see how long it took you to get to the bookstore.

As an example, refer to Figure 2.11. In the example, the timer count is recorded at the beginning and end of the pulse. Then the timer counts are used along with the system clock to determine the actual pulse width.

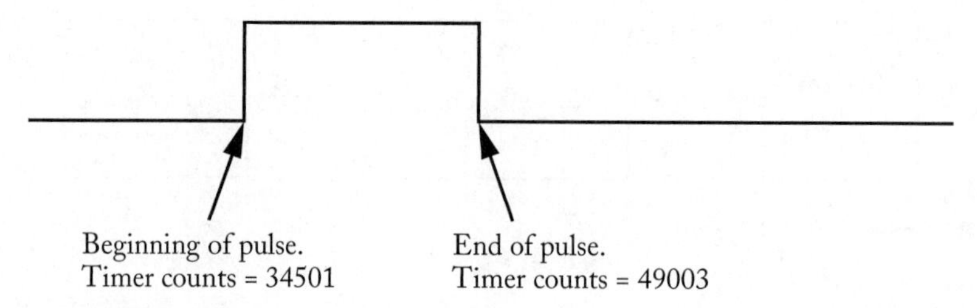

Figure 2.11 *Pulse-Width Measurement*

In the example shown in Figure 2.11, the pulse width would be calculated as follows:

1. Calculate the time per clock for the signal applied to the timer. For example, assume a 10 MHz clock for the timer:

   $Time_{clock} = 1 / F_{clock} = 1 / 10\text{ MHz} = 0.1$ microsecond

2. Determine the number of clock pulses that occurred during the pulse:

   #clocks = ending count – starting count = 49003 – 34501 = 14502

3. Determine the pulse width by multiplying the # clocks by the time per clock:

   Pulse width = #clocks x time/clock = .1 microsecond x 14502 = 1.45 milliseconds

The pulse in the example, then, was 1.45 milliseconds in width. If the object were to calculate the frequency of the waveform (assuming a square wave), then the period would be multiplied by 2 (to allow for both halves of the waveform), to get the period of the waveform, and the result would be inverted to get the frequency. In the case of asymmetrical waveforms, the time at the beginning and ending of the entire waveform would be timed and used to calculate the frequency.

2.7.1.3 Timers as Output Devices

Timers used to produce outputs such as periodic waveforms are, in essence, variations of timers used as ticks to produce an output waveform; a tick is created that is half of the period of the output waveform. Each time the tick occurs, a port bit is complemented to produce the output waveform. In this way a continuous waveform is produced at a port bit. In order to change the frequency of the output waveform, the reset number for the timer may be changed to vary the tick length. With a little creativity, the tick length can be changed to play a song, such as your school fight song.

Pulse-width modulation (PWM) is one of a number of methods of providing digital to analog conversion. PWM is the scheme in which the duty cycle of a square wave output from the microcontroller is varied to provide a varying DC output by filtering the actual output waveform to get the average DC. Figure 2.12 illustrates this principle.

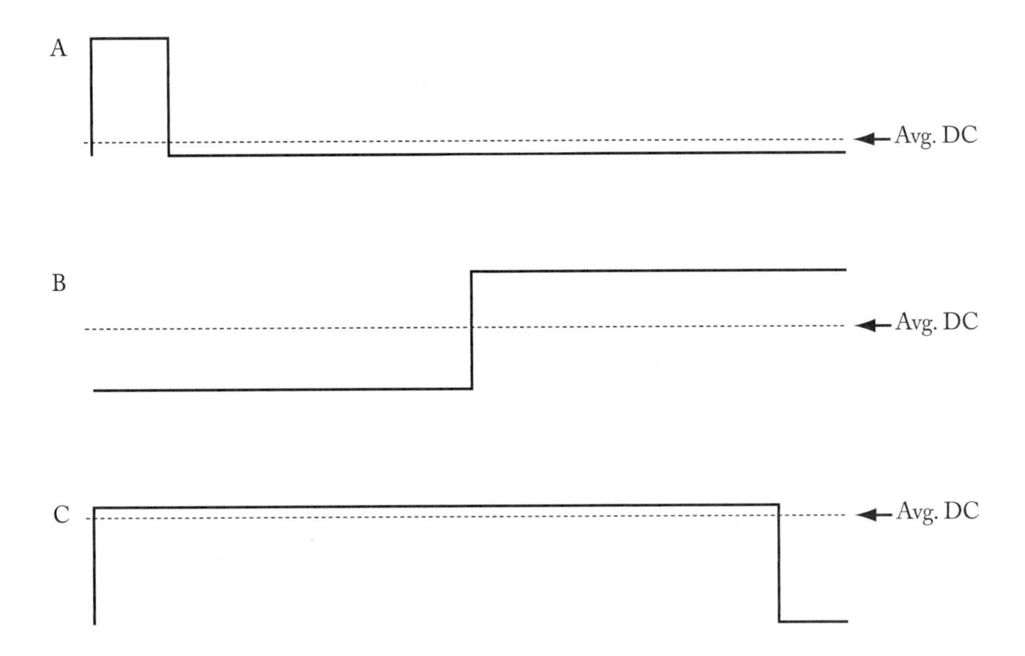

Figure 2.12 *Pulse-Width Modulation (PWM)*

As is shown in Figure 2.12, varying the duty cycle (or proportion of the cycle that is high) will vary the average DC voltage of the waveform. The waveform is then filtered and used to control analog devices, creating a digital-to-analog converter (DAC). Examples of PWM control schemes are shown in Figure 2.13.

In Figure 2.13A, the RC circuit provides the filtering. The time constant of the RC circuit must be significantly longer than the period of the PWM waveform. Figure 2.13B shows an LED whose brightness is controlled by the PWM waveform (note that in this example, a logic 0 will turn the LED on and so the brightness will be inversely proportional to the PWM). In this case, our eyes provide the filtering because we cannot distinguish frequencies above about 42 Hz, which is sometimes called the "flicker rate." In this case, the frequency of the PWM waveform must exceed 42 Hz, or we will see the LED blink.

The final example, Figure 2.13C, shows a DC motor control using PWM. The filtering in this circuit is largely a combination of the mechanical inertia of the DC motor and the inductance of the windings. It simply cannot physically change speed fast enough to keep up with the waveform. The capacitor also adds some additional filtering, and the diode is important to suppress voltage spikes caused by switching the current on and off in the inductive motor.

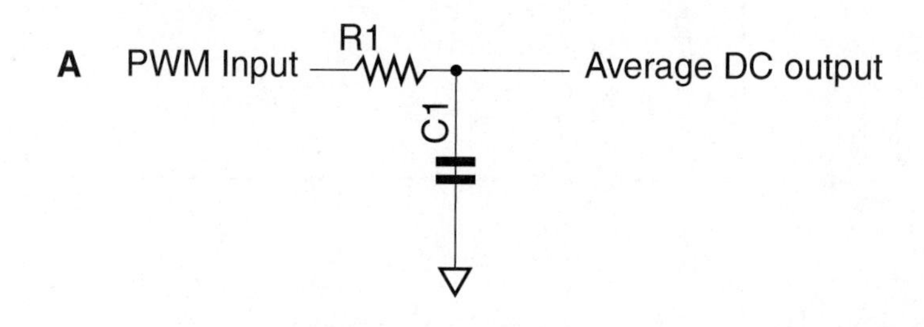

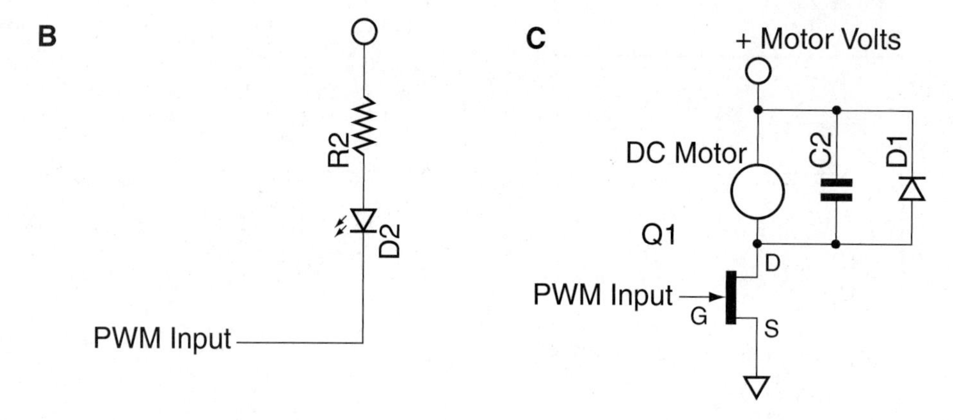

Figure 2.13 *PWM Examples*

PWM is achieved manually by changing the reload number of a timer, creating ticks for every half cycle of the output. Often the total of the high time and low time is set to 100, so that the PWM can be controlled as a percentage of the high time.

2.7.2 TIMER0

Timer0 is usually an 8-bit timer/counter created around an 8-bit counter unit. Hence its count range is restricted to 0–255. It is most often used as a time base or tick. The control registers associated with Timer0 include the OPTION_REG register, the INTCON register, and the TMR0 register.

The TMR0 register is the Timer0 8-bit counter. The Timer0-relevent bits of the OPTION_REG and INTCON registers are shown in Figure 2.14.

Figure 2.15 and Figure 2.16 are examples of a project that uses Timer0 as a time base. In this case, the project is alternately blinking two LEDs at a 0.1-second rate. The hardware and software for the project are shown in Figure 2.15 and Figure 2.16, respectively.

The examples in Figure 2.15 and Figure 2.16 show Timer0 being used as a time base for a program. The timer is initialized by the *setup_counters* built-in function as described in

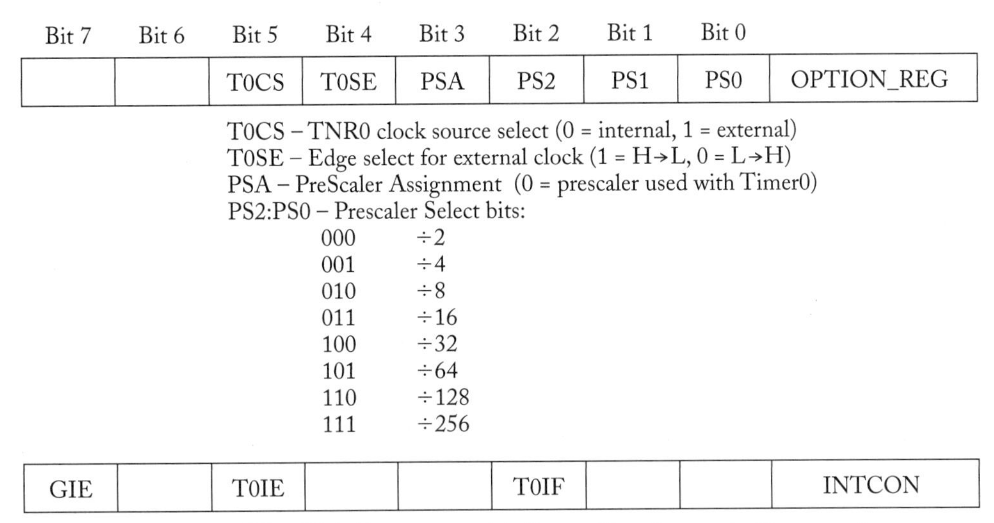

Figure 2.14 *Timer0 Control Registers*

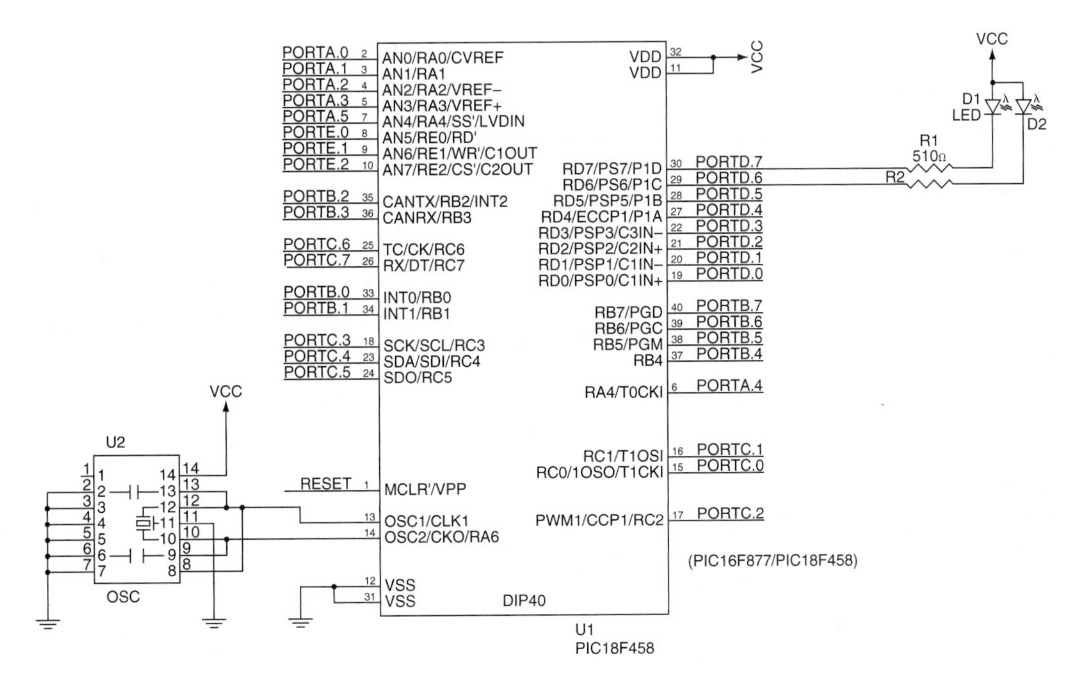

Figure 2.15 *Timer0 Time Base Example Hardware*

```
#include <16F877.h>
#use delay(clock=10000000)
#fuses HS,NOWDT
int millisecs, outs;

#int_TIMER0
void TIMER0_isr(void)
{
   set_timer0(100);    //reset for 1 ms overflow
   if (millisecs++ == 100)
   {
      millisecs = 0;       //reset for next .1 second interval
      outs = outs ^ 0xc0;   //swap states of upper 2 bits
      output_d(outs); //write data to port
   }
}

void main()
{
   setup_counters(RTCC_INTERNAL,RTCC_DIV_16);
   enable_interrupts(INT_TIMER0);
   enable_interrupts(global);
   outs = 0x80;   //set upper two bits to opposite states
   while(1)
      ;         //main loop does nothing
}
```

Figure 2.16 *Timer0 Time Base Example Software*

Chapter 4. Essentially, what this setup command does is to set the bits of OPTION_REG appropriately to use the internal clock (T0CS = 0) and a divide-by-16 prescaler (PS2:PS0 = 0b011). You, the programmer, could set these bits manually, but the built-in function is virtually always easier.

Likewise the *enable_interrupts* built-in function unmasks the Timer0 overflow bit (T0IE = 1) in INTCON to allow the Timer0 overflow to generate an interrupt that functions as the tick.

Within the Timer0 ISR, the set_timer0(100) built-in function resets the timer unit to 100_{10} each time the interrupt occurs. This is supposed to create a 1-millisecond tick time. The clock signal applied to the timer is F_{clk} / 4 divided by any prescaler (in this case, 16). Checking the math:

$$Time_{clock} = 1/\ (10\ MHz\ /\ (4 * 16)) = 6.4\ microseconds$$

Number clocks to roll over: Counts from 100 (loaded value) up to 256 (8-bit counter) before it rolls, so #clocks = 256 – 100 = 156

Time between ticks = 156 * 6.4 microseconds = 998.6 microseconds

Well, it is *almost* 1 millisecond. The result is actually 0.6 % low. This demonstrates that, often, it is not possible to hit an exact time with prescaler ratios, etc. Whether or not the difference is appreciable will depend on the specific situation.

One of the situations in which slight time errors will matter a lot is when you are trying to keep track of real time (e.g., creating an alarm clock or other time-based system). For instance, if you were to use the system above to keep time, the error would be 0.6%. This translates to 0.36 second over a minute (0.6% * 60 seconds), and over a day (0.6% * 60 seconds/minute * 60 minutes/hour * 24 hours/day), 518 seconds or 8.6 minutes per day of error. This clock would be 1 hour off (8.6 minutes/day * 7 days/week = 60.48 minutes) every week. This is definitely unsatisfactory for a clock. In the next section we will see how to create a time-correct clock.

2.7.3 TIMER1

Timer1 is a 16-bit timer that can be used in exactly the same manner as Timer0. However, Timer1 is actually designed to be used in an entirely different fashion. Where Timer0 usually accomplishes its tasks by reloading or resetting the timer as need to produce the delays or 'ticks', Timer1 is designed to keep running and never be reset or reloaded.

Timer1 may be used alone or with the capture/compare/PWM module. The PWM function works the same with either Timer1 or Timer2. The capture and compare function will be discussed here and PWM will be discussed with Timer2.

Timer1 is designed to keep running continuously in order to allow the pulse capture and output compare modules to work correctly. Pulse widths (or frequencies) are measured by noting the Timer1 count at the beginning and at the end of the pulse, and the pulse width is determined by subtracting the beginning time from the ending time. Similarly, the output compare module provides output by using the running count on Timer1 to determine when to assert an output signal.

Timer1 is composed of two registers: TMR1H and TMR1L. The Timer1-relevent bits of other registers that control Timer1 are shown in Figure 2.17.

Timer1 also provides a method to keep accurate time and to implement a real-time clock by providing an external input from a 32.768 watch crystal as the clock for Timer1. This is accomplished by setting TMR1CS high in T1CON and enabling Timer1. The examples in Figure 2.18 and Figure 2.19 implement a real-time clock (RTC) to blink and LED at exactly a 2-second rate (i.e., changes state every 2 seconds).

The RTC software is relatively self-explanatory (refer to the comments). An extra "0x8" was added to the *setup_timer0()* function to set the T1OSCEN bit in T1CON to enable the oscil-

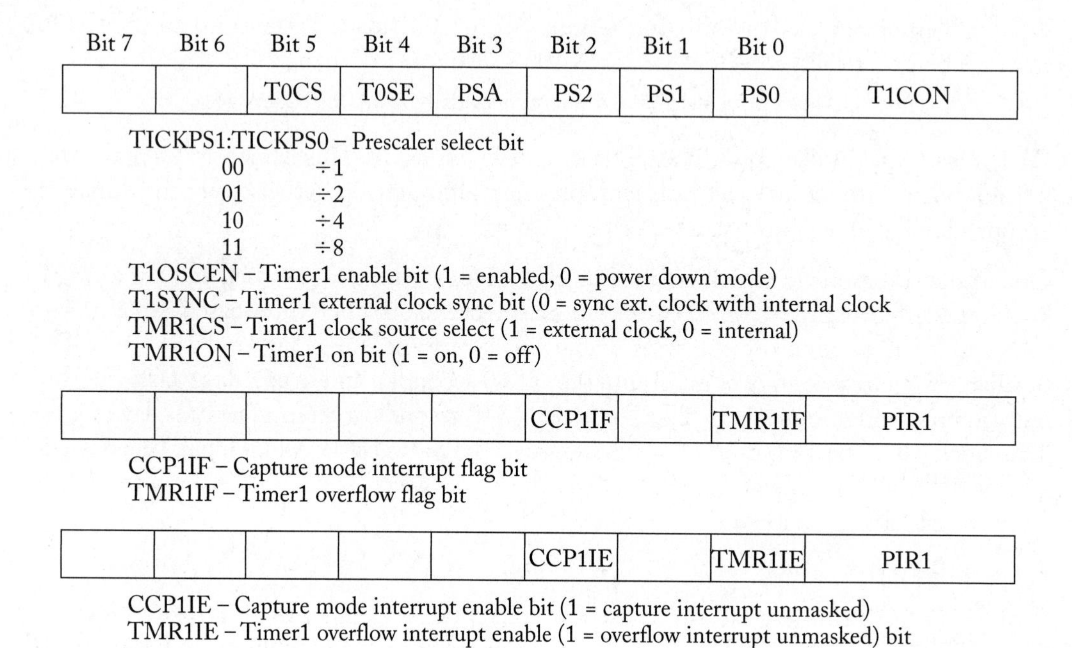

Figure 2.17 *Timer1 Register Bits*

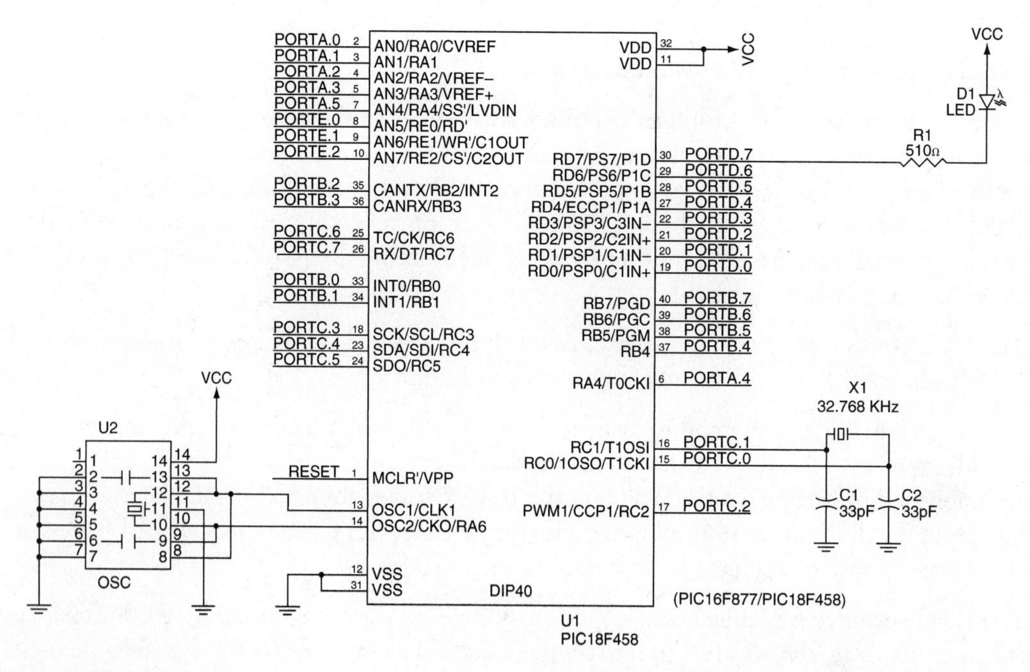

Figure 2.18 *RTC Hardware*

```
#include <16F877.h>
#use delay(clock=10000000)
#fuses HS,NOWDT
int outs;

#int_TIMER1
TIMER1_isr()
{                               //interrupt occurs every 2 seconds
   outs = outs ^ 0x1;           //toggle output bit
   output_bit(PIN_D7, outs);    //replace this line with code to keep
                                // track of seconds, minutes, hours, days,
                                //etc.  to implement a real clock.
}

void main()
{
   setup_timer_1(T1_EXTERNAL|T1_DIV_BY_1|0x8);
              //enable T1 for external clock & oscillator, divide by 1
   enable_interrupts(INT_TIMER1); //unmask timer1 overflow interrupt
   enable_interrupts(global);     //enable all unmasked interrupts
   outs = 0x1;                    //preset output variable
   while(1)
       ;                          //main does nothing
}
```

Figure 2.19 *RTC Software*

lator for the 32.768 KHz crystal. This was done as an excellent example of manually setting control register bits. The CCS_PICC wizard provides this function, but under the name of T1_CLK_OUT, which may or may not seem to relate to enabling the external oscillator. Understanding the control register bits allows you to set the peripherals as you wish.

The RTC overflows every 2 seconds because the crystal frequency is 32,768 Hz and it takes 65,536 counts (2* 32768) to overflow the counter.

2.7.3.1 Capture and Compare Modules

Timer1 also is involved with the capture and compare modes of operation. Capture mode is used to capture the contents of Timer1 when an external event occurs on Port C bit 2 (RC2). The event may be either a rising or a falling edge, and a prescaler is provided to capture every 1st, 4th, or 16th event. The timer values are used to calculate the period or frequency of the events.

Compare mode is a related function in that it uses the contents of the Timer1 registers to determine when to cause an output event on RC2. Because RC2 is used as the input pin for capture mode and the output pin for compare mode, only one mode can be used at any given instant in the PIC16F877. Pulse-width modulation (PWM) is a variation of compare mode.

The capture and compare operation is controlled by the capture and compare control register CCP1CON. The relevant bits of CCP1CON are shown in Figure 2.20.

2.7.3.1.1 Capture Operation

Capture operation involves the Timer1 registers and the capture and compare register (CCPR1), which is composed of two 8-bit registers, CCPR1H and CCPR1L. CCPR1 captures the contents of Timer1 when the trigger event occurs. An interrupt, if enabled, will also occur when the time is captured.

When the trigger event occurs, the program must store the contents of CCPR1 until the next event has taken place. Once the second event has occurred, the two time values are used to calculate the total time in timer ticks. The timer ticks, in conjunction with the clock frequency, may be used to calculate the frequency or period of the trigger events.

The hardware and software in Figure 2.21 and Figure 2.22 comprise an engine control monitor and alarm. The purpose is to monitor the tachometer signals from an engine. If the period of the pulses drops below 1 millisecond, the engine is running too fast and a red LED lights up (the alarm).

The engine monitor example in Figure 2.21 and Figure 2.22 relies on the capture interrupt event to do all of the work. Following are the steps it is executing:

1. Reading the capture register and storing the result as *end_time*. CCP_1 is a CCS_PICC variable that is provided as a convenient way to read the 16-bit capture register.

Bit 7	Bit 6	Bit 5	Bit 4	Bit 3	Bit 2	Bit 1	Bit 0	
		CCP1X	CCP1Y	CCP1M3	CCP1M2	CCP1M1	CCP1M0	CCP1CON

CCP1X,CCP1Y – Two LSB bits of the duty cycle in PWM mode
CCP1M3:CCP1M0 – Capture/compare mode select bits
0000 – Capture/compare disabled
0100 – Capture mode, capture every falling edge
0101 – Capture mode, capture every rising edge
0110 – Capture mode, capture every 4th rising edge
0111 – Capture mode, capture every 16th rising edge
1000 – Compare mode, set output pin on match
1001 – Compare mode, clear output pin on match
1010 – Compare mode, generate interrupt on match (output pin not affected)
1011 – Compare mode. Special—see specification for details.
11xx – PWM mode

Figure 2.20 *CCP1CON Bit Definitions*

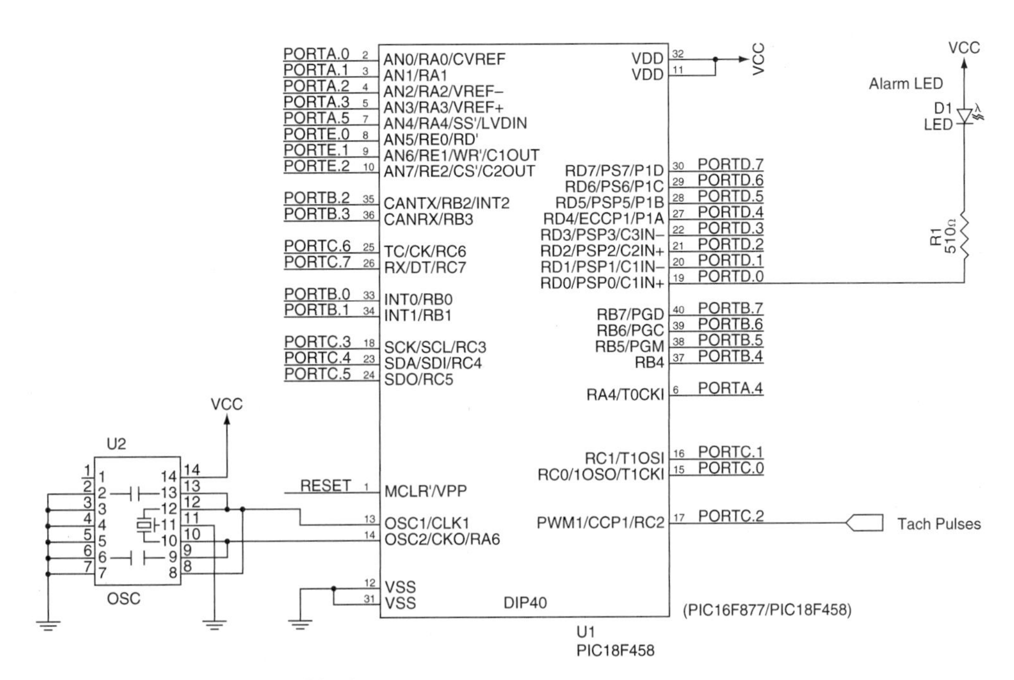

Figure 2.21 *Engine Monitor Hardware*

2. Using *end_time* (the Timer1 count at the end of the period) and a previously stored beginning Timer1 count (*start_time*), the actual period of the wave is calculated in timer ticks. The Timer1 overflow interrupt value is used to account for any counter overflow that may have taken place during the measurement period. The calculation for the number of ticks is:

   Pulse_ticks = overflow_count* 0x10000 – start_time + end_time

3. Comparing the actual *pulse_ticks* to 2500 to see if the engine is running too fast. The limit is when actual time of the waveform's periods is less than 1 millisecond, indicating a high engine speed problem. The 2500 limit was calculated this way:

   ticks for 1 millisecond = 1 ms / time per tick = 1 ms / (1/(4 MHz/4)) = 2500

4. Adjusting the output LED according to the current reading.
5. Finally, saving the *end_time* for the current pulse as the *start_time* for the next pulse and clearing the *overflow_count* for the next pulse calculation.

2.7.3.1.2 Compare Operation

Compare operation is almost the corollary of capture mode. In compare mode, the programmer loads CCPR1, and an interrupt occurs when the count in Timer1 matches the value loaded into CCPR1. At this point the program executes whatever was supposed to happen when CCPR1 and Timer1 match and reloads CCPR1 for the next match time.

```
#include <16F877.h>
#use delay(clock=10000000)
#fuses HS,NOWDT
int overflow_count;
unsigned int32 start_time, end_time;
int32 pulse_ticks;

#int_TIMER1
TIMER1_isr()
{
   ++overflow_count; //increment whenever an overflow occurs
}

#int_CCP1
CCP1_isr()
{
   end_time = CCP_1;    //read captured timer ticks
   //check for pulse time (in ticks) accounting for any overflow
   //that may have occurred
   pulse_ticks = (0x10000 * overflow_count) - start_time + end_time;
   if (pulse_ticks < 2500) output_low(PIN_D0); //light LED if too fast
   else output_high(PIN_D0); //extinguish LED if speed is OK
   //now set up for the next pulse
   start_time = end_time;  //end time of this pulse is the start time
                           //for the next one
   overflow_count = 0;  //clear overflow counts
}

void main()
{
   setup_ccp1(CCP_CAPTURE_RE);
   setup_timer_1(T1_INTERNAL|T1_DIV_BY_1);//set timer1 to run at system
   clock/4
   enable_interrupts(INT_TIMER1);  //unmask Timer1 overflow interrupt
   enable_interrupts(INT_CCP1);    //unmask capture event interrupt
   enable_interrupts(global);      //enable all unmasked interrupts
   while(1);  //as usual, the main() function does nothing
}
```

Figure 2.22 *Engine Monitor Software*

As an example, consider the case in which the microcontroller is supposed to output a 1 KHz signal. The first step is to determine how many clock pulses will be applied to the timer during half of the period of the desired output signal (assuming the output is to be symmetrical). In this example, the 1 KHz wave has a period of 1 millisecond and so the output will need to change states every 500 microseconds to create the signal. The desired output signal is shown graphically in Figure 2.23.

Using a system clock of 10 MHz and a prescaler of 1, the clock (usually referred to as the *instruction* clock) applied to the timer would be:

10 MHz / 4 = 2.5 MHz

The time per clock of the 2.5 MHz clock is:

1 / 2.5 MHz = 0.4 microsecond

So the number of clock pulses that will occur in 500 microseconds is:

500 microseconds / .4 microsecond per clock pulse = 1250 clock pulses

Therefore, if the program waits 1250 clock pulses between toggling an output port bit, a 1 KHz signal will be created. In order to do this with the capture register, the port bit is toggled each time a match occurs and the capture register is incremented by 1250 counts for the next match point. Figure 2.24 and Figure 2.25 show the hardware and software that produce a 1 KHz signal on Port D bit 0.

Figure 2.25 shows the method that enables the capture mode (`setup_ccp1(CCP_COMPARE_INT)`) to create an interrupt and not affect the output pin associated with CCP1 (Port C.2). Because the only automatic functions are to set or clear the bit, and the bit needs to be toggled, it is necessary to toggle the bit in the ISR associated with the match event.

The compare ISR also reloads the compare register for the next desired match event (`CCP_1 = CCP_1 + 1250;`) by reading it and incrementing the value by 1250. One additional important point relative to the output compare registers, and specifically relative to the calculation of the next match point number, is to consider the situation in which adding the interval number to the current contents of the compare register results in a number bigger than 16 bits. For example, if the output compare register contains a number 65,000, and the interval is 1000, then:

65,000 + 1000 = 66,000 (a number greater than 65,535)

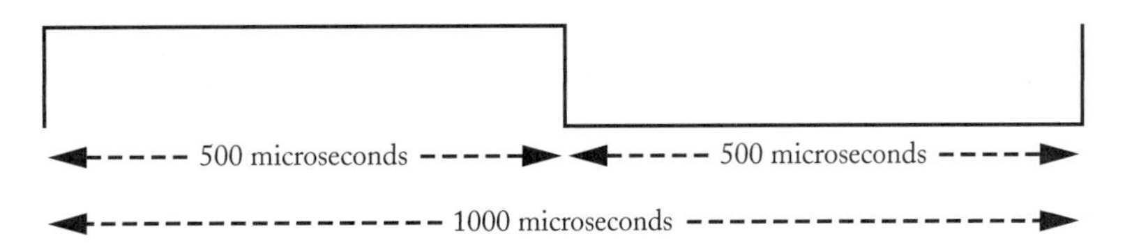

Figure 2.23 *1 KHz Waveform*

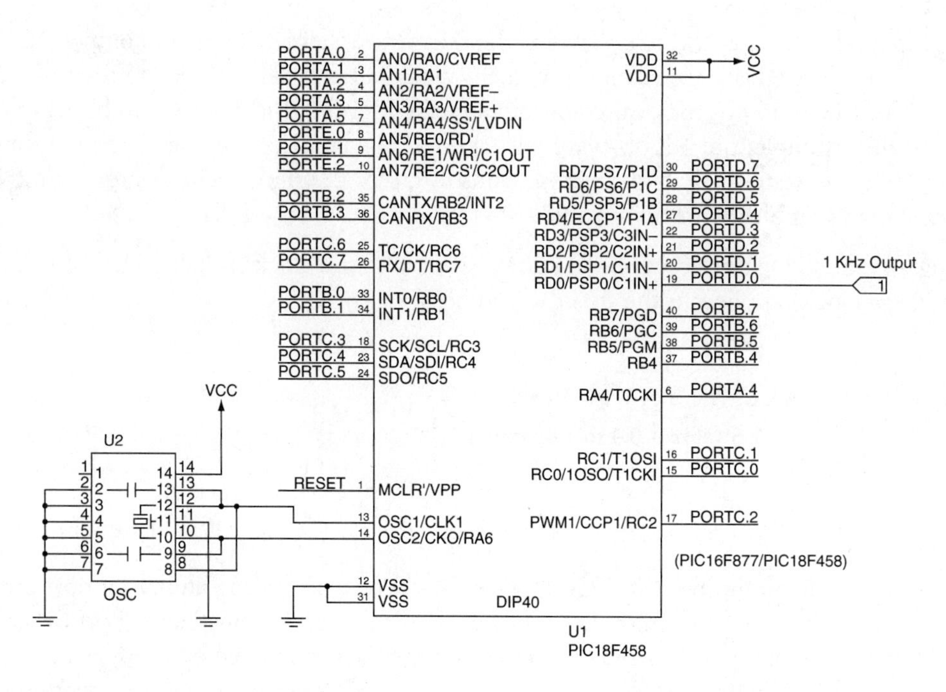

Figure 2.24 *1 KHz Generator*

As long as unsigned integers are used for this calculation (CCP_1 is unsigned), those bits greater than 16 will be truncated and so the actual result will be:

65,000 + 1000 = 464 (drop the 17th bit from 66,000 to get 464)

This will work out perfectly because the *output compare register* is a 16-bit register and the timer/counter is a 16-bit device as well. The timer/counter will count from 65,000 up to 65,536 (a total of 536 counts) and then an additional 464 counts to reach the match point. The 536 counts plus the 464 counts is exactly 1000 counts as desired. In other words, in both the timer/counter and the compare register, rollover occurs at the same point, and as long as unsigned integers are used for the related math, rollover is not a problem.

```
#include <16F877.h>
#use delay(clock=10000000)
#fuses HS,NOWDT
int outs = 1;   //preset to start process

#int_CCP1
CCP1_isr()
{
   CCP_1 = CCP_1 + 1250;        //increment compare cpounter
   outs ^= 1;                   //exclusive OR 'outs' to toggle bit
   output_bit(PIN_D0, outs);    //output bit to pin
}

void main()
{
   setup_timer_1(T1_INTERNAL|T1_DIV_BY_1); //timer1 enabled, instruction
clock / 1
   setup_ccp1(CCP_COMPARE_INT); //set CCP1 to cause an interrupt on
match
   enable_interrupts(INT_CCP1);   //unmask compare1 match interrupt
   enable_interrupts(global);     //enabled all unmasked interrupts
   while(1)
       ;     //main does zilch
}
```

Figure 2.25 / *KHZ Generator Program*

EXAMPLE

Chapter 2 Example Project – Part B

This is the second part of this chapter's example program to create a system to collect operational data for a stock car racer. In the first part, it was determined that the system had sufficient memory resources to do the task. In this part, the overall code structure and user interface will be created.

It is important to determine which peripheral resources are to be used for each measurement, input, and output. If you were creating the data collection system from scratch, you could determine the resources to use, and, as you assigned the resources, you would be sure that the processor had sufficient resources. In this case, however, you need to determine which peripherals are being used for each measurement so you can write the program accordingly.

Investigation (and perhaps a little circuit tracing) has shown that the system connections are as follows:

The "start" button is connected to the INT1 (External Interrupt1) pin on the

PIC18F458). Pressing it pulls INT1 low.

The "upload" button is connected to Port B, bit 0. Pressing it pulls the bit low.

The engine rpm pulses are connected to the CCP1 (capture/control) pin.

The drive shaft rpm pulses are connected to the INT0 (External Interrupt 0) pin.

The engine temperature signal is connected to the AN1 (analog to digital converter input #1) pin 3.

For now all that you need to be concerned with will be the start and upload buttons. The other signals will be handled in other parts of the example project.

Now you will want to plan the overall structure of the program. Thinking back to the section on "Real-Time Methods" in Chapter 1, it would make sense to apply those methods to this project. In general, this program is supposed to record data at specific intervals and eventually upload the data to a PC for analysis. Using real-time programming methods allows us to keep the measurements running in real time (using interrupts) and then simply record the current values when the 1-second interval has elapsed. In other words, the measurements are kept as current as possible all the time, whether they are being recorded or not. When the time comes to record the data, the recording function simply grabs the latest data and records it.

So, eventually, interrupt service routines will be needed for the two rpm measurements and the one temperature measurement. Additionally, you must handle the start and upload buttons.

A simple algorithm that will handle the job is for the program to record 120 sets of data then stop; pressing the start button simply clears the data set counter so it records another 120 sets of data at the time you want. The advantage here is that pressing the start button will always start a new set of data in case someone starts the recording process errantly. The upload button would send the data whenever it is pressed if the 120 sets were complete.

The program skeleton in Figure 2E.1 seems to fit the needs and will be filled in as you proceed with this example:

This code structure sets up the operation of the entire system. INT1 is connected to the start button; when the button is pressed it resets the data set counter to 0. Timer0 will be set (details in the next section) to produce an interrupt that will provide a 1-second interval. Within the Timer0 interrupt service routine, the data is recorded at a 1-second interval anytime the data counter is less than 120. So the scheme is: pressing the *start* button sets the data counter to 0, and data is then recorded until 120 sets have been recorded. In the meantime, all of the data is kept up-to-date in a real-time fashion by the various interrupts. The last piece, uploading to the PC, is handled in the *while(1)* loop in *main()*. When the upload button is pressed, the data is sent out serially.

More pieces of the example program will be added in the latter sections of this chapter.

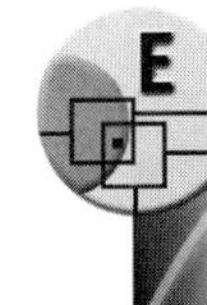


```
{
   setup_adc_ports(A_ANALOG);                    //set up A/D to use internal
                                                 //reference

   setup_adc(ADC_CLOCK_DIV_32);                  //code to set up A/D clock
   setup_wdt(WDT_OFF);                           //WDT off
   setup_timer_0(RTCC_DIV_256);                  //Timer0 enabled with a 256
                                                 //prescaler
   setup_timer_1(T1_INTERNAL|T1_DIV_BY_1);//Timer1 enabled using system
                                                 //clock
   setup_ccp1(CCP_CAPTURE_RE);                   //enable capture mode
                                                 //operation
   enable_interrupts(INT_TIMER0);                //unmask timer0 overflow
                                                 //interrupt
   enable_interrupts(INT_TIMER1);                //unmask timer1 overflow
                                                 //interrupt
   enable_interrupts(INT_EXT);                   //unmask external interrupt 0
   enable_interrupts(INT_EXT1);                  //unmask external interrupt 1
   enable_interrupts(INT_AD);                    //unmask A/D interrupt
   enable_interrupts(INT_CCP1);                  //unmask capture 1 interrupt
   enable_interrupts(global);                    //enable all unmasked
                                                 //interrupts
   while(1)
   {
        //place code here to monitor port B.1 to upload data
   }
}
```

Figure 2E.1 *Skeleton Code Structure*

2.7.4 TIMER2

Timer2 is configured to provide a time base for several functions. In this regard it works in a manner similar to Timer0. The output of Timer2's time base functions may be used to provide the time base for the pulse-width modulation function, the baud rate for the serial port, or the ticks for program operation. A basic block diagram for Timer2 is shown in Figure 2.26.

Figure 2.26 shows that the clock applied to Timer2 is the instruction clock (system clock / 4), which may be further prescaled by ratios of 1, 4, or 16. Each time that the Timer2 count matches the count in PR2, Timer2 is automatically reset. The reset or match signal is used to provide a baud rate clock for the serial module. Notice that Timer2 counts up to the match point and is then *reset to 0* as opposed to the count-up-to-rollover method used with Timer0.

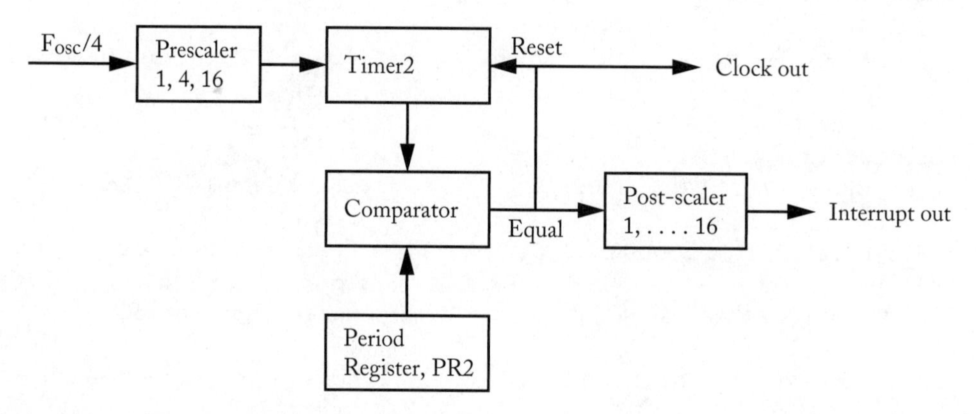

Figure 2.26 *Timer2 Block Diagram*

Each match may also provide an interrupt. The interrupt can be post-scaled to actually provide the interrupt every 1, 2, 3, 4, 5, and so forth, up to 16 resets. This is useful for creating lengthier delays than possible with a normal 8-bit timer.

The control register that is associated with Timer2 is shown in Figure 2.27.

In order to use Timer2 effectively, a formula that shows the relationship of the various prescalers, system clocks, and the like is needed.

Inasmuch as the limiting item in the equation is the 8-bit counter (Timer2), it seems reasonable to determine a formula that allows calculation of the PR2 value needed for any given output. In this way, prescaler and post-scaler values may be tried until a reasonable value for PR2 is achieved. The larger the value of PR2, the better it will be to achieve the best resolution and accuracy.

Starting with the clock applied:

$$\text{Timer2 clock} = F_{osc}/4 \ / \ \text{prescaler} = F_{osc} \ / \ (\text{prescaler} * 4)$$

Bit 7	Bit 6	Bit 5	Bit 4	Bit 3	Bit 2	Bit 1	Bit 0	
	TOUTPS3	TOUTPS2	TOUTPS1	TOUTPS0	TMR2ON	T2CPS1	T2CPS0	T2CON

T2PS1:T2PS0 – Prescaler control bits (00 = 1, 01 = 4, 1x = 16)
TMR2ON – On/off control bit for power consumption control (1 = on)
TOUTPS3:TOUTPS0 – Post-scaler control bits
0000 – 1:1 post-scale
0100 – 1:2 post-scale
.
.
1111 – 1:16 post-scale

Figure 2.27 *Timer2 Control Register*

The time period for the clock pulses is the inverse of the clock signal:

$$\text{Timer2 clock period} = \text{time-per-tick} = (4 * \text{prescaler}) / F_{osc}$$

The time it will take for Timer2 to produce an interrupt (*period*) is the number given by PR2 multiplied by the time-per-tick and the post-scaler:

$$\text{Period} = \text{time-per-tick} * \text{PR2} * \text{post-scaler} = ((4* \text{prescaler}) / F_{osc}\text{PR2} * \text{post-scaler}$$

A bit of algebra reduces this formula to:

$$\text{Period} = (4 * \text{prescaler} * \text{post-scaler} * \text{PR2}) / F_{osc}$$

Finally, solving to PR2 yields the following useful formula:

$$\text{PR2} = (\text{period} * F_{osc} / (4 * \text{prescaler} * \text{post-scaler})$$

This formula is useful for a trial-and-error approach to picking prescaler and post-scaler values to meet design goals. For instance, taking the example from Figure 2.24 and Figure 2.25 that is designed to produce a 1 KHz waveform, the period of the match events is 500 microseconds. A little trial-and-error with the possible prescaler and post-scaler values gives the following:

$$\text{PR2} = \text{period}*F_{osc} /(4* \text{prescaler} * \text{post-scaler}) = 500\text{ E-6} * 10\text{E6} / (4 * 1 * 5) = 250$$

That is, a prescaler of 1 and a post-scaler of 5 means that setting PR2 to 250 will yield interrupts at a 500-microsecond rate. Setting PR2 to 250 is using most of the range of the timer and is, therefore, likely to produce the best accuracy. Figure 2.28 shows the software to produce the 1 KHz waveform using Timer2.

The program in Figure 2.28 is virtually the same as the one in Figure 2.25, except that Timer2 is being used and PR2 does not need to be reloaded each time, saving programming effort and wasted time during execution.

2.7.4.1 Pulse-Width Modulation (PWM)

PWM is the technique of controlling duty cycle to control the average DC applied to a device as was explained in the preceding "general" section.

To create the PWM signal, Timer2 is used to create the clock that makes the PWM signal; that is, Timer2 sets the frequency of the PWM signal. This is done in the usual manner for Timer2 by setting PR2. When Timer2 counts up to the value that matches PR2, Timer2 is reset and the process restarts.

When the Timer2 reset occurs, the PWM output bit (Port C bit 2) is set high. As Timer2 counts up, it will eventually match the value placed into CCPR1L (the low 8 bits of CCPR1) at which time the PWM output will be cleared to 0. Then Timer2 counts up until it matches PR2 and the whole thing starts over again. In this manner, the PWM output is created.

```
#include <16F877.h>

#use delay(clock=10000000)
#fuses HS,NOWDT
int outs = 1;  //preset to start process

#int_TIMER2
TIMER2_isr()
{
   outs ^= 1;  //exclusive OR outs to toggle bit
   output_bit(PIN_D0, outs);  //output bit to pin
}

void main()
{
   setup_timer_2(T2_DIV_BY_1,250,5);
        //Timer2 enabled, prescaler of 1, PR2 = 250, post-scaler = 5
   enable_interrupts(INT_TIMER2);
              //unmask timer2 interrupt to respond to timer2 match
   enable_interrupts(global);
              //enable all unmasked interrupts
   while(1)
       ;      //main serves no function
}
```

Figure 2.28 *1KHz Waveform Program*

For example, setting PR2 to 100 and CCPR1L to 45 would create a 45% duty cycle. Timer2 would start at 0, and the PWM output would be set high. When Timer2 counts up to 45, the match between Timer2 and CCPR1L will cause the PWM output to go low for the duration of the time that Timer2 counts up to 100. In this way, the output is high for 45 out of 100 counts, creating a 45% duty cycle. Setting CCPR1L to 0 will force a zero duty cycle, and setting CCPR1L (plus the two extra bits if in 10-bit mode) to a value higher than Timer2 can reach will force 100% duty cycle.

In this example, the output would have a resolution of 1% (1 part out of 100). In many cases, this is insufficient and so two extra bits are used with CCPR1L to create a 10-bit duty cycle number. The two extra bits are bits 4 and 5 of CCP1CON (refer to Figure 2.20) and two extra least significant bits are used with Timer2 to make it all work. The two extra bits are the two LSBs of the counter that does the divide-by-4 to reduce the system clock to the instruction clock.

As you have likely considered, loading the two LSBs of the duty cycle into bits 4 and 5 of CCP1CON sounds somewhat like a challenge (actually it is just tedious, not really a challenge). Fortunately, CCS-PICC handles this task directly for you through the *SET_PWMx_DUTY(value)* function. "X" is either 1 or 2 depending on the source of the PWM, and the value may be either 8 or 10 bits. The size of the variable will determine whether 10-bit resolution or 8-bit resolution PWM is used.

As an example, take the case of a child's electric wheelchair. Assume that the child has switches to go forward, turn (two ways), and reverse. Depending on the abilities of the child, it is prudent to set the maximum speed of the wheelchair, and as the child's abilities change, you need to be able to set the speed to match the abilities. The wheelchair motors are to be driven in a manner similar to the motor shown in Figure 2.13C.

Step 1 determines the frequency (period) of the PWM signal and the value for PR2 to create the correct frequency. From the PIC specification, the formula for PWM period is:

$$\text{PWM period} = (\text{PR2} + 1) * 4 * T_{osc} * \text{TMR2 prescaler}$$

Algebraically, the formula may be rearranged to solve for PR2 (the frequency for the system clock also is included rather than the period for convenience):

$$\text{PR2} = ((\text{PWM\_period} * F_{osc}) / (4 * \text{prescale})) - 1$$

In this example, it is desirable to set the PWM frequency above the range of human hearing so that people are not bothered by the noise from the motors. The frequency chosen is, therefore, 25 KHz (period = 1/25 KHz = 40 microseconds). It is reasonable to set the prescaler to 1 because the frequency needs to be as high as possible. Hence, the formula works out as:

$$\text{PR2} = ((40 \text{ microseconds} * 10 \text{ MHz}) / 4) - 1 = 99$$

This result means that the project can use 8-bit resolution PWM and still have almost 1% resolution.

The next step is to determine the values of PWM that need to be produced. In this example, pick 35% (the slowest speed), 60% (the mid-range speed), and 95% (the highest speed). The speeds are to be jumper-selectable. It is now possible to calculate the duty cycle values to load into CCPR1 to control the speed of the motors. The calculations are:

$$\text{CCPR1}_{35\%} = .35 * 99 = 35$$

$$\text{CCPR1}_{60\%} = .60 * 99 = 59$$

$$\text{CCPR1}_{95\%} = .95 * 99 = 94$$

Figure 2.29 and Figure 2.30 show the basic PWM portion of the wheelchair controller example. In this program, a jumper on Port B directly selects the PWM output. The PWM output is only present if a jumper to ground is present on one of the least significant 3 bits of Port B. This portion only verifies the operation of the PWM drive. The balance of the wheelchair controller is left to your imagination.

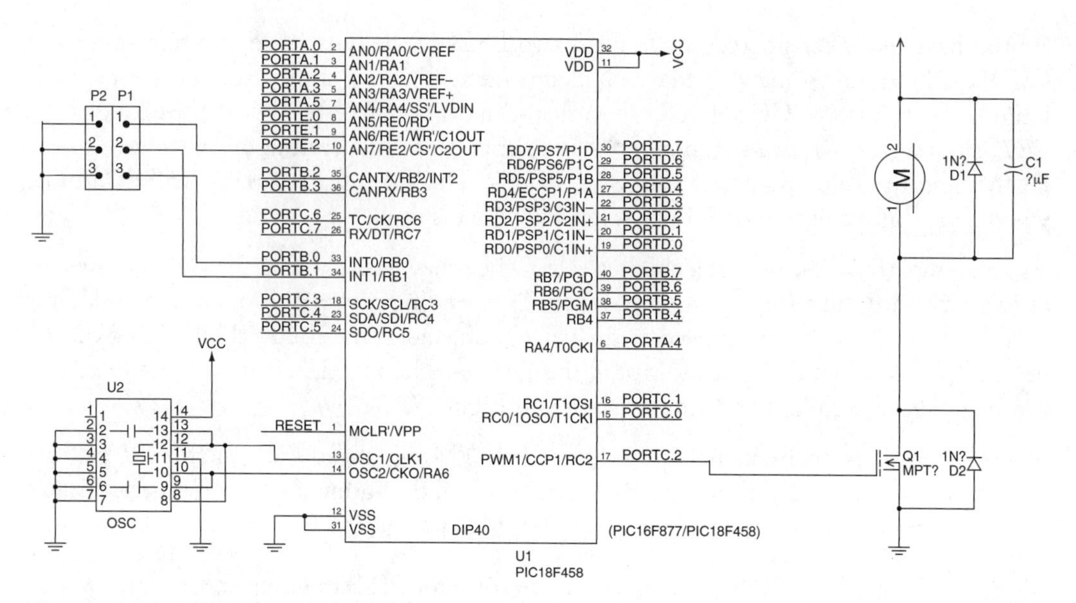

Figure 2.29 *PWM Example*

```
#include <16F877.h>
#use delay(clock=10000000)
#fuses HS,NOWDT
int duty_cycle, control_bits, hold_value;

void main()
{
   port_b_pullups(TRUE);     //engage pull ups on port B
   setup_timer_2(T2_DIV_BY_1,99,1); //enable Timer2, PR2=99, prescaler=1
   setup_ccp1(CCP_PWM);  //enable PWM mode
   while (1)
   {  //main actually does something!
      //read control port and wait for change
      hold_value = (~(input_b()) & 0x7); //get pin value
      while ((control_bits = (~(input_b()) & 0x7)) == hold_value);
      //wait for a change
      switch (control_bits)
      {
         case 0x1: duty_cycle = 35; //35% PWM
                   break;
         case 0x2: duty_cycle = 59;  //60% PWM
                   break;
```

Figure 2.30 *PWM Example Program (continues)*

```
            case 0x4: duty_cycle = 94;    //94% PWM
                      break;
            default: duty_cycle = 0;
         }
         CCP_1 = duty_cycle;    //set duty cycle
      }
}
```

Figure 2.30 *PWM Example Program (continued)*

The program in Figure 2.30 demonstrates the use of the control registers to effect PWM operation. One thing to notice is the fact that the program waits for a change on the input pins before selecting a new pulse-width duty cycle and reloading CCPR1. Although it technically does not hurt to reload CCPR1 constantly, it is bad practice and a waste of processor time.

2.7.5 WATCHDOG TIMER

The watchdog timer (WDT) is essentially a safety device. It monitors the operation of the microcontroller and resets the microcontroller if the program were to go awry. This operation is especially important where incorrect operation of the device can cause personal harm or extensive (read: expensive) damage.

The WDT begins to count up as soon as it is enabled. If the count is allowed to reach its maximum, then the processor is reset and operation starts anew. It is the job of the programmer to keep reloading the WDT so it is not allowed to time out and cause the reset to occur. The theory is that if the microcontroller has "lost its mind," a reset will bring it back into correct operation.

The program in Figure 2.31 is strictly to demonstrate the WDT function. `#fuses HS,WDT` enables the WDT and `setup_wdt(WDT_72MS)` sets up the WDT for a 72-millisecond timeout. The `restart_wdt()` is used to reset the WDT each time the main *while(1)* loop is executed.

The program lights the LED (connected to Port D.0) for 2 seconds every time it restarts. As long as Port B.0 is high the main loop continues to execute, the WDT is regularly reloaded, and the LED flashes at a 0.2-second rate. When Port B.0 is pulled low, the *while(!input(PIN_B0))* line holds the program and stops the execution of the while(1) loop. In this case the WDT times out (it isn't being reloaded) because the inner *while()* loop holds the program from executing the outer *while(1)* loop and the WDT is not being reloaded. When the WDT times out, the processor is reset and the LED once again lights for 2 seconds. Assuming Port B.0 is again allowed to go high, the LED once again starts blinking at the faster rate.

```
#include <16F877.h>
#use delay(clock=10000000, RESTART_WDT)
#fuses HS,WDT

void main()
{
   port_b_pullups(TRUE);   //pull ups on input port
   setup_wdt(WDT_72MS);    //setup and start wdt
   output_bit(PIN_D0,0);   //light the LED
   delay_ms(2000);        //wait 2 seconds
   output_bit(PIN_D0,1);   //LED off
   while(1)
   {
      restart_wdt();  //restart WDT time out period
      while (!input(PIN_B0));   //hold here if port b.0 = 0;
      output_bit(PIN_D0,0);   //light the LED
      delay_ms(200);         //wait .2 seconds
      output_bit(PIN_D0,1);   //LED off
      delay_ms(200);         //wait .2 seconds
   }
}
```

Figure 2.31 *WDT Example*

The *#use delay(clock=10000000, RESTART_WDT* line is used to tell the compiler to insert an appropriate code to automatically reload the WDT during delay functions to prevent errant timeouts.

2.8 SERIAL I/O

Serial communication is the process of sending multiple bits of data over a single wire. It is an offshoot of the original telegraph in which the bits were the dots and dashes of Morse code. The bits of a serial byte are separated by time, so that the receiving device can determine the logic levels of each bit.

Microcontrollers usually use either asynchronous or synchronous serial communication. Asynchronous communication does not require a common communication clock between the two devices, but does require an additional *start* and *stop* bit in addition to the 8-bit data word being transmitted or received. Asynchronous communication is common between PCs or other devices separated by distance. Asynchronous communication is supported in the PIC devices through the Universal Synchronous/Asynchronous Receiver Transmitter (USART) peripheral and in some devices such as the PIC18F458 via the CAN bus.

Synchronous communication does not require start and stop bits, but does require a common clock signal for the two devices that are communicating. Synchronous communication, therefore, requires an additional signal wire to carry the clock signal. Synchronous communication is more often applied within a piece of equipment because of the need for the additional wire. Synchronous communication is supported in the PIC device through the Master Synchronous Serial Port (MSSP) peripheral.

2.8.1 ASYNCHRONOUS SERIAL PORT (USART)

The USART is used to communicate from the microcontroller to various other devices. Examples of such devices include the *serial port monitor* tool in CCS-PICC (often used for troubleshooting and program debugging), other microcontrollers that need to communicate to control a system, or a PC that is communicating with the microcontroller to complete a task.

The usual form of serial communication, and the form being discussed here, is *asynchronous* serial communication. It is asynchronous in the sense that a common clock signal is not required at both the transmitter and the receiver in order to synchronize the data detection. Asynchronous serial communication uses a start bit and a stop bit added to the data byte to allow the receiver to determine the timing of each bit.

Figure 2.32 shows the elements of a standard asynchronous serial communication byte. This figure shows both the waveform and a definition for each bit of the serial word. The serial transmit line idles at a logic 1 and drops to 0 to indicate the beginning of the start bit. The start bit takes 1 full bit time and is followed by the 8 bits of the data byte that appear on the serial line in inverse order (i.e., the least significant bit appears first, whereas the most significant bit appears last). The stop bit follows the most significant data bit and is a logic 1, the same level as the idle state.

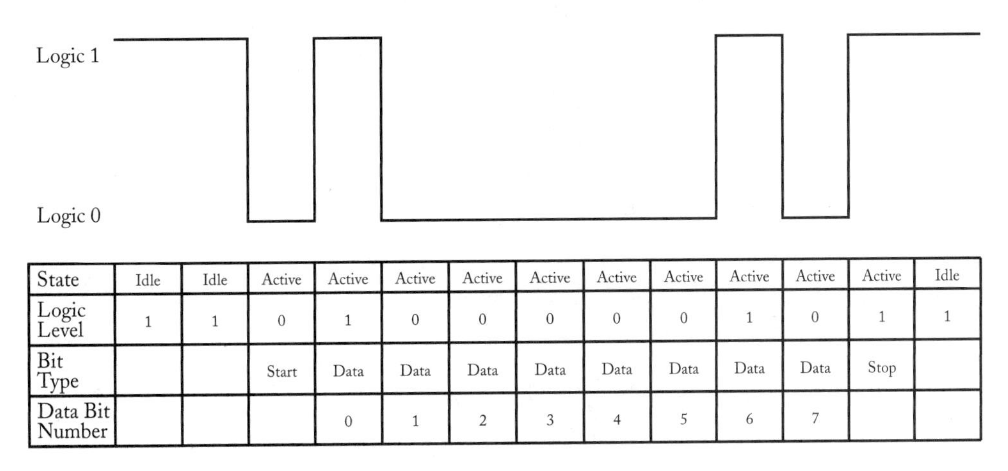

State	Idle	Idle	Active	Active	Active	Active	Active	Active	Active	Active	Active	Active	Idle
Logic Level	1	1	0	1	0	0	0	0	0	1	0	1	1
Bit Type			Start	Data	Data	Data	Data	Data	Data	Data	Data	Stop	
Data Bit Number				0	1	2	3	4	5	6	7		

Figure 2.32 *Serial Waveform*

The falling edge of the start bit begins the timing sequence in the serial receiver. Beginning from the falling edge of the start bit, the receiver waits 1 1/2 bit times before sampling the receiving line to get the first data bit. After that the receiver waits 1 bit time per bit, thereby sampling each successive data bit in the center of its time period for maximum reliability.

Fortunately, all of the timing relative to the serial byte formatting, the sampling of the serial bits, and the addition of the start stop bits are handled automatically by the *Universal Synchronous/Asynchronous Receiver Transmitter*, the USART (pronounced “you sart”). None of these issues directly affects the programmer.

The only issue confronting the programmer is to ensure that the serial communication parameters of both the transmitter and the receiver match. Specifically, this includes setting the correct number of data bits (usually 8), determining whether or not to include a parity bit (usually not), and setting the baud rate. The baud rate is the speed of serial communication and determines the timing of the bits. Baud rate is defined as the inverse of the time per bit (baud rate = 1/bit time).

The serial waveform and other parameters relating to the serial communication discussed above are all involved with the *information* being transmitted. The serial information is independent of the *media* being used to carry the serial information. The *media* may consist of a wire, an infrared beam, a radio link, or other means.

The most common media is defined as RS-232. It was developed in order to provide a greater distance for reliable serial communication using wire to carry the signal. RS-232 is an inverted scheme in that a logic 1 is represented by a negative voltage more negative than -3 V, and a logic 0 is represented by a positive voltage more positive than + 3 V. Using a different nonzero voltage for each logic level allows some hardware error checking, because a broken line will present 0 V at the receiver and so may be detected. Most microcontroller-to-PC communication is RS-232.

CCS-PICC, like most C language compilers, provides built-in library functions to handle the common serial communication tasks. These tasks usually involve communicating with a terminal device (e.g., a PC executing a terminal program to transmit serially the characters typed on the keyboard and to display in a window the characters received serially from the microcontroller). CCS_PICC also provides a built-in terminal emulator (called the serial port monitor tool) in the development environment as a convenience to the programmer. For more information on the standard library functions, see Chapter 3.

The PIC USART communication speed (the baud rate) is controlled by the *USART baud rate generator* (SPBRG). The SPBRG is a free-running 8-bit timer. Essentially, the desired baud rate is set by loading the SPBRG with the proper value to provide the desired baud rate. The baud rate is also affected by the BRGH bit in the TXSTA register (discussed further later). BRGH determines whether high-speed or low-speed communications are desired. The SPBRG calculation is as follows:

$$SPRRG = ((F_{osc} * 4^{BRGH}) / (64 * \text{baud rate})) - 1$$

For example, calculating the SPBRG value for 9600 baud communication using a PIC with a 10 MHz clock would yield this:

$$SPBRG = ((10E6 * 4^{-1}) / (64 * 9600)) - 1 = (4E6 / 614.4E3) - 1 = 64.003$$

In the example above, 64 would be loaded into SPBRG to provide the desired baud rate. Where possible (i.e., it results in a SPBRG value less than 256 so it will fit into the 8-bit register), using high-speed mode (BRGH = 1) will always yield actual baud rates closer to the desired rate than low-speed mode will. Hence using high-speed mode is desirable.

The major control registers associated with the USART are transmit status and control (TXSTA) and receive status and control (RCSTA). These registers are shown in Figure 2.33.

The PIC microcontrollers require specific sequences to set up the control registers. Fortunately, CCS_PICC handles all of these issues via the **#use** directive. Figure 2.34 and Figure 2.35 show an example of serial port use. The example program makes use of the built-in functions to receive a character via the serial port, print a response that echoes the character sent, and displays the ASCII code for the character sent on the Port D LEDs. American Standard Codes for Information Interchange (ASCII) codes are standard codes that allow the exchange of data between devices; for example, the ASCII code for 'A' is 0x41. When I type an 'A', 0x41 is sent to the other device (for instance, your PC). If your computer thought 0x41 meant 'B', we could never exchange e-mail because it would never make sense.

The example in Figure 2.34 and Figure 2.35 shows the use of the **#use** statement to set up the USART and the use of the built-in functions *getchar()* and *printf()*. Note that the getchar() function waits until a character is received before allowing the program to continue

Bit 7	Bit 6	Bit 5	Bit 4	Bit 3	Bit 2	Bit 1	Bit 0	
	TX9	TXEN	SYNC		BRGH	TRMT	TX9D	TXSTA

TX9 – Enable 9th bit transmission mode (1 = enabled)
SYNC – Selects asynchronous or synchronous mode (0 = asyncronous mode)
BRGH – High speed select for baud rate generator (1 = high speed mode)
TRMT – Transmit shift register status (1 = TSR empty)
TX9D – Ninth data bit in 9-bit transmission mode
TMR1ON – Timer1 on bit (1 = on, 0 = off)

SPEN	RX9		CREN	ADDEN	FERR	OERR	RX9D	RXSTA

SPEN – Serial port enable bit (1 = use standard pins—port C.6 and C.7, 0 = disabled)
RX9 – Enable 9th bit mode (1 = enabled)
CREN – Enables continuous receive (1 = enabled)
ADDEN – Address detect enable for 9th bit mode
FERR – Framing error detected (1 = error detected)
OERR – Overrun error detected (1 = detected)

Figure 2.33 *USART Control Registers*

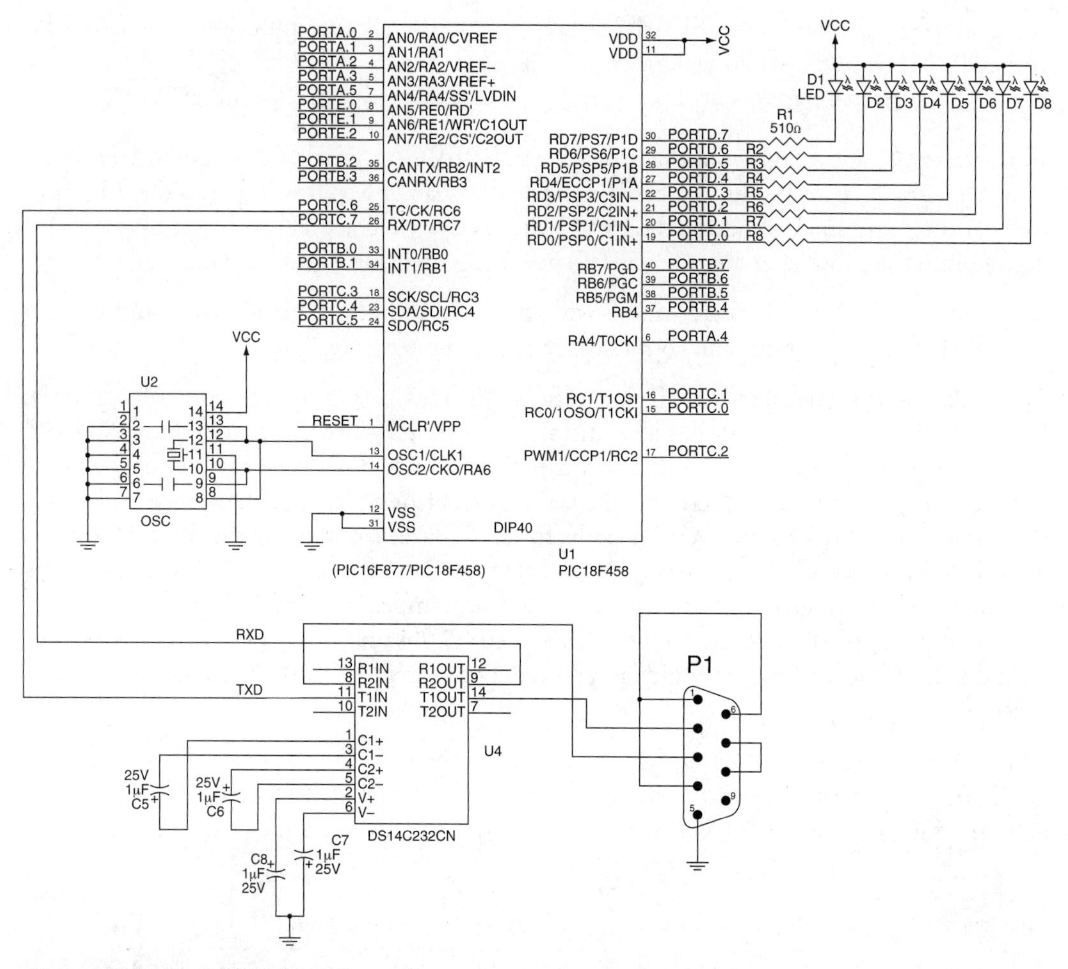

Figure 2.34 *Serial Hardware Example*

executing. Use of the *kbhit()* function would allow the program to use the time that it was waiting for serial character. An equivalent program using *kbhit()* is shown in Figure 2.36. More details on the built-in functions may be found in Chapter 3.

Figure 2.36 is different in that the *kbhit()* function only returns a true value if a serial character has been received. In this way, the *getchar()*, *output_bit()*, and *printf()* functions are only executed when a character is received. Consequently, the program can be doing other things (in this case blinking an LED) rather than waiting for a serial character in the *getchar()* function. This is actually much better programming practice than the example shown in Figure 2.35. It is also possible to use the USART on an interrupt basis where receipt of a serial character causes an interrupt. The ISR would take care of reading the serial character and the appropriate action. In this example, the action would be to output the character code to the LEDs and to send the response message.

```
#include <16F877.h>
#use delay(clock=10000000)
#fuses HS,NOWDT
#use rs232(baud=9600,parity=N,xmit=PIN_C6,rcv=PIN_C7,stream=,bits=8)
#include <stdio.h>


void main()
{
   char ch;  //character variable
   while(1)
   {
      ch = getchar();   //wait for and get serial character
      output_D (~ch);    //display on LEDs
      printf ("\n\r You pressed %c",ch);  //echo it with comment
   }
}
```

Figure 2.35 *Serial Example Software.*

```
#include <16F877.h>
#use delay(clock=10000000)
#fuses HS,NOWDT
#use rs232(baud=9600,parity=N,xmit=PIN_C6,rcv=PIN_C7,stream=,bits=8)
#include <stdio.h>


void main()
{
   char ch;  //character variable
   while(1)
   {
      if (kbhit())
      {  //only get the character if one has been recieved.
         ch = getchar();   //wait for and get serial character
         output_D (~ch);    //display on LEDs
         printf ("\n\r You pressed %c",ch);  //echo it with comment
      }
      output_bit(PIN_C2,0);    //light LED on Port C.2
      delay_ms(100);  //wait .1 seconds
      output_bit(PIN_C2,1);    //extinguish LED on Port C.2
      delay_ms(100);  //wait .1 seconds
```

Figure 2.36 *KBHIT() Example (continues)*

```
        //more program steps could be placed here that would
        //be executed while the program was waiting for a serial
        //character.
    }
}
```

Figure 2.36 *KBHIT() Example (continued)*

E

EXAMPLE

Chapter 2 Example Project – Part C

This is the third part of this chapter's example system to collect operational data for a stock car. This part will be concerned with using the timers to measure the 1-second recording interval and the two rpm measurements.

One-Second Recording Interval Using Timer0

Using a timer to produce a 'tick' or delay interval has been discussed. In this case, you are trying to produce a long (in microcontroller terms) time interval, and you have other interrupt-dependent processes running as well. So it would be well to use as slow a clock as is practical for the 'tick' clock so as to minimize the amount of time used in the tick routine, as long as accuracy of recording interval can be maintained.

Further investigation of the PIC18F458 system you have been assigned to use shows that the system clock is 10 MHz. The Timer0 may be configured in this processor as either an 8- or 16-bit timer/counter. The Timer0 prescaler allows division ratios of 1, 2, 4, 8, 16, 32, 64, 128, and 256. Choosing a prescaler value of 256 produces a 9.766 KHz clock applied to Timer0:

10 MHz / 4 / 256 = 9.766 KHz

Now you can figure out the number of clock pulses that must elapse for 1 second to have passed. In this case, it is easy and 9766 clock pulses must occur for 1 second to have elapsed. Timer0 is a 16-bit counter and so it will need to be reloaded to 55,770:

65536 – 9766 = 55770

Now the question might be accuracy. At that reload rate, what will the real time per sample be and how far off will it be at the end of 2 minutes?

Time per sample: 9766 * actual tick time = 9766 * 4 * 256 / 10 MHz = 1.0000384 seconds

This is an error of about 0.004%. Over the period of 2 minutes the actual time will be:

120 * 1.0000384 seconds = 120.0046 seconds

Your boss agrees that this is accurate enough for his purposes.

The following code sets up the data arrays and the interrupt service routine for the Timer0 overflow:

```
int16 e_rpm[120], s_rpm[120];    //array for rpm data
int16 current_e_rpm, current_s_rpm;  //current values of RPM
int temp[120];                   //array for temp data
int current_temp;                //current temperature value
int data_set_cntr;  //counter for 120 sets of data

#int_TIMER0
TIMER0_isr()
{
   set_timer0(55770);    //reload for the next second
   if (data_set_cntr < 120)
   {
      e_rpm[data_set_cntr] = current_e_rpm;    //record engine rpm
      s_rpm[data_set_cntr] = current_s_rpm;    //record shaft rpm
      temp[data_set_cntr++] = current_temp;    //record engine temp
   }
}
```

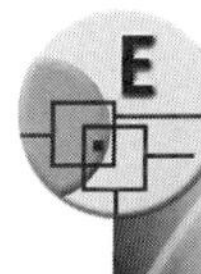

EXAMPLE

Figure 2E.2 *Timer0 ISR Code*

Notice that the *data_set_cntr* variable is post-incremented as a part of the storing of the current temperature. Adding *data_set_cntr = 0* to the *main()* loop just above the *while(1)* will ensure that the data starts recording when the program starts.

The following line is temporarily added to the Timer0 ISR for test purposes:

output_d(~data_set_cntr);

This will allow the programmer to check both the correctness of the counter and the timing of the record loop. Starting the program should cause current-sinking LEDs attached to Port D to count up to 120 over a 2-minute period.

Engine rpm Measurement Using Capture/Compare Module 1

The capture feature is used to capture the period of the engine rpm pulses. The period is used to calculate the rpm. The project assumes a sensor that provides 1 pulse per revolution.

The system clock is 10 MHz in this system. The system clock is divided by 4 to provide the clocking pulses for Timer1. These clock pulses drive Timer1, and the *Capture/Compare/PWM Register 1 (CCPR1)* will be used to capture the Timer1 count each time a rising edge occurs on the CCP1 pin of the PIC18F458. When this occurs, an interrupt will be generated. In the interrupt service routine, the current captured count is compared to the previously captured time to determine the period of the pulses, which is used to calculate the engine rpm.

The code in Figure 2E.3 makes up the ISR for the capture event:

```
TIMER1_isr()
{
   ++T1_rolls;     //count Timer1 roll overs for Engine RPM
}

int16 previous_capture_time;
#int_CCP1
CCP1_isr()
{
   int16 current_capture_time;
   int32 period;     //variables for calculations
   current_capture_time = CCP_1;
   //CCP_1 is from CCS-PICC for the capture value
   period = (T1_rolls * (int32)0xFFFF) - (int32)previous_capture_time +
                                      (int32)current_capture_time;
   current_e_rpm = (int32)75E6 / (int32)period;  //calculate RPM
   previous_capture_time = current_capture_time;
   //use for next calc
   T1_rolls = 0;    //clear rollovers for next calculation
}
```

Figure 2E.3 *RPM Capture Code*

In this code, a global variable, *previous_capture_time*, is initialized to retain the value from the previous capture. The ISR function reads the current capture time and uses it, along with the previous captured time, to calculate the rpm. The calculation takes into account any possible *Timer1 rollover.* This is necessary to allow for the case in which the 16-bit timer/counter rolls over from 0xFFFF to 0x0000 during the elapsed period. The last statement in the ISR saves the *current_capture_time* for use as the *previous_capture_time* the next time a pulse occurs and clears the Timer1 rollover counter for the next period measurement.

The rpm equation is derived as follows:

rps (revolutions per second) = rpm / 60

Period for the rps = 1 / rps = 1 / (rpm/60) = 60 / rpm

Timer counts for any rpm = period / .8 usec. (.8 usec is the period of the T1 clock)

= (60 / rpm) / .8 usec.

rpm = (60 / counts) / .8 usec. = 75×10^6 / counts

Note that casting is used to ensure accuracy with the large numbers.

Drive Shaft rpm Measurement Using Timer1

This measurement gets a little bit more involved because the shaft rpm signal is connected to INT0. You can create your own capture register within the INT0 ISR by reading the

Timer1 count when the interrupt occurs. Then the rest of the function works very much like the one above for the engine rpm:

```
int16 previous_shaft_time;   //variable to save shaft start time
int shaft_rolls;             //roll over counter for shaft RPM
#int_EXT
EXT_isr()
{
   int16 current_shaft_time;
   int32 period; //local variables

   current_shaft_time = get_timer1();   //read current timer1 count
   period = (int32)Shaft_rolls * (int32)0xffff -
             (int32)current_shaft_time
             + (int32)previous_shaft_time;
   current_s_rpm = (int32)75E6 / period; //calculate shaft RPM
   previous_shaft_time = current_shaft_time;
                             //save value for next calculation
   shaft_rolls = 0;          //clear timer1 roll over counter
}
```

Figure 2E.4 *Shaft RPM Code*

The calculation is the same as the one for the engine rpm. An additional variable is incremented in the Timer1 rollover ISR (*++shaft_rolls*). The relatively slow shaft rpm makes it very important to cast variables to larger sizes. Also note that some of the local variables are of the same name as in the engine rpm calculations. Because they are local, this is acceptable, but may provide some debugging grief later when you are trying to track the values in two different variables with the same name.

2.8.2 CAN BUS MODULE

The USART asynchronous communication presented in the preceding section is a very simple communications system. Implementing requires only two devices: a transmitter and a receiver. The transmitter sends information and the receiver receives it. There is very little opportunity for errors to creep into the system. If there is a chance of errors, simple error checking scheme such as parity bits and checksums can protect the communications.

When multiple devices are present and communicating via a single bus, all sorts of possibilities for errors exist and, in addition, the whole problem of addressing becomes important. In a two-device system, one talks and one listens—very simple. With multiple devices it must be defined which device is sending data to which other device on the bus. These needs call for a communications system *protocol.*

A protocol defines the addressing method, the error checking, and the general data format for a bus used by multiple devices. One such protocol is the Controller Area Network (CAN) bus. The CAN specification and entire communications scheme is well beyond the scope of this text. Happily, though, for those who use the CAN bus, the PIC18 family supports it (check the specification for individual PIC devices to see which ones support CAN) and CCS-PICC provides drivers that makes using CAN convenient. Using the built-in CAN drivers eliminates the need for detailed bit control via registers.

CCS-PICC and the PIC controllers support CAN 2.0 A and B protocols. For more information, refer to the BOSCH CAN specification.

As with most peripherals, the use of the CAN peripheral amounts to getting the module initialized and then using the built-in functions to communicate via the CAN bus. Figure 2.37 is a working CAN test program. The program transmits eight sequential numbers every second via the CAN bus. The numbers transmitted range from 0 to 255 and repeat. At the same time, the program monitors the CAN bus for transmissions sent to it and displays those via the USART.

The comments in Figure 2.37 explain the program clearly. CAN is also used in the scooter project in Chapter 5.

2.8.3 SYNCHRONOUS SERIAL PORT (MSSP)

The MSSP is the synchronous corollary to the asynchronous port described earlier. The major difference is that the synchronous port requires a common clock and, hence, does not need the start and stop bits to be added to each serial byte. The PIC MSSP is capable of operating in two modes: Synchronous Peripheral Interface (SPI) and Inter-integrated Circuit (I^2C). SPI (pronounced "spy") and I^2C (developed by Philips/Signetics for internal communication of TV ICs that has become an industry standard) are covered in the sections that follow.

2.8.3.1 SPI Bus

The SPI bus is a synchronous serial communication bus, meaning that the transmitter and receiver involved in SPI communication must use the same clock to synchronize (hence, the term *synchronous* communication) the detection of the bits at the receiver. Normally, the SPI bus is used for very short distance communication with peripherals or other microcontrollers that are located on the same circuit board or at least within the same piece of hardware. This is different from the USART, which is used for communication over longer distances, such as between units or between a microcontroller and PC. The SPI bus was developed to provide relatively high-speed, short-distance communications using a minimum number of microcontroller pins.

SPI communication involves a master and a slave. Both the master and a slave send and receive data simultaneously, but the master is responsible to provide the clock signal for the data

```
#include <18F458.h>
#use delay(clock=10000000)
#fuses HS,BROWNOUT,WRT,NOWDT,WDT1
//set for immediate I/O
#use fast_io(B)
//use the USART ti report CAN status and results
#use rs232(baud=19200,parity=N,xmit=PIN_C6,rcv=PIN_C7)
//include the CCS-PICC CAN libraries and drivers
#include <can-18xxx8.c>

int16 ms;  //seconds counter
//variables needed by the program for received information and data
//struct rx_stat rxstat;
int32 rx_id;
int in_data[8];    //received data array
int rx_len;

//transmit variables
int out_data[8];  //transmit data array
int32 tx_id=24;   //transmit to CAN id 24
int1 tx_rtr=0;
int1 tx_ext=0;
int tx_len=8;     //transmit 8 bytes
int tx_pri=3;

#int_timer2
void isr_timer2(void)
{
   ms++; //keep a running timer that increments every milli-second
}

void main()
{
   int i,j;
   setup_adc_ports(NO_ANALOGS);
   setup_adc(ADC_CLOCK_DIV_2);
   setup_psp(PSP_DISABLED);
   setup_spi(FALSE);
   setup_wdt(WDT_OFF);
   setup_timer_0(RTCC_INTERNAL);
   setup_timer_1(T1_DISABLED);
   setup_timer_3(T3_INTERNAL|T3_DIV_BY_1);
   setup_ccp1(CCP_PWM);
```

Figure 2.37 *CAN Example (continues)*

```
setup_ccp2(CCP_PWM);
setup_comparator(FALSE);

for (i=0;i<8;i++)    //clear both transmit and recieve arrays
{
   out_data[i]=0;
   in_data[i]=0;
}

setup_timer_2(T2_DIV_BY_4,79,16);    //setup up timer2 to interrupt
                                     //every 2ms

can_init();                         //This function dinitializes the CAN
                                    //bus module

enable_interrupts(INT_TIMER2);      //enable timer2 interrupt for ticks
enable_interrupts(GLOBAL);          //enable all interrupts (else timer2
                                    //wont happen)

while(TRUE)                         //main while(1) loop
{

   if ( can_kbhit() )    //if data has been received via CAN
   {
      if(can_getd(rx_id, in_data, rx_len, rxstat))
      { //...then get data from buffer and print report
            printf("\r\nGOT: BUFF=%U ID=%LU LEN=%U OVF=%U ",
            rxstat.buffer, rx_id,
                  rx_len, rxstat.err_ovfl);
         printf("FILT=%U RTR=%U EXT=%U INV=%U", rxstat.filthit,
         rxstat.rtr,
                  rxstat.ext, rxstat.inv);
         printf("\r\n    DATA = ");

         for (i=0;i<rx_len;i++)        //get number of data bytes
                                       //specified by message
            printf("%X ",in_data[i]);//print the data as it is
                                       //retrieved
         printf("\r\n");               //next line
      }
      else
      {
         printf("\r\nFAIL on GETD\r\n");  //report a failure to get
         data
```

Figure 2.37 *CAN Example (continues)*

```
      }
   }

   //every second, send new data if transmit buffer is empty
   if (can_tbe() && (ms > 500))  //can_tbe() checks the CAN transmit
                                 //buffer
   {
      ms=0;                      //reset ms for next 1 second period
      for (i=0;i<8;i++)          //load transmit array with next 8
                                 //sequential numbers
            out_data[i] = j++;

      i=can_putd(tx_id, out_data, tx_len,tx_pri,tx_ext,tx_rtr);
                                   //put data in transmit buffer
      if (i != 0xFF)  //i contains value returned from transmit
                      //function
      { //success, a transmit buffer was open so print report
         printf("\r\nPUT %U: ID=%LU LEN=%U ", i, tx_id, tx_len);
         printf("PRI=%U EXT=%U RTR=%U\r\n   DATA = ", tx_pri, tx_ext,
         tx_rtr);
         for (i=0;i<tx_len;i++)
         {
            printf("%X ",out_data[i]);  //show data transmitted
         }
         printf("\r\n"); //newline
      }
      else
      { //fail, transmit error occurred
         printf("\r\nFAIL on PUTD\r\n");  //so report failure
      }
   }
   else    //nothing received
   {
      if(ms > 1000)  //if no transmit occurs and nothing is recieved
                     //for 2 seconds,
                     //report status for troubleshooting
      {
         printf("\r\n\nCOMSTAT = %02X",COMSTATUS);
         printf("\r\nTXB0CON = %02X",TXB0CONR);
         printf("\r\nTXB1CON = %02X",TXB1CONR);
         printf("\r\nTXB2CON = %02X",TXB2CONR);
         printf("\r\nRXERRCNT = %U",RXERRCNT);
         printf("\r\nTXERRCNT = %U",RXERRCNT);
```

Figure 2.37 *CAN Example (continues)*

```
                ms = 0;
            }
        }
    }
}
```

Figure 2.37 *CAN Example (continued)*

transfer. In this way, the master has control of the speed of data transfer and is, therefore, *in control* of the data transfer.

Figure 2.38 shows the connections between the master and the slave units for SPI communication. The master supplies the clock on the SCLK pin and 8 bits of data, which are shifted out of the *serial data out* (SDO) pin. The same 8 bits are shifted into the slave (1 bit per clock pulse) on its *serial data in* (SDI) line. As the 8 bits are shifted out of the master and into the slave, 8 bits also are shifted out of the slave on its *SDO* line and into the master on its SDI pin. SPI communication, then, is essentially a circle in which 8 bits flow from the master to the slave and a different 8 bits flow from the slave to the master. In this way, the master and a slave can exchange data in a single communication.

It is entirely possible to connect many different devices together using an SPI bus, because all of the SDO pins and all of the SDI pins could be hooked together. In this instance, a fourth pin, slave select (SS), is used to determine which device is the slave that will communicate with the master. Usually the SS pins from the slaves are connected to either a parallel port on the master or to a decoder that determines which device will be the active slave.

The MSSP is controlled by three registers: Sync Serial Status Register (SSPSTAT), Sync Serial Port Control Register (SSPCON), and Sync Serial Port Control Register 2 (SSPCON2). SSPCON2 is useful only in I²C mode and, therefore, is not shown. These registers are shown in Figure 2.39. The bits in these registers have different functions in I²C mode; consult your processor specification for details.

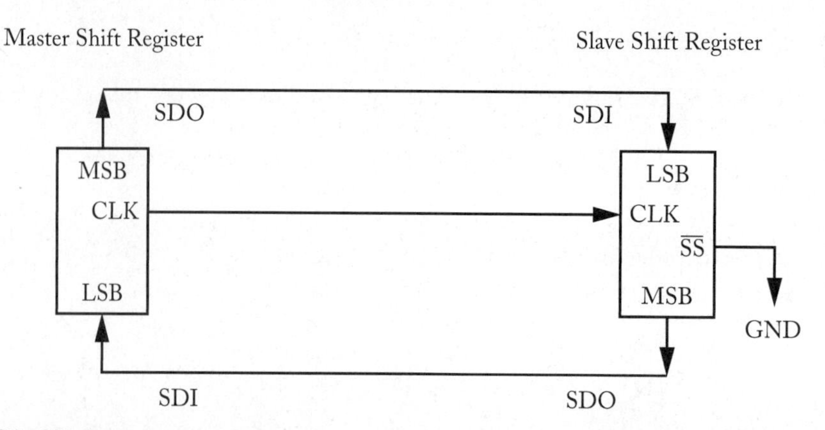

Figure 2.38 *SPI Communication Scheme*

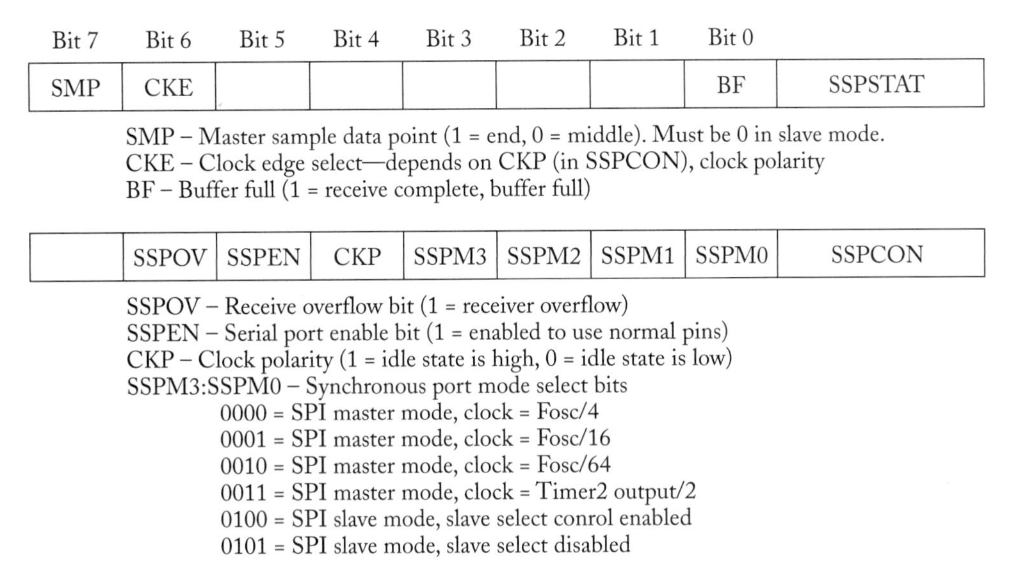

Bit 7	Bit 6	Bit 5	Bit 4	Bit 3	Bit 2	Bit 1	Bit 0	
SMP	CKE						BF	SSPSTAT

SMP – Master sample data point (1 = end, 0 = middle). Must be 0 in slave mode.
CKE – Clock edge select—depends on CKP (in SSPCON), clock polarity
BF – Buffer full (1 = receive complete, buffer full)

	SSPOV	SSPEN	CKP	SSPM3	SSPM2	SSPM1	SSPM0	SSPCON

SSPOV – Receive overflow bit (1 = receiver overflow)
SSPEN – Serial port enable bit (1 = enabled to use normal pins)
CKP – Clock polarity (1 = idle state is high, 0 = idle state is low)
SSPM3:SSPM0 – Synchronous port mode select bits
0000 = SPI master mode, clock = Fosc/4
0001 = SPI master mode, clock = Fosc/16
0010 = SPI master mode, clock = Fosc/64
0011 = SPI master mode, clock = Timer2 output/2
0100 = SPI slave mode, slave select conrol enabled
0101 = SPI slave mode, slave select disabled

Figure 2.39 *SPI Control Registers*

As a simple example of the use of the SPI bus, Figure 2.40 and Figure 2.41 show an example oversimplified system. In this example, the SDO pin of a PIC16F877 is connected directly to its SDI pin. Consequently, anything transmitted on the SPI bus will also be received

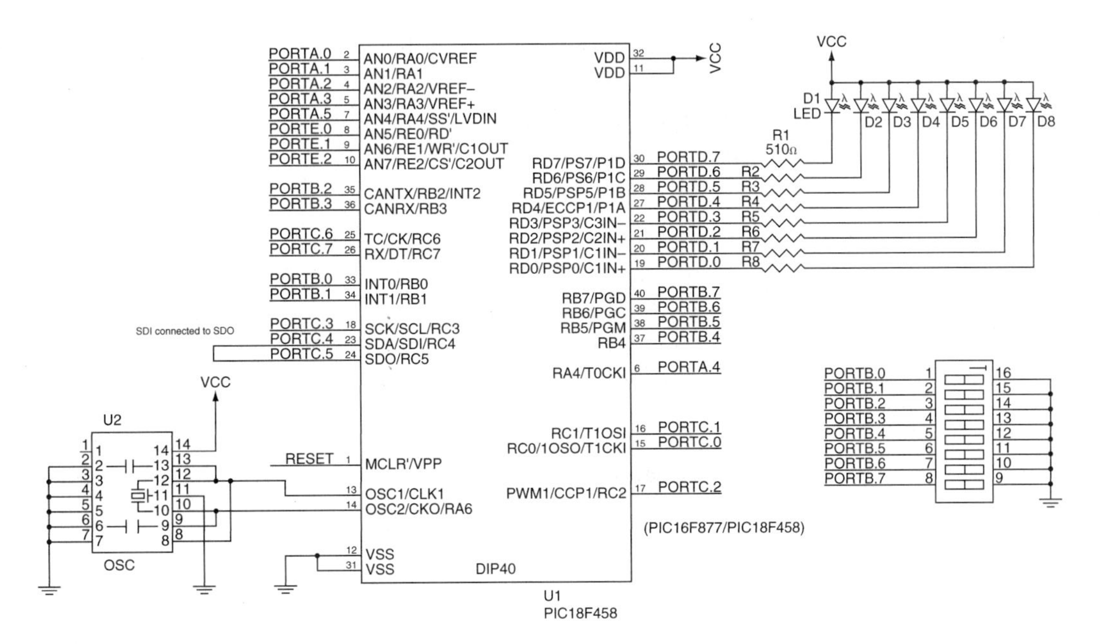

Figure 2.40 *SPI Hardware Example*

```
#use delay(clock=10000000)
#fuses HS,NOWDT

void main()
{
   setup_spi(SPI_MASTER|SPI_L_TO_H|SPI_CLK_DIV_16);
   port_b_pullups(TRUE);  //pull ups active on Port B
   spi_write(0x00);  //get new data and send it out the SPI
   while(1)
   {
      if (spi_data_is_in())
      {
         output_D(~(spi_read()));         //show data on LED's
         spi_write(input_b());  //get new data and send it out the SPI
      }
      output_bit(PIN_C2,0);     //light LED on Port C.2
      delay_ms(50);             //wait .1 seconds
      output_bit(PIN_C2,1);     //extinguish LED on Port C.2
      delay_ms(50);             //wait .1 seconds
   }
}
```

Figure 2.41 *SPI Software Example*

by the SPI receiver. This example then is really showing how to set up and use the MSSP in SPI mode. Configuring a slave processor to work with this system would only involve slight changes to the setup of the slave (i.e., make it a slave by changing the mode in SSPCON). For an example using two processors, refer to Appendix A under the *spi_read()* function description.

The example program in Figure 2.41 shows the appropriate method to set up and use the SPI port of the MSSP module in a manual-operating mode. The built-in functions are used to read and write data to the SPI port. It is also possible to set up the SPI to operate on an interrupt basis in order to prevent the program from having to either wait on the SPI or to need to poll the SPI port as the example does.

2.8.3.2 I^{2}C Bus

Inter-integrated Circuit (I^{2}C) communication is a protocol-oriented synchronous communication system much as CAN is to asynchronous communication. It is designed to allow multiple devices to share the same I^{2}C bus. I^{2}C is different from SPI communication in two ways. The first is the protocol and the second is that, with I^{2}C there is only one data wire and one clock wire. The lines are called synchronous clock (SCL) and synchronous data (SDA) for the clock and data, respectively. For the built-in PIC MSSP, these lines are typically Port C.3

and Port C.4, respectively. These are the same pins used by the MSSP in SPI mode.

Using only two wires requires a different approach to data flow. In SPI, data flows out of the master into the slave and *simultaneously* out of the slave and into the master. I^2C uses a *sequential* system in which data flows from the master to the slave on the SDA and then, if necessary (in the case of the master reading data from the slave), data flows from the slave to the master on the SDA. In either case, the master always supplies to the clock as appropriate to a master/slave system.

I^2C is not as complex a protocol as the CAN bus. The first byte sent by the master is the address byte, and the slave whose address matches wakes up and receives any following message bytes. Or the address may contain a 'read' bit indicating that the next transfer is to send data to the master. In this case, the software in the slave responds and loads the I^2C write buffer with the requested data. When the master next sends out clock pulses, the data is sent from the slave's buffer to the master on the SDA. For further details on the I^2C protocol, refer to the Bosch I^2C specification.

Figure 2.42, Figure 2.43, and Figure 2.44 show an example system that implements two-way communication via the I^2C bus. Figure 2.42 is a diagram of two sets of the hardware shown in Figure 2.40 with slight changes to the connections to Port C to implement I^2C. Using I^2C, the data on the input to the master's Port B (switches) is transferred to and displayed on the slave's Port D (LEDs) and vice versa.

As can be seen in the example I^2C software, using the built-in functions of CCS-PICC make the job much easier, rather than attempting to directly control the communication through the register bits.

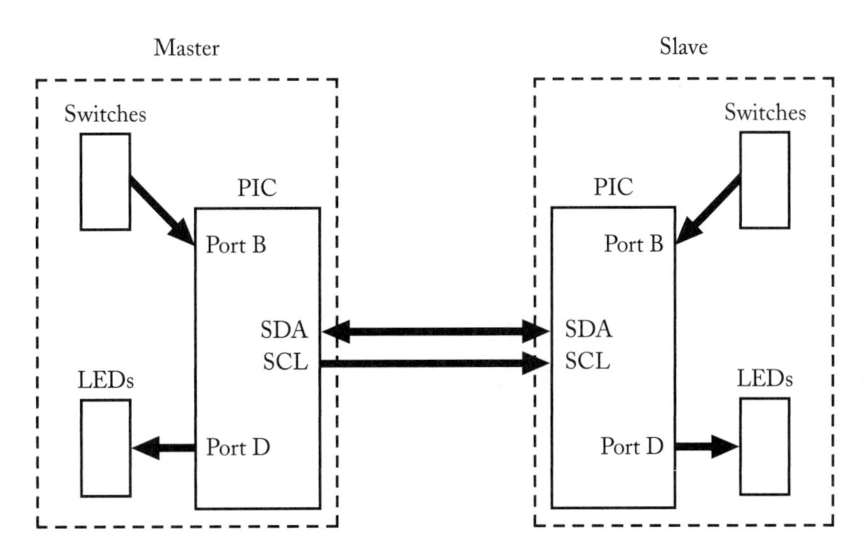

Figure 2.42 *I^2C Example System*

```
#include <16F877.h>
#use delay(clock=10000000)
#fuses HS,WDT
//set up i2c peripheral and use hardware SSP
#use i2c(Master,sda=PIN_C4,scl=PIN_C3,RESTART_WDT,FORCE_HW)

#int_TIMER1
TIMER1_isr()
{   //communicate every 50 ms
   i2c_start();                   //set i2c to start condition
   i2c_write(0xa0);               //send address
   i2c_write(input_b());          //send master data from port B bits
   i2c_start();                   //restart to change data direction
   i2c_write(0xa1);               //address with direction bit changed
   output_d(~i2c_read(0));        //get slave data and display
   i2c_stop();                    //stop i2c activity
}

void main()
{
   setup_counters(RTCC_INTERNAL,WDT_1152MS);
   setup_timer_1(T1_INTERNAL|T1_DIV_BY_1);
   port_b_pullups(TRUE);          //pull ups for input pins
   enable_interrupts(INT_TIMER1);
   enable_interrupts(global);
   while(1)
   {
      //use WDT in case i2c 'hangs up' due to faulty slave
      restart_wdt();
   }
}
```

Figure 2.43 *Master I²C Software*

The general scheme of the I²C protocol is:

1. The master asserts the start conditions onto the bus. This alerts the slaves that something is coming and synchronizes the data flow.
2. The master then sends the address of the slave that is to be active. All of the slaves receive this byte, and the slave whose address matches actively receives and interprets the data while the others ignore the communication until another start pulse occurs. The address consists of the upper 4 bits of the address byte. The lowest bit controls the direction of the next data transfer (1 or more bytes). A '0' indicates a master-to-slave transfer, and a '1' indicates the opposite transfer.

```
#include <16F877.h>
#use delay(clock=10000000)
#fuses HS,NOWDT
#use i2c(SLAVE,sda=PIN_C4,scl=PIN_C3,address=0xa0)

enum {address,data};
int mode;

#int_SSP
SSP_isr()
{
   int ch;
   //check for valid char received.  The interrupt occurs, but no
   //valid character is received when setting up for the read
   if (i2c_poll())
   {
      ch = i2c_read();              //get data received
      //ch contains address on first transmission of the interchange
      if (mode == address)  mode = data;  //so ignore and set for data
                                          //byte
      else //must be data byte
      {
         output_D(~ch);  //output master data on LEDs
         mode = address;  //go back to waiting for address
      }
   }
   //get to 'else' only if second command byte of interchange, so load
   //i2c output register for read operation
   else i2c_write(input_b());     //load for read
}


void main()
{
   port_b_pullups(TRUE);         //pullups on input bits
   enable_interrupts(INT_SSP);   //interrupt mode required
   enable_interrupts(global);
   mode = address;               //set starting mode
   while(1)                      //main does nothing
      ;
}
```

Figure 2.44 *Slave I²C Software*

3. Data is then transferred as indicated by the address byte.
4. Data may be transferred in both directions during an interchange by resending the start pulse with the direction bit changed as in the example software.
5. Finally, the master must issue a stop condition on the bus to conclude the transfer.

Figure 2.43 shows the I^2C software for the master PIC. The scheme is based on a 50-millisecond tick created when Timer1 overflows. The I^2C communications occurs in the Timer1 overflow ISR. The following steps occur to do the I^2C transfer:

1. The master asserts the *start* condition using *ic2_start()*.
2. The master sends the address of the slave using *i2c_write(0xa0)*. The slave's address is 0xa, which is placed on the upper 4 bits of the address byte. The least significant bit of the address is set to '0' to indicate a master-to-slave transfer.
3. The master reads its own Port B and sends the result again using *i2c_write(input_b())* to both read the switches and send the data.
4. The master asserts another *start* condition to set up changing the data flow direction. Actually, CCS_PICC modifies such a request into a *restart* condition appropriate to the I^2C protocol.
5. The master then resends the slave address with the write bit set high to indicate a slave-to-master transfer using *i2c_write(0xa1)*.
6. The master uses *i2c_read()*, which causes the master to create the clock pulses for the slave to use in sending its data. The received data is inverted (those pesky current-sinking LEDs again) and written to the master's LEDs. This is accomplished in one step using *output_d(~i2c_read())*.
7. Finally, the master asserts the stop condition using *i2c_stop()* to end the communication cycle.

The slave software is similar but a bit trickier. Figure 2.44 is the slave software, which is based on an interrupt and a state machine. Each time a character is received, the SSP_isr() is executed (the start and stop conditions are not complete characters and are handled by the I^2C peripheral without creating an interrupt). The order of events in the slave is as follows:

1. The slave starts up in the 'address' state (*mode = address in main()*).
2. Each time a byte (*valid* or not) is received, the SSP interrupt occurs. The *i2c_poll()* function is used to determine if the character is one that the I^2C software needs to evaluate (a valid character).
3. The first valid character received is the address byte sent by the master. This byte causes the slave to become active in the communication sequence but is of no real use as a data byte, so the receipt of the address byte is used to simply change the software into the *data* state.
4. The second valid byte is the data to be displayed on the LEDs. When this character is received, it is displayed on the LEDs (with inversion to account for the current-sink nature of the LEDs). Also the mode is changed back to *address*.

5. While in address mode, the restart condition occurs and the address byte with the read bit set (0xa1) is received. This byte causes an interrupt but is not considered a valid character and so *i2c_poll()* returns a FALSE value. The FALSE value makes the *if (i2c_poll())* fail, and the *else* is used to write the data to be read by the master into the I^2C write buffer (*i2c_write(input_b())*).
6. The slave sends the data to the master as the master provides the clock pulses for the transfer.

The complete communications cycle is thus completed. Depending on the software, the transfer may be multiple bytes or single bytes in whatever direction is required.

I^2C is often used to communicate with serial EEPROMs, A/D converters, LCD displays, etc. In this case, the slave end (the EEPROM or any of the others) will determine the number of bytes and direction of transfer, and the master must conform because the EEPROMs and the like are factory-programmed as to their I^2C needs.

2.9 ANALOG TO DIGITAL I/O

2.9.1 ANALOG TO DIGITAL BACKGROUND

Analog to digital conversion (and digital to analog conversion) is largely a matter of *proportion*. That is, the digital number provided by the A/D converter relates to the proportion that the input voltage is of the full voltage range of the converter. For instance, applying 2 V to the input of an A/D converter with a full-scale range of 5 V will result in a digital output that is 40% of the full range of the digital output (2 V / 5 V = 0.4).

A/D converters with a variety of input voltage ranges and output digital ranges are available. The output digital ranges are usually expressed in terms of bits, such as 8 bits or 10 bits. The number of bits at the output determines the range of numbers that may be read from the output of the converter. An 8-bit converter will provide outputs from 0 up to 2^8-1, or 255; or a 10-bit converter will provide outputs from 0 up to $2^{10}-1$, or 1023.

In the previous example in which 2 V was applied to a converter with a full-scale range of 5 V, an 8-bit converter would read 40% of 255, or 102. The proportion/conversion factor is summarized in the following formula:

$$\frac{V_{in}}{V_{fullscale}} = \frac{X}{2^n-1}$$

In the formula above, "X" represents the digital output and "n" represents the number of bits in the digital output. Using this formula, and solving for X, you can calculate the digital number read by the computer for any given input voltage. Using the formula within a program where you have X, you can use the formula to solve for the voltage being applied. This is useful when you might be trying to display the actual voltage on an LCD readout.

An important issue buried in this formula is that of resolution. The resolution of measurement (or the finest increment that can be measured) is calculated as follows:

$$V_{resolution} = \frac{V_{fullscale}}{2^n - 1}$$

For an 8-bit converter that has a full-scale input range of 5 V, the resolution would be:

$$\text{Vres.} = 5\text{V}/(2^8-1) = 5\text{V}/255 = 20 \text{ mv (approx.)}$$

Therefore, the finest voltage increment that can be measured, in this situation, is 20 mv. It would be inappropriate to attempt to make measurements that are, for example, accurate to within 5 mv with this converter.

2.9.2 ANALOG TO DIGITAL MODULE

The A/D converter in the PIC microcontrollers is a 10-bit converter that can be used to convert analog voltages at one of eight inputs (or five inputs on the physically smaller devices). Four registers are used to control/read the A/D results. The A/D result high register (ADRESH) contains the upper 2 bits of the A/D result in 10-bit mode, or to contain the upper 8 bits of the result in 8-bit mode. ADRESL (A/D result register low) contains the balance of the bits in either mode. The control registers for the A/D converter are shown in Figure 2.45.

Implementing the A/D converter involves:

A. Choosing the A/D configuration

B. Choosing the clock and channel to convert (loading ADCON)

C. Starting the conversion (Setting GO high in ADCON)

D. Waiting for the GO bit in ADCON to go low, signaling the end of the conversion

E. Reading the result from the correct result register(s)

F. Repeat steps B thru E as needed to do additional conversions.

The A/D clock must be chosen to have a maximum speed of 625 KHz. In as much as the A/D clock is derived from the system clock, the ADCS1:ADCS0 bits must be chosen to select a divider that divides the system clock to a speed less than 625 KHz. To choose the appropriate divisor, divide the system clock by 625 KHz and then use the next higher divisor. For example, the divisor for a system with a clock of 10 MHz would be chosen as follows:

$$\text{Divisor} = F_{osc} / 625 \text{ KHz} = 10 \text{ MHz} / 625 \text{ KHz} = 16$$

The next higher divisor is 32 and so it must be used.

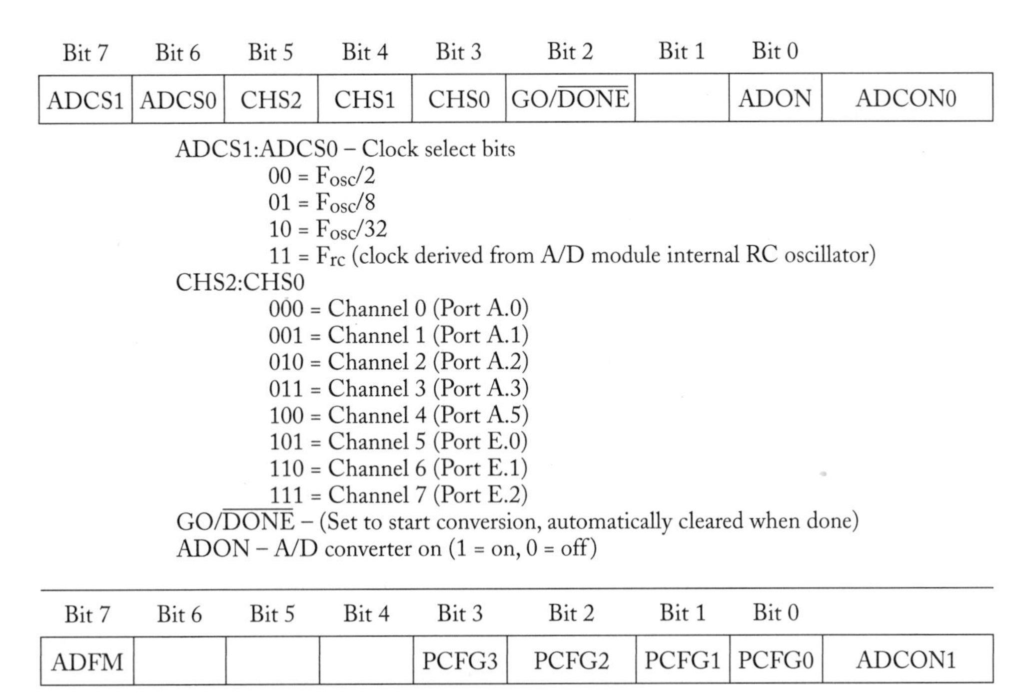

Bit 7	Bit 6	Bit 5	Bit 4	Bit 3	Bit 2	Bit 1	Bit 0	
ADCS1	ADCS0	CHS2	CHS1	CHS0	GO/$\overline{\text{DONE}}$		ADON	ADCON0

ADCS1:ADCS0 – Clock select bits
- 00 = $F_{osc}/2$
- 01 = $F_{osc}/8$
- 10 = $F_{osc}/32$
- 11 = F_{rc} (clock derived from A/D module internal RC oscillator)

CHS2:CHS0
- 000 = Channel 0 (Port A.0)
- 001 = Channel 1 (Port A.1)
- 010 = Channel 2 (Port A.2)
- 011 = Channel 3 (Port A.3)
- 100 = Channel 4 (Port A.5)
- 101 = Channel 5 (Port E.0)
- 110 = Channel 6 (Port E.1)
- 111 = Channel 7 (Port E.2)

GO/$\overline{\text{DONE}}$ – (Set to start conversion, automatically cleared when done)
ADON – A/D converter on (1 = on, 0 = off)

Bit 7	Bit 6	Bit 5	Bit 4	Bit 3	Bit 2	Bit 1	Bit 0	
ADFM				PCFG3	PCFG2	PCFG1	PCFG0	ADCON1

ADFM – A/D format (1 = 10-bit mode, 0 = 8-bit mode—refer to text)
PFCG3:PCFG0 – Configuration bits
(These bits determine which bits of the A/D port are analog inputs, which are digital inputs and which bits are used for V_{ref+} and V_{ref-}. The most common configuration is 0000 which sets all 8 analog inputs as analog inputs and uses V_{dd} and V_{ss} as the reference voltages. For other choices, refer to the part specification.)

Figure 2.45 *A/D Control Registers*

Use of the internal A/D clock is not recommended for systems with clocks above 1 MHz.

An example of A/D converter use is shown in Figure 2.46 and Figure 2.47. This example compares two analog voltages to see if the voltages are within 1 V of each other or, if not, which is higher. The desired output is:

Input	Output
$V_{ch1} > V_{ch2}$ by more than 1 V	Red LED (Port D.0) on
$V_{ch2} > V_{ch1}$ by more than 1 V	Green LED (Port D.7) on
$V_{ch1} = V_{ch2}$ within 1 V	Yellow LED (Port D.2) on

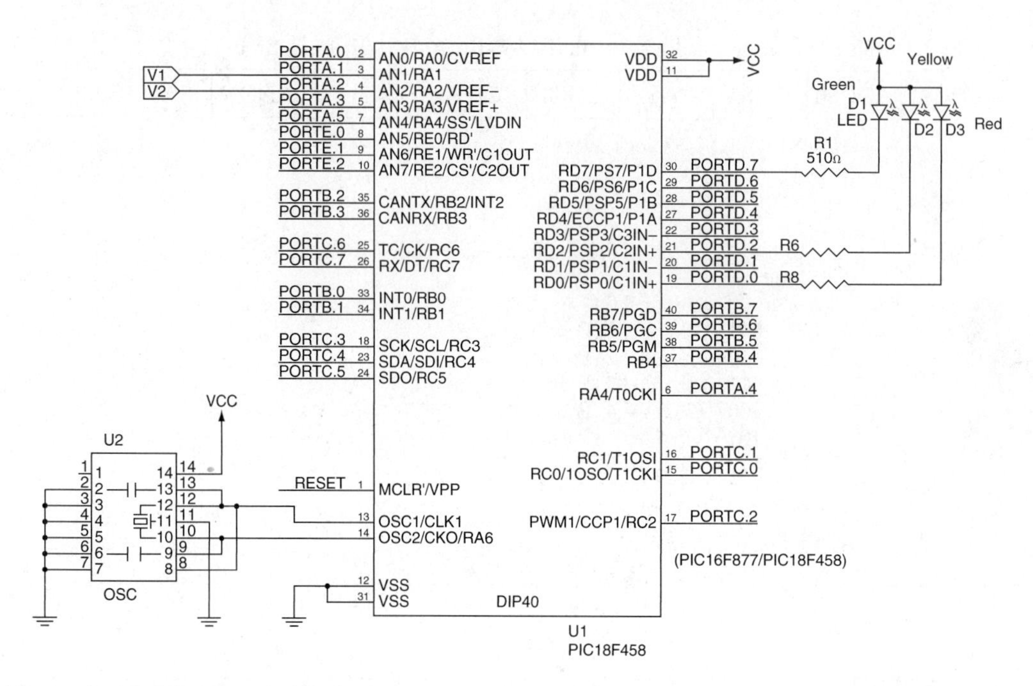

Figure 2.46 *A/D Example Hardware*

```
#device ADC=10  //set for 10 bit A/D
#use delay(clock=10000000)
#fuses HS,NOWDT


//define the A/D control register
#BYTE ADCON0 = 0x1F
int16 ch1_result, ch2_result;  //variables to hold A/D readings

#int_AD
AD_isr()
{
   //Pick out channel number to determine where to store result
   //  and which channel to read next
   if (((ADCON0 & 0x38) >> 3) == 1)
   {   //must be channel 1
      ch1_result = read_adc();    //get 10-bit reading for ch 1
      set_adc_channel(2);         //set to read channel 2 next
   }
   else
```

Figure 2.47 *A/D Example Software (continues)*

```
    {     //else must ch 2
       ch2_result = read_adc();     //get 10-bit reading for ch 1
       set_adc_channel(1);          //set to read channel 2 next
    }
    ADCON0 |= 0x4;    //restart  A/D
}

void main()
{
    setup_adc_ports(ALL_ANALOG);  //use internal reference all analog
                                  //channels active
    setup_adc(ADC_CLOCK_DIV_32);  //divide Fosc by 32 to get below clock
                                  //maximum
    enable_interrupts(INT_AD);  //unmask A/D interrupt
    enable_interrupts(global);  //enable all unmasked interrupts
    set_adc_channel(1);        //choose channel 1
    ADCON0 |= 0x4;             //start A/D process
    while(1)
    {
       if (ch1_result > (ch2_result + 205)) output_d(~0x1); //red LED on
       else if (ch2_result > (ch1_result + 205)) output_d(~0x80);//green
LED on
       else output_d(~0x4); //yellow LED on
    }

}
```

Figure 2.47 *A/D Example Software (continued)*

The program in Figure 2.47 uses the A/D in interrupt mode to accomplish the task. The A/D continues to read and record results while the *main()* loop looks at the results and adjusts the output display. This is a normal operating mode for the A/D. The A/D interrupt keeps all the analog readings up-to-date, and the rest of the program simply uses the readings when they are needed.

There are several unusual aspects to this particular example that need clarification:

1. The A/D and interrupts are set up at the beginning of *main()* as usual. However, the A/D channel is set and the A/D started before entering the *while(1)* loop. It is necessary to start a conversion in order for an interrupt to occur at the *end* of a conversion. This starts the whole process.

2. A *#device ADC=10* statement is required to set the ADC mode to 10 bit. This is only required so that the built-in *read_adc()* function works correctly. Watch the format on the *#device* statement carefully—no extra spaces are allowed!

3. There is no built-in function to start the A/D converter. Therefore, a *#BYTE* statement is used to assign a variable name to the ADCON0 register. Naming it "ADCON0" is simply for clarity; it could have been named "Fred" or "Dingbat" or whatever makes sense to you. Once the variable name is assigned, the programmer can manipulate the bits as required. In this case, *ADCON0 |= 0x4* sets the GO bit to start the conversion. If the program were not interrupt-based, the GO bit could be interrogated to determine when the conversion was completed.

4. Resolution also raises its ugly head here. Using 205 as the number to represent 1 V is close but not quite accurate. The actual counts that represent 1 V are:

$$\frac{V_{in}}{V_{fullscale}} = \frac{X}{2^n - 1} \Rightarrow \frac{1\,V}{5\,V} = \frac{X}{1023} \Rightarrow X = 204.6$$

Therefore, 205 is the nearest integer value that should be used to represent 1 V. Working the formula backwards using 205 and solving for V_{in}, shows that 205 gives an actual voltage of 1.002 V. If we were to use 204, the resulting voltage would be 0.997 V. This is not a big problem here, but does point out the issue of resolution. It is not always possible to get exact results using A/D or, for that matter, D/A, because of limited resolution.

Chapter 2 Example Project – Part D

This is fourth and final part of this chapter's example using a stock car data collection system. This portion is concerned with measuring the engine temperature and sending the collected data to the PC.

Measuring Engine Measurement Using the A/D Converter (ADC)

Previous investigation showed that the temperature signal is connected to AN1, the ADC input on Port A, pin 3. Some further investigation using the specification for the RTD thermocouple and the circuitry conditioning the RTD signal shows that the measurement range is, happily, exactly the range your boss wants, 100°F to 250°F. Using the 10-bit measurement mode on the ADC means that the resulting measured values will be:

$$100°F = 0x000 = 0_{10}$$

$$250°F = 0x3FF = 1023_{10}$$

This sets the conversion formula to be:

$$Temp = (150°F * ADC\ reading) / 1023 + 100°F$$

In this example, the rollover interrupt on Timer1 is used to keep restarting the ADC. In this way, the temperature will be kept up-to-date so that when the data is stored on the 1-second interval, the most recent value will be recorded.

The ADC ISR has the job of reading the current conversion value and converting it into temperature:

```
#int_AD
AD_isr()
{
   int16 temp_adc;  //local variable for raw temp reading

   set_adc_channel(1);     //set A/D to channel 1
   temp_adc = read_adc();  //get raw A/D reading
   //calculate last actual temperature
   current_temp = (((int32)150 * (int32)temp_adc) / (int32)1023) +
                        (int32)100;
   output_d(~current_temp);
}
```

Figure 2E.5 *A/D ISR Code*

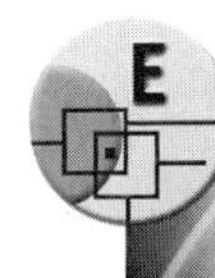

EXAMPLE

Sending Collected Data to the PC

The collected data is already converted to units of rpm and °F so this function needs only to send the data to a PC. This involves initializing the USART and using built-in functions to format and send the data. The USART is initialized by:

```
#use rs232(baud=9600,parity=N,xmit=PIN_C6,rcv=PIN_C7,bits=9)
```

The data sent to the PC is going to be analyzed using a Microsoft Excel® spreadsheet. Investigation shows that data can be input directly using a comma-delimited (CSV) format. This means that commas separate the data on each line of the spreadsheet, and CR separates each line; for example, data sent as:

```
data1, data2, data3
data4, data5, data6
```

will appear in the spreadsheet exactly as shown above occupying a space 2 cells high by 3 cells wide. In this case, you decide that the first column in the spreadsheet will contain the engine rpm, the second will contain the shaft rpm, and the third will contain the engine temperature. Each line of the spreadsheet will contain one set of data, so each line will contain data taken one second after the data in the previous line. The code in Figure 2E.6 is the *while(1)* from the *main()* function of the code:

This routine sends the 120 sets of data (and a line of column titles) using the built-in *printf()* function. A *newline* character ('\n') is included with each set of data to cause the next set to appear in the following line of the spreadsheet.

```
while (1)
{
      if (!input_pin(PIN_B0))
     {
           //note: switch must be released before data is all sent
           unsigned char x;   //temporary counter variable

           //print column titles into the spreadsheet in the first row
           printf ("%s , %s , %s \n","Engine RPM", "Shaft RPM",
                   "Temperature");
           for (x = 0; x < 120; x++)
           {
              // print one set of data into one line on the spreadsheet
              printf ("%d , %d , %d \n",e_rpm[x], s_rpm[x], temp[x]);
           }
      }
   }
```

Figure 2E.6 *Upload Code*

2.10 POWER DOWN (SLEEP) MODES

Many microcontroller-based projects are powered by batteries or by solar power or by other devices in which power conservation is important. PIC microcontrollers are capable of entering a sleep mode when they are not busy to conserve power. In most cases, an interrupt or a reset is required to wake the processor up again.

Consider, for instance, a clock. Recall the example in Section 2.7.3, "Timer1," in which a clock was implemented using Timer1 and a 32 KHz crystal. In this example, the timer overflowed every 2 seconds, which could be used to cause an interrupt. At the instant that the interrupt occurs, a number of registers will need to be updated. This would be accomplished in microseconds, but the rest of the time would be wasted. Because most clocks are based on LCD readouts (very low-power-consumption devices), the LCD could be reloaded before the processor went into sleep mode and it would continue to display the time until the next interrupt.

Placing a *sleep()* command function at the point in the program at which the processor needs to reduce power will implement the sleep mode. Then any interrupt will reawaken the processor, and execution will continue after the interrupt has been serviced.

During the period of the sleep mode operation, most of the main functions and clocks of the microcontroller are shut off, greatly reducing power consumption.

2.11 ASSEMBLY LANGUAGE

The C programming language is defined as a high-level language. A high-level language is one that is easy to read for humans, and allows complex operations to be performed in a single line of code. Microcontrollers, on the other hand, can actually only execute what is called *machine code*. Machine code consists solely of numbers and is a very low-level language.

Assembly language is very close to machine code in that it translates very directly into machine code. Assembly language is considered to be a low-level language because it is very close to machine language.

C language compilers *compile* the C language statements into the assembly code, which is then *assembled* into the machine code. The compilers accomplish this by using libraries that contain assembly language functions that implement the C language statements. The C language compiler looks at the C language statements and selects those library functions that implement the statements, compiling them into an assembly language program. Finally, the C language compiler uses an assembler to convert the assembly language into machine code.

The efficiency (and sheer number) of the assembly code routines in a compiler's library affect the size and speed of the resulting machine code significantly. If, for instance, a programmer were to use a simple while loop to create a delay, the code could either run relatively slowly or very quickly. If the C compiler had only one routine to implement a while loop and it had to be capable of handling any size numbers from 8 bits to 32 bits, then the routine would be relatively slow to execute even when it was counting an 8-bit size number. On the other hand, a compiler might have one while loop for 8-bit numbers, another for 16-bit numbers, and yet another for 32-bit size numbers. Then the compiler could select the appropriate function to match the number size and execute at maximum speed.

There are actually two additional steps buried in the compiling and assembling process: linking and locating. *Linking* is the process of integrating library functions and other program code into the program being compiled. *Locating* is the process of assigning actual code memory addresses to each of the executable instructions.

CCS-PICC adds a further step in that it converts the executable machine code into an Intel-formatted text file for downloading into a PIC microcontroller using an appropriate programmer.

The compiling process is summarized in Figure 2.48 using a line from the A/D example program.

Figure 2.48 details the process of turning a line of C code into machine code. The last step in the compiling process is to turn the code into an Intel-formatted HEX file so that the code can be transferred into the PIC microcontroller. Figure 2.49 details the format of Intel HEX format.

Step 1: Write the C code:

```
if (((ADCON0 & 0x38) >> 3) == 1){    //must be channel 1
```

Step 2: Compile:
The compiler uses its libraries to compile the C code into following assembly code (comments were added to the assembly code by the author for clarity):

```
0035:  MOVF   1F,W     ;move ADCON0 into a file register W
0036:  ANDLW  38       ;And 0x38 with the contents of
                       register W.
                       ;The step above AND's ADCON0 with
                       0x38.
0037:  MOVWF  77       ;Move data from W to register 0x77
0038:  RRF    77,F     ;rotate the contents of 0x77 right
                       1 place
0039:  RRF    77,F     ;rotate again
003A:  RRF    77,F     ;rotate the third time to finish
                       ;The three steps above shift the
                       data 3 places right
003B:  MOVLW  1F       ;Move contents of ADCON0 into reg-
                       ister W
003C:  ANDWF  77,F     ;And the contents of W with con-
                       tents of 0x77
003D:  MOVF   77,W     ;Move the contents of register 0x77
                       back into W
003E:  SUBLW  01       ;subtract 1 from W to see if the
                       result was = to 1
003F:  BTFSS  03.2     ;test for equality by checking the
                       Z flag
0040:  GOTO   051      ;jump if not equal to 1
                       ;The next instruction would be exe-
                       cuted if
                       ;the result was 1
```

Step 3: Assemble:
The assembler produces the machine code shown in the center column. The left-hand column is the code memory address that will hold the instruction word from the center column and the assembly language instructions are shown on the right for clarity.

Figure 2.48 *Compiling Process (continues)*

```
0039:   28      RRF     77,F
003A:   0C      RRF     77,F
003B:   1B      MOVLW   1F
003C:   32      ANDWF   77,F
003D:   28      MOVF    77,W
003E:   22      SUBLW   01
003F:   08      BTFSS   03.2
0040:   84      GOTO    051
```

```
Step 4: Convert to Hex for downloading:
This line Intel Hex code contains the machine code (see emphasized sec-
tions):

        :1000300083128C308400001F1F280C1B32282208DA
        :1000400084002308F7002408F8002508F900260892
```

Figure 2.48 *Compiling Process (continued)*

The Intel hex format (refer to Figure 2.49) requires that each line begin with a ":" to denote the beginning of a line of hex code. This is followed by a byte that indicates the number of data bytes contained in a line (remember, the numbers are in the hex, so a "10" means there are actually 16 data bytes on the line). The next two bytes are the address of the first data byte in the line (subsequent data bytes are placed in subsequent addresses). The type code indicates the nature of the data in most older, smaller processors. Type 00 is simply regular data to be loaded. A 01 type code is used to denote the last line in a HEX file.

Finally, the appropriate number of data bytes is included, followed by a checksum used for error checking.

At this point, it may appear that assembly code is relatively hard to use (especially when compared to a high-level language such as C), convoluted and obtuse, and you are probably won-

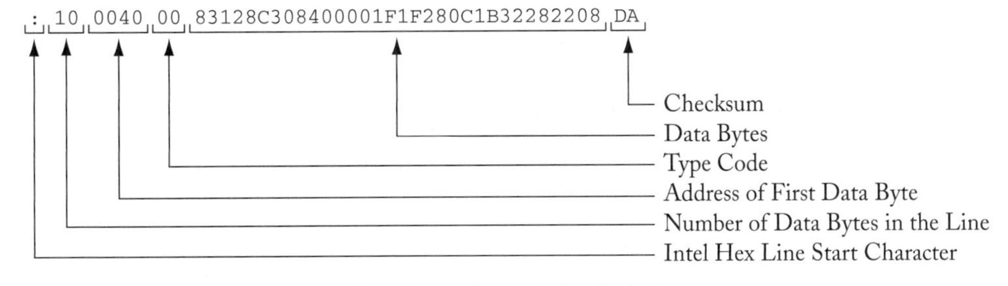

Figure 2.49 *Intel HEX Format*

dering why anyone would be concerned with the assembly code. Assembly code is useful when a programmer needs to do operations directly on the registers of the microprocessor or when the programmer needs to very critically control the speed of execution. Examples of the former would be writing advanced programs such as *Real-Time Operating Systems* (RTOSs) or other programs that must manipulate the microcontroller memory directly in order to operate. Likewise it is useful as a debugging tool. Errors in C programs often become apparent when the resulting assembly code is studied to determine what the processor is actually being told to do.

The second reason to use assembly code is to very tightly control the time of execution. C language compilers use a library of functions to implement the C language. Depending on the thoroughness of the compiler, there may be only a few functions that attempt to cover every instance, or there may be many functions that cover every particular instance of a programming command very efficiently. If there are only a few functions that cover all situations, the chances are very good that these functions use much more execution time than is necessary in order to cover all the scenario possibilities. Better C compilers, such as CCS-PICC, have large libraries of functions to cover most situations very efficiently, reducing the need for assembly code programming.

Should the need for assembly language programming arise, one relatively good way to handle it is to first write the function in C and then, using the assembly code files generated by the compiler, look at the assembly code the compiler proposes to use for your function. You can then analyze the assembly code by looking up each instruction in the function (the instructions are detailed in the appendix) and determining which, if any, instructions are really not necessary. Experience has shown that even well-optimized code can usually be speeded up somewhat by careful analysis of the assembly code.

2.12 CHAPTER SUMMARY

This chapter has provided you with the basic hardware background necessary to choose and apply a member of the PIC RISC microcontroller family.

Specifically, the architecture of the microcontrollers, including their memory system design and internal workings, has been presented. Each of the usual peripheral devices has been described, its use discussed, and an example has been presented showing how to use the peripheral. Should you encounter other peripherals not specifically covered in this chapter, their use should be readily accomplished because they all work very similarly.

As a student, you should now be ready to meet each of the objectives of this chapter.

Chapter 2 Example Project – Part E

This is the summary of the example project presented in pieces throughout the chapter. The complete program is (with code added to the INT1 ISR to clear the *data_counter*) shown in Figure 2E.7.

```
#include <18F458.h>
#device ADC=10
#use delay(clock=10000000)
#fuses HS,NOWDT,WDT1
#use rs232(baud=9600,parity=N,xmit=PIN_C6,rcv=PIN_C7,bits=9)

int16 e_rpm[120], s_rpm[120];    //array for rpm data
int16 current_e_rpm, current_s_rpm;  //current values of RPM
int temp[120];                   //array for temp data
int current_temp;                //current temperature value

int data_set_cntr;  //counter for 120 sets of data
#int_TIMER0
TIMER0_isr()
{
   set_timer0(55770);    //reload for the next second
   if (data_set_cntr < 120)
   {
      e_rpm[data_set_cntr] = current_e_rpm;    //record engine rpm
      s_rpm[data_set_cntr] = current_s_rpm;    //record shaft rpm
      temp[data_set_cntr++] = current_temp;    //record engine temp
   }
}

int16 previous_shaft_time;  //variable to save shaft start time
int shaft_rolls;        //roll over counter for shaft RPM
#int_EXT
EXT_isr()
{
   int32 current_shaft_time;
   int32 period; //local variables
   current_shaft_time = get_timer1();  //read current timer1 count
   period = ((int32)Shaft_rolls * (int32)0xffff) +
         (int32)current_shaft_time - (int32)previous_shaft_time;
   current_s_rpm = (int32)75E6 / period; //calculate shaft RPM
   previous_shaft_time = current_shaft_time;
                              //save value for next calculation
   shaft_rolls = 0;     //cear timer1 roll over counter
}
```

Figure 2E.7 *Complete Example Code (continues)*

```
#int_EXT1
EXT1_isr()
{
   data_set_cntr = 0;     //restarts data collection
}

#int_AD
AD_isr()
{
   int16 temp_adc;  //local variable for raw temp reading
   set_adc_channel(1);     //set A/D to channel 1
   temp_adc = read_adc();  //get raw A/D reading
   //calculate last actual temperature
   current_temp = (((int32)150 * (int32)temp_adc) / (int32)1023) +
                     (int32)100;
   output_d(~current_temp);
}

int T1_rolls;    //variable to track timer1 roll-overs
#int_TIMER1
TIMER1_isr()
{
   ++T1_rolls;     //count Timer1 roll overs for Engine rpm
   ++shaft_rolls; //count rollovers for shaft rpm
   read_adc();     //start A/D
}

int16 previous_capture_time;
#int_CCP1
CCP1_isr()
{
   int16 current_capture_time;
   int32 period;     //variables for calculations
   current_capture_time = CCP_1;
                     //CCP_1 is from CCS-PICC for the capture value
   period = (T1_rolls * (int32)0xFFFF) - (int32)previous_capture_time +
                                    (int32)current_capture_time;
   current_e_rpm = (int32)75E6 / (int32)period;  //calculate RPM
   previous_capture_time = current_capture_time;
                     //use current time for next calc
   T1_rolls = 0;     //clear rollovers for next calculation
}
```

Figure 2E.7 *Complete Example Code (continues)*

```
void main()
{
   setup_adc_ports(ALL_ANALOG);      //set A/D to use internal reference
   setup_adc(ADC_CLOCK_DIV_32);      //code to set up A/D clock
   setup_wdt(WDT_OFF);               //WDT off
   setup_timer_0(RTCC_DIV_256);      //Timer0 enabled with a 256 prescaler
   setup_timer_1(T1_INTERNAL|T1_DIV_BY_2);
                                     //Timer1 enabled using system
clock/4/2
   setup_ccp1(CCP_CAPTURE_RE);       //enable capture mode/ rising edge
   enable_interrupts(INT_TIMER0);    //unmask timer0 overflow interrupt
   enable_interrupts(INT_TIMER1);    //unmask timer1 overflow interrupt
   enable_interrupts(INT_EXT);       //unmask external interrupt 0
   enable_interrupts(INT_EXT1);      //unmask external interrupt 1
   enable_interrupts(INT_AD);        //unmask A/D interrupt
   enable_interrupts(INT_CCP1);      //unmask capture 1 interrupt
   enable_interrupts(global);        //enable all unmasked interrupts
   data_set_cntr = 0;                //start recording data sets at 0
   while(1)
   {
      if (!input(PIN_B0))
      {
            //note: switch must be released before data is all sent
         unsigned char x;  //temporary counter variable

         //print column titles into the spreadsheet in the first row
         printf ("%s , %s , %s \n","Engine RPM", "Shaft RPM",
            "Temperature");
         for (x = 0; x < 120; x++)
         {
            // print one set of data into one line on the spreadsheet
            printf ("%d , %d , %d \n",e_rpm[x], s_rpm[x], temp[x]);
         }
      }
   }
}
```

Figure 2E.7 *Complete Example Code (continued)*

In summary, this program is but one way to accomplish the task; there are many others that would work equally well. And there are some enhancements that would improve system operation. Some of the more likely variations and enhancements are:

1. Store the rpm and temperature data as raw data, not converted to rpm or temperature. In this way, time would be saved during each parameter measurement. Each piece of raw data would need to be converted in the routine that sends the data to the PC as the data is sent.
2. The rpm measurements can be acquired by actually counting the pulses for 1 second and multiplying by 60 to get the rpm. In this case, the 1-second routine would record the count and clear it for the next 1-second count.
3. There is no protection to prevent bumping the 'Start' switch after data is recorded and before it is transferred to the PC. Data could be lost in this way. One choice might be to disallow further recording until data is sent to the PC. The problem with this is that if the driver starts the recording process inadvertently, he or she has no way to recover and transfer the right data. Another choice might be to provide a 'Data Lock' toggle switch. With this switch set one way, data could still be recorded whenever the start switch is pressed. Setting this switch to the 'lock' position after the good data is recorded would prevent further recording until the data had been saved off to the PC.

There are many more variations and/or enhancements possible. Having completed this chapter you should be able to add several yourself.

EXERCISE

2.13 WRITTEN EXERCISES

1. Explain the differences between a microcontroller and a microcomputer (Section 2.2).

*2. Describe the following memory types and delineate their uses (Section 2.4):

   A. FLASH Code Memory

   B. Data Memory

3. Explain the function and purpose of the processor *stack* (Section 2.4).
4. Explain how an interrupt finds its way to the interrupt service routine to execute it when the interrupt occurs (Section 2.5).

*5. Write a fragment of C language code to initialize External Interrupt 1 to activate on a falling edge applied to the external interrupt pin (Section 2.5).

6. What is a watchdog timer and why is it used (Section 2.7.5)?

*7. Write a fragment of C language code to initialize the Port D pins so that the upper nibble may be used for input and the lower nibble may be used for output (Section 2.6).

8. Write a fragment of C code to initialize Port B so that the least significant bit of Port B is input and the balance of Port B is output (Section 2.6).

9. Name and describe at least two distinctly different functions that can be performed by a timer/counter (Section 2.7).

*10. Sketch the waveform appearing at the output of the USART when it transmits an 'H' at 9600 baud. The sketch should show voltage levels and the bit durations in addition to the waveform (Section 2.8).

11. Sketch the waveform appearing at the output of the UART when it transmits an 'F' at 1200 baud. The sketch should show voltage levels and the bit durations in addition to the waveform (Section 2.8).

*12. Compute the missing values to complete the following table that relates to analog to digital conversion (Section 2.9):

V_{in}=	$V_{fullscale}$	Digital Out	# of bits
4.2 V	10 V		8
1.6 V	5 V		10
	5 V	123	10
	10 V	223	8

13. Describe the differences between the USART, I^2C, and SPI serial communications (Section 2.8).

14. Detail the steps in the compiling process (Section 2.11).

"*" by an exercise number indicates that the question is answered or partially answered in the appendix. Don't be peeking until you try the problems yourself!

2.14 LABORATORY EXERCISES

1. Create a program that will turn on an LED when a falling edge occurs on external interrupt 0 and will turn it off when a rising edge occurs on external interrupt 1.

2. Create a program that will demonstrate how a watchdog timer resets the processor if the program hangs up in an infinite loop. (Don't use the one from Section 2.7.5—do your own).

3. Create a program that will read the data on all 8 bits of Port B, swap the nibbles of that data, and output the result to Port D.

4. Create a simulated engine speed monitor that will light a yellow LED if the motor 'speed' (a TTL-level square wave) drops below 2000 Hz, a red LED if the speed exceeds 4000 Hz, and a green LED when the speed is between these two limits.

5. Create a program to vary the speed of a small motor (or the brightness of an LED) in 16 steps, from full off to full on, using PWM control. Control the speed or brightness via logic levels applied to the lower nibble of Port B.

6. Create a program to output the ASCII character 'G' every 50 milliseconds via the USART at 9600 baud. Observe both the TTL level signal on the microcontroller's TXD line and the RS-232 signal at the output of the media driver using an oscilloscope. On each waveform, identify the start bit, the stop bit, and the data bits of the transmitted signal as well as the voltage levels of the waveform.
7. Modify the program above so that the 'G' is also transmitted from the SPI bus. Use the oscilloscope to display both the transmitted SPI signal and the SPI clock, and identify the data bits. Repeat for I^2C.
8. Use the A/D converter to create a simple voltmeter that measures voltage in the range of 0 to 5 V to an accuracy of 0.1 V. Display the results either on the terminal using serial communications or on a port where the upper nibble would be the volts digit and the lower nibble would be the tenths of volts digit.

CHAPTER 3

Standard I/O and Preprocessor Directives

3.1 OBJECTIVES

At the conclusion of this chapter, you should be able to:

- Use standard I/O functions to yield data from your programs for informational as well as debugging purposes.
- Use the standard input functions to read user data into your programs.
- Understand the use of standard output formatting to simplify program coding and get professional-looking results.
- Use the **#define** statement to declare constants as well as redefine functions.
- Use the **#include** statement to bring information into your program from external files.
- Use the **#pragma** statement to optimize or customize your program during the compilation process.
- Use the device specification directives to correctly select a device for your program.
- Use the **#use** statements to direct the compiler in generating code for the I/O and communication operations in a specific manner.

3.2 INTRODUCTION

The standard C language input/output (I/O) functions are a method of sending information to and gaining information from your program as it executes. These functions are built from simple character input and output functions that are easily tailored to meet system hardware requirements. Some of the higher-level functions associated with the standard C language I/O include the ability to format numeric and text output, as well as to process and store numeric or text data input.

The standard C language I/O functions outlined here were adapted to work on embedded microcontrollers with limited resources. The CCS-PICC compiler uses preprocessor commands and a database of microcontroller information to guide the generation of the software associated with these functions.

In general, preprocessor commands are used to add readability and expand the functionality of your program and to provide more detail in the definition and construction of the program. These commands include **#define, #include, #pragma** as well as other conditional commands such as **#ifdef, #else**, and **#endif**. These commands are used by the compiler and provide information about the program before it is actually compiled. In addition, the CCS-PICC compiler relies on preprocessor commands such as **#use, #device**, and **#int_xxx** for direction in generating Microchip PIC device-specific output.

3.3 CHARACTER INPUT/OUTPUT FUNCTIONS – *GETCHAR()* AND *PUTCHAR()*

At the lowest level, the input/output functions are *getchar()* and *putchar()*. These functions provide a method of getting a single character into a program or sending a single character out. These functions are generally associated with the Universal Synchronous/Asynchronous Receiver Transmitter (USART) or serial communications port of the microcontroller. The higher-level standard C language I/O functions use these "primitives" as a baseline for their operation. The standard form of these functions is:

```
char getchar(void);     // returns a character received
                        // by the USART, using polling.
void putchar(char c);   // transmits the character c using
                        // the USART, using polling.
```

Typically, prior to using these "stock" functions in an embedded system, you must:

1. Initialize the UART baud rate.
2. Enable the UART transmitter.
3. Enable the UART receiver.

The CCS-PICC compiler provides the **#use delay** and the **#use rs232** preprocessor commands to accomplish these tasks. Calls to these preprocessor commands are required before calling *getchar()* or *putchar()*. For example, in the code below, the program defines the clock frequency and the desired RS232 settings by calling **#use delay** and **#use rs232**, respectively. The compiler uses the information provided in these two preprocessor commands to initialize the USART. The program then sits in a **while** loop until power is removed. While in this loop, the program calls *getchar()* to wait for a character to be received. Once a character is received it is echoed back by calling the *putchar()* function:

```
#include <16F877.h>       // register definition file for
#fuses HS,NOWDT           // a Microchip PIC16F877
#use delay(clock=10000000)
#use rs232(baud=9600,parity=N,xmit=PIN_C6,rcv=PIN_C7,stream=RS232,bits=8)
```

```
void main(void)
{
        char k;

        while (1)
        {
             k=getchar(); /* receive the character */
             putchar(k);  /* and echo it back */
        }
}
```

When *getchar()* or *putchar()* is called, the CCS-PICC compiler uses the last **#use rs232** preprocessor command without a stream identifier to determine what serial communication port of the microcontroller to use. If the pin definitions in the RS232 initialization match a hardware USART of the microcontroller, the hardware USART is used. If the pin definitions do not match a hardware USART, then the compiler generates a software USART, resulting in a larger program and a more limited USART. If a hardware USART is used, then the hardware can buffer up to three characters and generate an interrupt on transmit and receive complete. However, if a software USART is used, then the function *getchar()* must be active while the character is being received by the microcontroller. For either a hardware or a software USART, a compilation error is generated if the specified baud rate cannot be achieved with the given clock frequency.

It is possible to define more than one serial communication port for the microcontroller. The CSS-PICC compiler refers to the serial communication port as a *stream.* In the call to **#use rs232**, the parameter, **stream=**, is used to associate an identifier with a communication port. The functions *fgetc()* and *fputc()* utilize this identifier. The prototypes for these functions are:

```
char fgetc(stream x); // returns a character received
                      // from stream x, using polling.

void fputc(char c, stream x);
                      // transmits the character c using
                      // stream x, using polling.
```

Using a stream identifier is useful when more than one communication port is in use. Consider the following example where communications are received from a GPS on one port and sent to a host PC on another port. A third port is used to send debugging information to a monitoring program.

```
#include <16F877.h>      // register definition file for
#fuses HS,NOWDT          // a Microchip PIC16F877
#use delay(clock=10000000)
#use rs232(baud=9600,xmit=PIN_C6,rcv=PIN_C7,stream=HOSTPC)
#use rs232(baud=1200,xmit=PIN_B1,rcv=PIN_B0,stream=GPS)
```

```
#use rs232(baud=9600,xmit=PIN_B3,stream=DEBUG_MONITOR)

void main(void)
{
   char k;
   fprintf(HOSTPC,"Running!");
   while (1)
   {
      k=fgetc(GPS);       /* receive the character from GPS*/
      fputc(k,HOSTPC);    /* echo it back to the host PC*/

      /* if a carriage return is encountered, tell the
        debug monitor about it */
      if (k == '\r')
         fputs("Got a CR!",DEBUG_MONITOR);
   }
}
```

When *getchar()* and *fgetc()* are called in the above examples, they force the microcontroller to wait until a character is received before allowing the program to go on. In some applications, it may not be desirable to sit and wait for a character to arrive. Instead, interrupts may be used to allow serial communications to be handled in the background while the main program continues to run. Interrupt driven serial communications allow characters to be placed into a buffer or queue as they come in, in the background. Then, when the main program is ready for the received characters, it reads the characters from the buffer and processes them.

The same is true for transmitting characters. To prevent waiting for a character to leave the serial port, interrupt driven transmitting can be used. As the transmit register of a USART empties, it can generate an interrupt that allows the next character from a transmit buffer to be sent. This process would repeat for as long as there were characters in the buffers.

In the example below, *bgetc()* and *bputc()* get characters from, and place characters into, the *Rx_Buffer* and *Tx_Buffer* character arrays, respectively. In this case, the *bgetc()* and *bputc()* functions and their support may appear as in Figure 3.1.

In this example, the variable *RX_Counter* contains the number of characters that have been placed into the receiver character array *RX_Buffer* by the interrupt routine. This prevents the program from calling *getchar()* and then sitting idle until a new character arrives. *RX_Counter* is tested in each pass of the **while** loop. If it is not zero, then characters exist in the *RX_Buffer* character array and a call to *bgetc()* will retrieve a character without hesitation.

Because *bputc()* is interrupt driven as well, it adds almost no time to the program execution. *bputc()* places a character into the transmit buffer and starts the transmit interrupts. From that point on, interrupts move the characters from the transmit buffer to the UART each time it becomes empty, until the transmit buffer is empty.

```
#include <16F877.h> // register definition file for
#fuses HS,NOWDT           // a Microchip PIC16F877
#use delay(clock=10000000)
#use rs232(baud=9600,parity=N,xmit=PIN_C6,rcv=PIN_C7,bits=8)

#define  RX_BUFFER_SIZE        24
char Rx_Buffer[RX_BUFFER_SIZE+1]; // character array (buffer)
char RX_Wr_Index = 0;//index of next char to put into the buffer
char RX_Rd_Index = 0;//index of next char to fetch from buffer
char RX_Counter = 0;  //a total count of characters in the buffer
int1 RX_Buffer_Overflow = 0;    // This flag is set on UART
                                   // Receiver buffer overflow

// UART Transmit buffer
#define  TX_BUFFER_SIZE        24
char TX_Buffer [TX_BUFFER_SIZE+1]; // character array (buffer)
char TX_Rd_Index = 0; //index of next char to put into the buffer
char TX_Wr_Index = 0; //index of next char to fetch from buffer
char TX_Counter = 0;  //a total count of characters in the buffer

// USART Receiver interrupt service routine
#int_rda    // preprocessor directive identifying the
            //following routine as an interrupt routine
void serial_rx_isr()
{
    // put received char in buffer
    Rx_Buffer[RX_Wr_Index] = getc();

    if(++RX_Wr_Index > RX_BUFFER_SIZE)// wrap the pointer
        RX_Wr_Index = 0;

    if(++RX_Counter > RX_BUFFER_SIZE) // keep a character count
    {                                 // overflow check..
        RX_Counter = RX_BUFFER_SIZE;  // if too many chars came
        RX_Buffer_Overflow = 1;    // in before they could be used
    }                              // that could cause an error!!
}

// Get a character from the UART Receiver buffer
char bgetc(void)
{
    char c;

    while(RX_Counter == 0)          // wait for a character...
```

Figure 3.1 *Interrupt Driven Transmit and Receive (continues)*

```
        ;

    c = Rx_Buffer[RX_Rd_Index];      // get one from the buffer..

    if(++RX_Rd_Index > RX_BUFFER_SIZE) // wrap the pointer
        RX_Rd_Index = 0;

    if(RX_Counter)
        RX_Counter--;          // keep a count (buffer size)

    return c;
}

// USART Transmitter interrupt service routine
#int_tbe            // preprocessor directive identifying the
                    //following routine as an interrupt routine
void serial_tx_isr()
{
    // if there are characters to be transmitted...
    if(TX_Counter != 0)
    {
        // send the character out the port
        putc(TX_Buffer[TX_Rd_Index]);

        // test and wrap the pointer
         if(++TX_Rd_Index > TX_BUFFER_SIZE)
            TX_Rd_Index = 0;

        TX_Counter--;       // keep track of the counter
        // if there are no more characters,
        // then disable the interrupt
        if (TX_Counter == 0)
            disable_interrupts(int_tbe);
    }
}

// write a character to the serial transmit buffer
void bputc(char c)
{
    char restart = 0;

    while(TX_Counter > (TX_BUFFER_SIZE-1))
        ;        // WAIT!! Buffer is getting full!!
```

Figure 3.1 *Interrupt Driven Transmit and Receive (continues)*

```
    if (TX_Counter == 0) // if buffer empty, setup for interrupt
        restart = 1;

    TX_Buffer[TX_Wr_Index++]=c; // jam the char in the buffer..

    if(TX_Wr_Index > TX_BUFFER_SIZE)    // wrap the pointer
        TX_Wr_Index = 0;

    // keep track of buffered chars
    TX_Counter++;

    // if there were not chars but are now,
    // re-enable the interrupts
    if(restart == 1)
        enable_interrupts(int_tbe);
}

void main(void)
{
    unsigned char k;

    enable_interrupts(global); // enable interrupts in general
    enable_interrupts(int_rda);// enable the receive interrupt

    // print message that we are up and running!
    printf(bputc,"\r\n\Running...\r\n");

    while(1)
    {
        // buffer up 5 characters before transmitting them!
        if(RX_Counter > 4)
        {
            // send the characters until there are none left!
            while (RX_Counter > 0)
            {
                k = bgetc();    // get the character
                bputc(k);              // and echo it back
            }
        }

        // since there is no waiting on getchar or putchar..
        // other things can be done!!
    }
}
```

Figure 3.1 *Interrupt Driven Transmit and Receive (continued)*

The *printf()* function defaults to using *putchar()* to output the formatted string. In this case, we want *printf()* to use our new interrupt driven *bputc()* instead. Some compilers accomplish this by allowing us to redefine *putchar()*. However, under CCS-PICC, *printf()* accepts a parameter, which is a function name. This function is then used by *printf* to output the formatted string. In the preceding example, we passed the function name *bputc()* to get interrupt driven output from *printf()*. This method of passing a function name could also be used to direct *printf()* to output a formatted string to other types of communications ports such as a *serial peripheral interface* (SPI) or a parallel port.

3.4 STANDARD OUTPUT FUNCTIONS

The output functions of a standard library include the put string, *puts()*; 'file' put string, *fputs()*; print formatted, *printf()*; and 'file' print formatted, *fprintf()*. The term 'file' refers to the requirement for a stream identifier with this function call. This allows the programmer to specify outputs other than the last defined serial port.

3.4.1 PUT STRING, *PUTS()*, AND FILE PUT STRING, *FPUTS()*

The standard form of *puts()* and *fputs()* are:

```
void puts(char *str);
void fputs(char *str, stream x);
```

The put string function, *puts()*, calls *putchar()* to output *str* to the most recently defined serial communication port that does not include a stream identifier. *str* may be either a pointer to a null terminated character array or a constant string. The file put string function, *fputs()*, calls *putchar()* to output *str* to the serial communication port associated with the **stream** *x*. Both *puts()* and *fputs()* append a carriage return ('\r') and a line-feed ('\n') to the end of the transmitted string. The following example calls *puts()* and *fputs()* to output the strings "Hello" and "World" to two separate serial communications ports. The strings are each followed by a carriage return and a line-feed character.

```
#include <16F877.h>     // register definition file for
#fuses HS,NOWDT         // a Microchip PIC16F877
#use delay(clock=10000000)
#use rs232(baud=9600,xmit=PIN_B3,stream=DEBUG_MONITOR)
#use rs232(baud=9600,xmit=PIN_C6,rcv=PIN_C7)

char s[6] = "Hello";

void main(void)
{
        while (1)
        {
                // first, print to PIN_C6 since it was the last
                // defined RS232 port without a stream identifier)
                puts(s);        // prints "Hello" followed by carriage
```

```
                              // return and a line-feed to last
                              // defined serial port

                fputs("World",DEBUG_MONITOR);
                              // prints "World" followed by carriage
                              // return to DEBUG_MONITOR port

        }
}
```

In this program, *s*, which is the address of the array *s[]*, is passed to the *puts()* function as a pointer to the array. "World", on the other hand, is a constant string. *DEBUG_MONITOR* is the identifier associated with the first serial port defined using the **#use rs232** preprocessor command. The call to *fputs()* prints "World" to the *DEBUG_MONITOR* serial port, which in this case happens to be the hardware port. The call to *puts()* uses the last **#use rs232** definition without a stream identifier to select the serial port to print to. In this example, the call to *puts()* prints to the **PIN_C6** and **PIN_C7** serial port.

3.4.2 PRINT FORMATTED, *PRINTF()*, AND FILE PRINT FORMATTED, *FPRINTF()*

The print formatted and the file print formatted functions have the standard forms of:

```
void printf([function_name, ] char flash *fmtstr
     [ , arg1, arg2, ...]);
void fprintf(stream x, char flash *fmtstr [, arg1, arg2, ...]);
```

The *printf()* function outputs the formatted text using the function *function_name* or, if *function_name* is blank, then it uses *putchar()* to print to the last defined RS232 port without a stream. The *fprintf()* function outputs the formatted text to the RS232 port identified by *stream x*.

For either print function, the text is formatted according to the specifications in the constant string *fmtstr*.

fmtstr can be made up of constant characters to be output directly as well as special format commands or specifications. If no variables and, therefore, no special format commands, are used, the string, *fmtstr*, may also be an array of characters. If format specifications are used, then *fmtstr* must be a constant string. *fmtstr* is processed by the *printf()* function. As *printf()* is processing the *fmtstr*, *printf()* outputs the characters and expands each argument according to the format specifications that may be embedded within the string. A percent sign (%) is used to indicate the beginning of a format specification. Each format specification is then related to an argument, *arg1*, *arg2*, and so on, in sequence. There should always be an argument for each format specification and vice versa.

The described implementation of the *printf()* format specifications is a reduced version of the standard C function. This is done to meet the minimum requirements of an embedded system. A full "ANSI-C" implementation would require a large amount of memory space, which, in most cases, would make the functions useless in an embedded application.

In the CCS-PICC library, the format specifications take the generic form, *%wt*. *w* is optional and may be **1-9** to specify how many characters are to be output or **01-09** to indicate leading zeros. For floating point, *w* may be **1.1-9.9**. *t* is the type and may be one of the types shown in Table 3.1.

Specification	Format of Argument
c	Outputs the next argument as an ASCII character
s	Outputs the next argument as a null terminated character string
u	Outputs the next argument as an unsigned decimal integer
x	Output the next argument as an unsigned hexadecimal integer using lowercase letters
X	Outputs the next argument as an unsigned hexadecimal integer using uppercase letters
d	Outputs the next argument as a decimal integer
e	Outputs the next argument in exponential format
f	Outputs the next argument as a floating point number
Lx	Outputs the next argument as an unsigned hexadecimal long integer using lowercase letters
LX	Outputs the next argument as an unsigned hexadecimal long integer using lowercase letters
lu	Outputs the next argument as an unsigned decimal long integer
ld	Outputs the next argument as a signed decimal long integer
%	Outputs the % character

Table 3.1 *printf - Format Specifications*

Here are some examples of how the format specifications affect the output. Pipe marks, '|' were used to help clarify the effect of the formatting by showing where the beginning and end of the text and white space are located. They are not required by the *printf()* function:

```
int i = 123;
int16 i16 = 1234;
char p[10] = "Hello";

printf("|%u|",i);              // prints |123|
printf("|%5u|",i);             // prints |00123|
printf("|%1u|",i);             // prints an undefined string
printf("|%05u|",i);            // prints |00123|
printf("|%x|",i);              // prints |7b|
printf("|%04X|",i);            // prints |007B|
printf("|%Lx|",i16);           // prints |04d2|
```

```
printf("|%1d|",i16);            // prints |1234|
printf("|%s:%04X|",p,i);        // prints |Hello:007B|
```

Following is the program that uses *printf()* to print a string and an integer value from the same statement:

```
#include <16F877.h>     // register definition file for
#fuses HS,NOWDT         // a Microchip PIC16F877
#use delay(clock=10000000)
#use rs232(baud=9600,xmit=PIN_C6,rcv=PIN_C7)

void main(void)
{
    int I;
    for(I=0; I<100; I++)
    {
        // print a formatted string and tell about I
        printf("printf example, I = %u\n\r",I);
    }

    while(1)
        ;
}
```

Running this program would result in the text string, *"printf example, I ="*, being printed 100 times. Each time the number 0 to 99 would be printed following the string. Note that the *fmtstr* contains all the text and the **%d**, indicating where the number *I* should appear as well as the end of line character. That one statement does a lot of work!!

The end of line character is denoted in C language as '\n' and informs programs displaying data to move the cursor to the next line. The carriage return character, '\r', is used to inform programs to move the cursor to the beginning of the current line. These two characters are often used together to start a new line—'\n' to move the cursor down and '\r' to move the cursor to the beginning of the new line.

You should be aware of the costs associated with the use of *printf()*. The *printf()* function is doing a lot of work and, therefore, needs code space. Using the *printf()* function adds flexible, powerful, formatted output, without many lines of "typed in" code. However, the standard library *printf()* function itself contains several hundred lines of assembly code that are automatically added to your program in order to support *printf()*. Once the function has been added to the code, each call to the function adds but a few lines of assembly.

3.5 STANDARD INPUT FUNCTIONS

The input functions of the CCS-PICC standard library include get string, *gets()*; and file get string, *fgets()*. Another function, *get_string()*, offers a third, more robust method of reading strings from standard input. (*get_string()* is not part of the standard library and is located in the "Drivers" folder.) Just as the root output function is *putchar()*, the root input function is *getchar()*. In all cases, *getchar()* is called to read the input data. So by changing the operation of *getchar()*, the data source of *gets()* and *fgets()* can easily be altered.

3.5.1 GET STRING FUNCTIONS – *GETS()* AND *FGETS()*

The standard forms of *gets()* and *fgets()* are:

```
void gets(char *str);
void fgets(char *str, stream x)
```

These functions use *getchar()* to input characters into the character string *str*, which is terminated by the new line character. The new line character is replaced with a null ('\0') by the *gets()* function. If *gets()* is used, then *getchar()* reads characters from the communication port defined by the last **#use rs232** preprocessor command without a stream identifier. If *fgets()* is used, then *getchar()* reads from the communication port identified by *stream x*.

These functions should be used with caution. If the character array *str* is not long enough to contain all of the characters received before the new line character is received, then the memory following the character array is corrupted. Consider the following example that uses *gets()* to read a string from the standard input and uses *puts()* to write the string to the standard output:

```
#include <16F877.h>     // register definition file for
#fuses HS,NOWDT         // a Microchip PIC16F877
#use delay(clock=10000000)
#use rs232(baud=9600,xmit=PIN_C6,rcv=PIN_C7)

char s[12];     // declare a place to put the string
char j[14] = "0123456789\n\r";
void main(void)
{
   while(1)
   {
        gets(s);    // get a string from the standard input
        puts(s);    // print the string to the standard output
        puts(j);    // print out j also!
   }
}
```

If the string "abcd" followed by a new line character is sent to the device as this program is executing, then the program copies "abcd" into the memory allocated for *s* and terminates the string with a null. Because the string is only 5 bytes long (four characters plus the null), it fits within the memory allocated for *s* and the value of *j* remains unchanged. After the program processes the string "abcd\n", the memory for *s* and *j* is shown in Table 3.2.

However, if the string "abcdefghijklmnopqrstuvwxy" followed by a new line character is sent to the device, then the program attempts to assign the entire input string to *s*. *s* is only allocated 12 bytes by its declaration. So, as *gets()* stores characters beyond the end of *s*, the memory after *s* gets overwritten. In our example, *j* is allocated the memory immediately following

	Array Index													
Variable	0	1	2	3	4	5	6	7	8	9	10	11	12	13
s	'a'	'b'	'c'	'd'	'\0'	?	?	?	?	?	?	?		
j	'0'	'1'	'2'	'3'	'4'	'5'	'6'	'7'	'8'	'9'	'\n'	'\r'	'\0'	?

Table 3.2 *Values for s and j After gets() Reads "abcd\n"*

s and is corrupted by *gets()*. After the program receives the entire string, *s* contains the first 12 characters or "abcdefghijkl" and *j* contains the next 13 characters or "mnopqrstuvwxy". Finally, *gets()* writes null to the memory location following the storage of 'y'—this happens to be the last byte of memory allocated for *j*. The memory for *s* and *j* after the program processes "abcdefghijklmnopqrstuvwxy\n" as the input is shown in Table 3.3.

	Array Index													
Variable	0	1	2	3	4	5	6	7	8	9	10	11	12	13
s	'a'	'b'	'c'	'd'	'e'	'f'	'g'	'h'	'i'	'j'	'k'	'l'		
j	'm'	'n'	'o'	'p'	'q'	'r'	's'	't'	'u'	'v'	'w'	'x'	'y'	'\0'

Table 3.3 *Values for s and j After gets() Reads "abcdefghijklmnopqrstuvwxyz\n"*

3.5.2 GET STRING FUNCTION – *GET_STRING()*

There is a third function available in CCS-PICC to read strings, *get_string()*. It is located in the "input.c" file of the "Drivers" folder. The standard form of this routine is as follows:

```
void get_string(char *s, int len)
```

The *get_string()* routine is more robust than the *gets()* function in that it accepts the maximum length of the string as a parameter. The maximum length of the string including the null terminator is *len*. If *len*-1 characters were read without encountering the new line character, then the string is terminated with a null and the function ends.

For example:

```
#include <16F877.h>    // register definition file for
#fuses HS,NOWDT // a Microchip PIC16F877
#use delay(clock=10000000)
#use rs232(baud=9600,xmit=PIN_C6,rcv=PIN_C7)
#include <input.c>

char s[12];     // declare a place to put the string
char j[14] = "0123456789\n\r";
void main(void)
{
```

```
    while(1)
    {
        get_string(s,11);   // get a string from the standard input
        puts(s);     // print the string to the standard output
        puts(j);     // print out j also!
    }
}
```

In this program utilizing *get_string()* instead of *gets()*, sending "abcd\n" to the device results in the same values for *s* and *j* as the example using *gets()* did. These results are shown in Table 3.2.

The real difference between the two functions can be noticed by sending the second string example, "abcdefghijklmnopqrstuvwxyz\n". *get_string()* stops storing characters to *s* when *len*-1 number of characters have been read. Here, *len* is 11, so *get_string()* truncates the input after *len*-1 characters, terminates the string with a null, and returns. This prevents the memory after the string *s* from being overwritten. The memory for *s* and *j* after the program received and processed the alphabet string is shown in Table 3.4.

	Array Index													
Variable	0	1	2	3	4	5	6	7	8	9	10	11	12	13
s	'a'	'b'	'c'	'd'	'e'	'f'	'g'	'h'	'i'	'j'	'\0'	?		
j	'0'	'1'	'2'	'3'	'4'	'5'	'6'	'7'	'8'	'9'	'\n'	'\r'	'\0'	?

Table 3.4 *Values for s and j After get_string() Reads "abcdefghijklmnopqrstuvwxyz\n"*

3.6 STANDARD PREPROCESSOR DIRECTIVES

Preprocessor directives are not actually part of the C language syntax but are accepted as such because of their use and familiarity. The preprocessor is a separate step from the actual compilation of a program and happens before the actual compilation begins. The most common directives are **#define** and **#include**. The preprocessor directives allow you to:

- Include text from other files, such as header files containing library and user function prototypes.
- Define macros that reduce programming effort and improve the legibility of the source code.
- Set up conditional compilation for debugging purposes and to improve program portability.
- Issue compiler specific directives to generalize and/or optimize the compilation.

3.6.1 THE #INCLUDE DIRECTIVE

The **#include** directive may be used to include another file in your source. There can be as many files included as needed, or as is allowed by the compiler, and includes may be nested (meaning that an included file may contain another, nonrecursive **#include**). Typically, there is a limit to the depth of the nesting of 8 to 16 files.

The standard form of the statement is:

```
#include <file_name>

            or

#include "file_name"
```

The less than (<) and greater than (>) sign delimiters indicate that the file to be included is part of a standard library or a set of library files. The compiler will typically look for the file in the "\inc" directory. When the file name is delimited with quotation marks, the compiler will first look in the same directory as the C file being compiled. If the file is not found there, then the compiler will look for the file in the default library directory ("\inc").

The **#include** statement may be located anywhere in the code but is typically used at the top of a program module to improve program readability.

3.6.2 THE #DEFINE DIRECTIVE

A **#define** directive should be thought of as a "text substitution," referred to as a "macro." The standard form of a **#define** is:

```
#define              NAME            Replacement_Text
```

Typically, **#define** directives are used to declare constants but the tag *NAME* and the *Replacement_Text* may also have parameters. The preprocessor will replace the tag *NAME* with the expansion, *Replacement_Text*, whenever it is detected during compilation. If *NAME* contains parameters, the real ones found in the program will replace the formal parameters written in the text *Replacement_Text*.

There are two simple rules that apply to macro definitions and the #define directive:

1. Always use closed comments (/* ... */) when commenting the line of a **#define**.
2. Remember that the end of the line is the end of the **#define**, and that the text on the left will be replaced by ***all*** of the text on the right.

For example, given the following macro definitions:

```
#define ALPHA 0xff          /* mask off lower bits */
#define SUM(a,b) a+b
```

The following expression:

```
int x = 0x3def & ALPHA;
```

would be replaced with:

```
int x = 0x3def & 0xff     /* mask off lower bits */ ;
```

And this expression:

```
int i=SUM(2,3);
```

would be replaced with:

```
int i=2+3;
```

The diversity and capability of the **#define** directive is somewhat dependent on the sophistication of the compiler's preprocessor. Most modern compilers have the ability to redefine functions and operations as well as sort out the parameters that go with them (as in the examples above). The CCS-PICC compiler is very capable in these areas. Lesser compilers may only allow the simplest operations such as defining constants and strings. Using a **#define** to alias functions, or create alternate calls for functions, can be a powerful and handy thing! For example, an existing function in a library may appear as:

```
int _write_SCI1(int c)        // this function sends a char to
{                             // an alternate comm. port
    . . .
  return c;
}
```

Giving the function an alias:

```
#define putchar2(c) _write_SCI1(c)
```

allows the function to be referred to in two ways:

```
putchar2('b');

      or

_write_SCI1('b');
```

This is quite useful when reutilizing code from another program, or combining several libraries in a large development. Instead of locating and renaming every call or reference, the function was simply given an alias, allowing it to be referred to in more than one way.

This use of **#define** can also enhance the readability of a program by allowing a function to be "redeclared." Let's assume a function called *set_row_col()* places the cursor at a position on a 4-row by 20-character LCD display based on two parameters, a *row* and a *column*:

```
void set_row_col(char row, char column)
{
      . . .
}
```

By using a macro to form a "redeclaration," the positions on the LCD display can be referred to as an *x* and *y* coordinate instead of a row and column:

```
#define  goto_xy(x,y)     set_row_col(y,x)
```

Another possibility is a redeclaration that can use a character position to set the cursor:

```
#define set_char_pos(p)    set_row_col((p/20),(p%20))
```

The row is calculated by dividing the character position by the number of columns, and the column position will be what is left, a modulus function of the width. This is all done as a function of "fancy text substitution." The compiler only sees what is on the right side of the **#define**, because the preprocessor did all the work ahead of time. These changes in reference potentially make the program more understandable to not only the programmer but to others reading the code as well.

The substitution process can even be more finely controlled. When defining macros, you can use the # operator to convert the macro parameter to a character string. For example:

```
#define print_MESSAGE(t)   printf(#t)
```

will substitute this expression:

```
print_MESSAGE(Hello);
```

with:

```
printf("Hello");
```

It is also possible to concatenate two parameters using the ## operator. In the example below, the formal parameters *a* and *b* are to be concatenated. In the substitution, the left side, *ALPHA(a,b)*, is effectively replaced with the right side, *ab*. In the program, the real parameters x and y are used. So, the macro definition:

```
#define alpha(a,b) a ## b
char alpha(x,y)=1;
```

will result in:

```
char xy=1;
```

Typically, a replacement text is defined on a single line along with the tag name. It is possible to extend the definition to a new line by using a backslash, '\', at the end of the line:

```
#define MESSAGE "This is a very \
long text..."
```

The backslash character will not appear in the expansion, and the new line character will be eliminated as well. The result is that *"This is a very long text..."* will be treated as a single, unbroken string.

A macro can be undefined using the **#undef** directive. This feature is primarily used in conditional compilation where a tag name is used as a flag:

```
#undef ALPHA
```

The **#undef** directive allows a previously defined macro to be redefined. For instance, if a macro was defined as:

```
#define A_CHAR      'A'
```

and you wanted to redefine it to a lowercase 'a':

```
#undef  A_CHAR
#define A_CHAR      'a'
```

3.6.3 THE #IFDEF, #IFNDEF, #ELSE, AND #ENDIF DIRECTIVES

The **#ifdef**, **#ifndef**, **#else**, and **#endif** directives may be used for conditional compilation.

The syntax is:

```
#ifdef macro_name
        [set of statements 1]
#else
        [set of statements 2]
#endif
```

If *macro_name* is a defined macro name, then the **#ifdef** expression evaluates to TRUE and the *set_of_statements_1* will be compiled. Otherwise, the *set_of_statements_2* will be compiled. The **#else** and *set_of_statements_2* are optional. If *macro_name* is not defined, the **#ifndef** expression evaluates to TRUE. The rest of the syntax is the same as that for **#ifdef**.

The **#if**, **#elif**, **#else**, and **#endif** directives may be used for conditional compilation.

```
#if expression1
  [set of statements 1]
#elif expression2
  [set of statements 2]
#else
  [set of statements 3]
#endif
```

These conditional compilation statements work very similar to standard C language **if** statements. The compiler evaluates each **#if** and **#elif** expression until one evaluates to true. The statements following the first true expression are compiled into the program and the subsequent expressions and statements are ignored until the **#endif**. If none of the **#if** or **#elif** statements evaluates to true and an **#else** statement is encountered before an **#endif**, then the statements following the **#else** are compiled into the program.

In the syntactical example above, if expression1 evaluates to TRUE, the *set_of_statements_1* will be compiled into the program. If *expression1* evaluates to FALSE and if *expression2* evaluates to TRUE, the *set_of_statements_2* will be compiled.

Otherwise, the *set_of_statements_3* will be compiled. The **#else** and *set_of_statements_3* are optional.

Conditional compilation is very useful in creating one program that runs in several configurations. This makes maintaining the program (or the product the program runs in) easier, because one source code can apply to several operating configurations.

Another common use of condition compilation is for debugging purposes. The combination of conditional compilation and standard I/O library functions can help you get a program up and running faster with less need for additional equipment such as in-circuit emulators and oscilloscopes.

The example state machine from Chapter 1 used to control an "imaginary" traffic light could be modified to add some debugging capability. Using conditional compilation, the debugging features added could be turned on and off as needed, trading between providing information and increased performance. The modified traffic light software appears in Figure 3.2.

```
#include <16F877.h>
#use delay(clock=10000000)
#fuses HS,WDT

//#define    DEBUGGING_ON

// Removing the comment will cause the DEBUGGING_ON "flag" to
// be enabled. This will cause the compiler to include the
// additional library and lines of code required allowing the
// programmer to watch the state machine run from the UART.

#ifdef DEBUGGING_ON
      // include the standard C I/O function definitions
      #use rs232(baud=9600,xmit=PIN_C6,rcv=PIN_C7)
#end if

#byte     PORTD = 0x08                  /* output port D definition */
#bit      EW_RED_LITE     = PORTD.5     /* definitions to actual outputs */
#bit      EW_YEL_LITE     = PORTD.6     /* used to control the lights */
#bit      EW_GRN_LITE     = PORTD.7
#bit      NS_RED_LITE     = PORTD.1
#bit      NS_YEL_LITE     = PORTD.2
#bit      NS_GRN_LITE     = PORTD.3

#byte     PINB  = 0x06                 /* input port B definition */
#bit      PED_XING_EW     = PINB.0     /* pedestrian crossing push button */
#bit      PED_XING_NS     = PINB.1     /* pedestrian crossing push button */
#bit      FOUR_WAY_STOP = PINB.3       /* switch input for 4-Way Stop */

char time_left;           // time in seconds spent in each state
int  current_state;       // current state of the lights
char flash_toggle;        // toggle used for FLASHER state

// This enumeration creates a simple way to add states to the machine
// by name. Enumerations generate an integer value for each name
// automatically, making the code easier to maintain.

enum { EW_MOVING , EW_WARNING , NS_MOVING , NS_WARNING , FLASHER };
```

Figure 3.2 *Expanded "Imaginary Traffic Light" Software (continues)*

```
// The actual state machine is here..
void Do_States(void)
{
      switch(current_state)
      {
            case  EW_MOVING:         // east-west has the green!!
                  EW_GRN_LITE = 0;
                  NS_GRN_LITE = 1;
                  NS_RED_LITE = 0;  // north-south has the red!!
                  EW_RED_LITE = 1;
                  EW_YEL_LITE = 1;
                  NS_YEL_LITE = 1;

                  if(PED_XING_EW || FOUR_WAY_STOP)
                  {     // pedestrian wishes to cross, or
                        // a 4-way stop is required
                        if(time_left > 10)
                              time_left = 10;   // shorten the time
                  }
                  if(time_left != 0)            // count down the time
                  {
                        --time_left;
                        return;                 // return to main
                  }                          // time expired, so..
                  time_left = 5;        // give 5 seconds to WARNING
                  current_state = EW_WARNING;
                                          // time expired, move
                  break;                  // to the next state

            case  EW_WARNING:
                  EW_GRN_LITE = 1;
                  NS_GRN_LITE = 1;
                  NS_RED_LITE = 0;  // north-south has the red..
                  EW_RED_LITE = 1;
                  EW_YEL_LITE = 0;  // and east-west has the yellow
                  NS_YEL_LITE = 1;

                  if(time_left != 0)            // count down the time
                  {
                        --time_left;
                        return                  // return to main
                  }               // time expired, so..
```

Figure 3.2 *Expanded "Imaginary Traffic Light" Software (continues)*

```
            if(FOUR_WAY_STOP) // if 4-way requested then start
                current_state = FLASHER; // the flasher
            else
            {                         // otherwise..
                time_left = 30;       // give 30 seconds to MOVING
                current_state = NS_MOVING;
            }                         // time expired, move
            break;                    // to the next state

        case    NS_MOVING:
            EW_GRN_LITE = 1;
            NS_GRN_LITE = 0;
            NS_RED_LITE = 1;  // north-south has the green!!
            EW_RED_LITE = 0;  // east-west has the red!!
            EW_YEL_LITE = 1;

            if(PED_XING_NS || FOUR_WAY_STOP)
            {     // if a pedestrian wishes to cross, or
                  // a 4-way stop is required..
                  if(time_left > 10)
                        time_left = 10;   // shorten the time
            }
            if(time_left != 0)            // count down the time
            {
                --time_left;
                return;                   // return to main
            }                 // time expired, so..
            time_left = 5;    // give 5 seconds to WARNING
            current_state = NS_WARNING;   // time expired, move
            break;                           // to the next state

        case    NS_WARNING:
            EW_GRN_LITE = 1;
            NS_GRN_LITE = 1;
            NS_RED_LITE = 1;  // north-south has the yellow..
            EW_RED_LITE = 0;
            EW_YEL_LITE = 1;  // and east-west has the red..
            NS_YEL_LITE = 0;

            if(time_left != 0)            // count down the time
            {
                --time_left;
                return;                   // return to main
            }                 // time expired, so..
```

Figure 3.2 *Expanded "Imaginary Traffic Light" Software (continues)*

```
            if(FOUR_WAY_STOP) // if 4-way requested then start
            current_state = FLASHER; // the flasher
            else
            {                    // otherwise..
             time_left = 30;    // give 30 seconds to MOVING
             current_state = EW_MOVING;
             }                 // time expired, move
             break;            // to the next state

        case  FLASHER:
            EW_GRN_LITE = 1;  // all yellow and
            NS_GRN_LITE = 1;  // green lights off
            EW_YEL_LITE = 1;
            NS_YEL_LITE = 1;

            flash_toggle ^= 1;       // toggle LSB..
            if(flash_toggle & 1)
            {
                  NS_RED_LITE = 0;  // blink red lights
                  EW_RED_LITE = 1;
            }
            else
            {
                  NS_RED_LITE = 1;  // alternately
                  EW_RED_LITE = 0;
            }
            if(!FOUR_WAY_STOP)       // if no longer a 4-way stop
                  current_state = EW_WARNING;
            break;                   // then return to normal

        default:
            current_state = NS_WARNING;
            break;       // set any unknown state to a good one!!
    }
    }

    void main(void)
    {
    port_b_pullups(TRUE);
 set_tris_b(0xFF); // port B is all input
    set_tris_d(0x00); // port D is all output

    current_state = NS_WARNING;       // initialize to a good starting
                                      // state (as safe as possible)
```

Figure 3.2 *Expanded "Imaginary Traffic Light" Software (continues)*

```
    while(1)
    {
        delay_ms(250);   // 1 second delay.. this time could
        delay_ms(250);   // be used for other needed processes
        delay_ms(250);
        delay_ms(250);

        Do_States();     // call the state machine, it knows
                         // where it is and what to do next
        #ifdef DEBUGGING_ON
            // send state and time data to serial port
            printf("Current State = %d :", current_state);
            printf("Time Left = %d  \r", (int)time_left);
        #endif

    }
}
```

Figure 3.2 *Expanded "Imaginary Traffic Light" Software (continued)*

3.6.4 THE #ERROR DIRECTIVE

The **#error** directive forces the compiler to generate an error when this directive is encountered during compilation of the source file. The command may be used to alert the user to an invalid compile time situation. The directive takes the following form:

```
#error text
```

The *text* may include macros that will be expanded for display. As such, the command may also be used to see macro expansions. Consider the following section of code:

```
#define BUFFER_SIZE 18
#if BUFFER_SIZE > 16
#error Buffer size is too large-Currently declared as BUFFER_SIZE
#endif
```

When the compiler encounters the **#if** directive, it evaluates *BUFFER_SIZE* and compares it to 16. Because *BUFFER_SIZE* is defined as 18, the compiler executes the code following the **#if** and generates an error message according to the text. Here, the generated error message reads, "Buffer size is too large-Currently declared as 18." *BUFFER_SIZE* is a macro in the text and is replaced by its defined value.

Here is an example that uses the **#error** directive to examine the results of redefining a function call using macros:

```
#define my_minimum(y,x) min(x,y)
#error Macro test: my_minimum(1,2)
```

Compiling this code results in the error message, "Macro test: min(2,1)." This shows that the function name is replaced and the variables are moved to their correct position for the new function call.

3.6.5 THE #PRAGMA DIRECTIVE

The previous preprocessor directives are fairly standard across compilers. However, you will find that most compilers also have compiler-specific preprocessor directives. Usually, the **#pragma** directive allows for these compiler-specific directives or switches. **#pragma** statements are very compiler dependent, so you should always refer to the compiler user's guide when using these controls.

The CCS-PICC compiler supports the **#pragma** directive but using it is optional. The preprocessor directives described in the following sections are CCS-PICC specific. Including the **#pragma** directive is optional when using them. For example, both of the following are acceptable under CCS-PICC:

```
#pragma device PIC16F877
#device PIC16F877
```

3.7 CCS-PICC FUNCTION-QUALIFYING DIRECTIVES

The preprocessor directives **#inline, #separate, #int_default, #int_global,** and **#int_xxx** are all function-qualifying directives specific to the CCS-PICC compiler. They allow the user to direct the compiler to handle some functions in a specific way. The **#inline** and **#separate** directives flags determine where the compiler places the compiled function. The **#int_default, #int_global,** and **#int_xxx** directives flag functions for handling interrupts. These directives are placed in the line directly preceding the function they are to affect. An example using the **#inline** directive is shown below:

```
#inline
void abd()
{
        .
        .
        .
}
```

3.7.1 THE #INLINE AND #SEPARATE DIRECTIVES

The **#inline** directive causes the compiler to place a duplicate copy of the specified function wherever it is called. Typically, the compiler makes the decision as to when it is necessary to make functions inline. Forcing functions inline may make the compiled code longer. However, when speed or stack space is critical, it is useful to have some functions placed inline.

The **#separate** directive is the opposite of the **#inline** directive in that it prevents the compiler from making the function inline ever. This makes the compiled code smaller but may use more stack space as more functions are called instead of pulled inline. The compiler will make all procedures marked **#separate**, separate, as requested, even if there is not enough stack space to execute.

3.7.2 THE #INT_DEFAULT, #INT_GLOBAL, AND #INT_XXX DIRECTIVES

The **#int_default**, **#int_global**, and **#int_xxx** preprocessor directives specify the interrupt routines to the compiler. Most of the PIC microcontrollers have only one interrupt vector and, as such, there is a single interrupt service routine called by any interrupt that occurs. By default, under CCS-PICC, that interrupt service routine is a dispatcher designed into the CCS-PICC compiler. The dispatcher routine calls the individual interrupt routines based on which interrupt flags are set and which interrupt routines were defined by the **#int_xxx** directive.

If the PIC triggers an interrupt and none of the expected interrupt flags are set, then the dispatcher routine calls the function specified by the **#int_default** directive. This can be useful for detecting interrupts that are not expected and not handled.

The **#int_global** directive causes the following function to replace the compiler interrupt dispatcher. The function is normally not required and should be used with great caution. The compiler interrupt dispatcher includes start-up code and cleanup code for the interrupts that saves and restores the registers. When **#int_global** is used, it is up to the user to write the code to handle the start-up and cleanup for the interrupt routines.

The **#int_xxx** directive specifies that the following function is an interrupt function. The compiler interrupt dispatcher calls the interrupt functions when the PIC triggers an interrupt. The actual interrupt function called is dependent on which interrupt flags are set in the PIC. The interrupt functions may not have any parameters. Not all directives may be used with all parts. See the devices *.h* file for all valid interrupts for the part. Table 3.5 shows the more common interrupts and their associated directives.

In some of the PIC microcontrollers, such as the PIC18F458, there are two interrupt vectors. This allows the prioritization of interrupts at a hardware level, even to the extent of a higher priority interrupt preempting an interrupt in process. The CCS-PICC compiler utilizes this feature of the hardware by using a flag, **FAST**, with the **#int_xxx** directive. Only one interrupt can be given this priority level under CCS-PICC because this flag is only allowed once in a program. The generic form of a fast interrupt is:

```
#int_xxx FAST
```

If the external interrupt 2 were to be made the fast interrupt for a program, its preprocessor directive would appear as:

```
#int_ext2 FAST
```

When an interrupt is used with the **FAST** flag, the interrupt does not save all the registers the way a normal interrupt does; it only saves the essential processor registers. The idea is to do something very simple or make sure you save and restore any registers you change during the interrupt.

Preprocessor Directive	Description of Interrupt
#int_ad	Analog to digital conversion complete
#int_adof	Analog to digital conversion timeout
#int_buscol	Bus collision
#int_button	Pushbutton
#int_ccp1	Capture or compare on unit 1
#int_ccp2	Capture or compare on unit 2
#int_comp	Comparator detect
#int_eeprom	Write complete
#int_ext	External interrupt
#int_ext1	External interrupt #1
#int_ext2	External interrupt #2
#int_i2c	I^2C interrupt (only on 14000)
#int_lcd	Activity
#int_lowvolt	Low voltage detected
#int_psp	Parallel slave port data in
#int_rb	Port B any change on B4-B7
#int_rc	Port C any change on C4-C7
#int_rda	RS232 receive data available
#int_rtcc	Timer0 (RTCC) overflow
#int_ssp	SPI or I^2C activity
#int_tbe	RS232 transmit buffer empty
#int_timer0	Timer0 (RTCC) overflow
#int_timer1	Timer1 overflow
#int_timer2	Timer2 overflow
#int_timer3	Timer3 overflow

Table 3.5 *Interrupt Preprocessor Directives*

3.8 CCS-PICC PREDEFINED IDENTIFIERS

Table 3.6 lists some predefined identifiers. These identifiers are specific to CCS-PICC but could be found in other compilers. These can be used to get information, such as the compile data and time, into the compiled program. They can also be used to conditionally compile based on device type or compiler type.

Predefined Identifier	Description
__DATE__	Date of compile in the form: 31-JAN-02
__DEVICE__	Base number of the current device*
__FILE__	File name of the file being compiled
__LINE__	Line number of the file being compiled
__PCB__	PCB compiler definition**
__PCM__	PCM compiler definition**
__PCH__	PCH compiler definition**
__TIME__	Time of compile in the form: hh:mm:ss
*Defined by the compiler from the information given by the #device directive. The base number is usually the number after the C in the part number. For example, the PIC16C622 has a base number of 622.	
**PCB, PCM, and PCH are all versions of the CCS-PICC compiler. The given compiler defines its associated identifier. #ifdef can be used to determine which compiler is performing the compilation when used in conjunction with the identifier. This would appear as, "#ifdef __PCB__."	

Table 3.6 *Predefined Identifiers*

3.9 CCS-PICC DEVICE SPECIFICATION DIRECTIVES

The device specification preprocessor directives tell the compiler important hardware specific information about the target device. These preprocessor directives include **#device**, **#fuses**, and **#id**.

3.9.1 THE #DEVICE DIRECTIVE

The **#device** preprocessor directive defines what microcontroller is the target device for the program. The **#device** directive takes the form,

```
#device chip options
```

chip is the name of a specific processor, such as *PIC16C74*.

options may effect the memory management, the A/D control, or the debugging capability. The memory management options allow the user to specify the number of bits used to contain a pointer to memory.

*=5	Use 5-bit pointers (for all parts)
*=8	Use 8-bit pointers (14- and 16-bit parts)
*=16	Use 16-bit pointers (for 14-bit parts)

A 5-bit pointer can only address up to 32 bytes of RAM but also requires very little memory to store the pointer. A 16-bit pointer, on the other hand, requires 2 bytes of memory to store the pointer but can address up to 65,536 bytes of RAM. Knowledge of the target device and the memory usage of the program are both required in order to effectively use these options.

The A/D option controls the number of bits read from the analog-to-digital converter by the internal function, *read_adc()*. This options is formatted in the following way:

ADC=x   x is the number of bits the read_adc() function should return

The final option is the debug control option. If specified, the code generated is compatible with Micorchip's ICD debugging software.

ICD=TRUE   Generate code compatible with Microchip's ICD debugging software

Both *chip* and *options* flags are optional, so multiple **#device** lines may be used to fully define the device. The **#device *chip*** line should always precede any **#device *options*** lines because the **#device *chip*** line clears all previous **#define** and **#fuse** settings. Also note that every program should have exactly one **#device** with a chip specification.

The following segments of code all select a PIC16C877 device with 8-bit pointers, 10-bit A/D, and no debugging output:

```
#device PIC16F877 *=16 ADC=10
```

-or-

```
#device PIC16F877
#device *=16   ADC=10
```

-or-

```
#device PIC16F877  ADC=10
#device *=16
```

3.9.2 THE #FUSE DIRECTIVE

This directive defines what fuses should be set in the part when it is programmed. This directive does not affect the compilation but places the information in the file for the programmers to read. The directive is formatted as follows:

#fuses options

The options available depend on the target device. To view a list of valid fuse options for particular device under CCS-PICC, select the **VIEW** pull-down menu. Then select **Valid Fuses**. For example, the valid options for the PIC16C877 are shown in Table 3.7.

An example of requesting the fuses to be set for a high-speed oscillator, brownout detection, and no watchdog timer follows:

```
#fuses HS,BROWNOUT,NOWDT
```

Some programmers require the information in a format other than the Microchip standard. If the fuses need to be in Parallax format, add a PAR option. Some device programmers

Fuse Option	**Description**
LP	Low power osc < 200 KHz
XT	Crystal osc <= 4 MHz
HS	High-speed Osc (> 4 MHz)
RC	Resistor/Capacitor Osc with CLKOUT
NOWDT	No watchdog timer
WDT	Watchdog timer enabled
NOPUT	No power-up timer
PUT	Power-up timer enabled
PROTECT	Code protected from reads
PROTECT_5%	Protect 5% of ROM
PROTECT_50%	Protect 50% of ROM
NOPROTECT	Code not protected from reading
NOBROWNOUT	No brownout reset
BROWNOUT	Reset when brownout detected
LVP	Low-voltage programming on B3
NOLVP	No low-voltage programming, B3 used for I/O
CPD	Protect EEPROM memory
NOCPD	No EEPROM memory protection
WRT	Allow programs to write to flash
NOWRT	Prevent programs from writing to flash

Table 3.7 *Fuse Options*

require the high and low bytes of non-program data in the *.hex* file to be swapped from the Microchip standard. The SWAP option performs this special function.

3.9.3 THE #ID DIRECTIVE

The PIC microcontrollers contain memory locations that are designated as ID locations. The size and accessibility of the ID memory is dependent on the device being used. For example, the PIC16F877 has 4 bytes reserved for its ID, and the ID is only accessible during program and verify operations. On the other hand, the PIC18F458 has 8 bytes reserved for its ID, and the ID is accessible during program, verify, and normal execution (through the **TBLRD** and **TBLWT** assembler instructions).

The **#id** preprocessor directive allows the user to place a specific ID in the output file to be written to the microcontroller in the ID locations. There are several methods of declaring an id. Their generic forms are:

#id *number16*

#id *number,number,number,number*

#id CHECKSUM

Here, *number16*, takes the 16-bit number and places a nibble into each of the four ID words. *number,number,number,number* specifies the exact value to be used in each of the four ID words. **CHECKSUM** is a keyword indicating that the checksum should be saved as the ID. Table 3.8 shows each in use and the resulting ID words.

Generic Form	Preprocessor Directive	ID Words
#id number16	`#id 0x1234`	0x0001,0x0002,0x0003,0x0004
#id number, number, number, number	`#id 0x1234,0x5678,0x90ab,0xcdef`	0x1234,0x5678,0x90ab,0xcdef
#id CHECKSUM	`#id CHECKSUM`	*0x0002,0x000A,0x0005,0x0005
*This is an example checksum. Actual checksum will be based on the compiled program.		

Table 3.8 *ID Preprocessor Directive*

3.10 CCS-PICC BUILT-IN LIBRARY PREPROCESSOR DIRECTIVES

The delay, I/O, and communications libraries in the CCS-PICC compiler require preprocessor directives for initialization and system parameters. These directives include **#use delay**, **#use fast_io**, **#use fixed_io**, **#use standard_io**, **#use rs232**, and **#use i2c**.

3.10.1 THE #USE DELAY DIRECTIVE

The **#use delay** directive takes the form,

#use delay(clock-speed[, restart_WDT])

This directive tells the compiler the speed of the processor and enables the use of the built-in functions *delay_ms()* and *delay_us()*. Speed is in cycles per second (Hz). An optional **restart_WDT** may be used to request the compiler to restart the watchdog timer while delaying. Two examples are shown below, one with a watchdog timer request and one without. Both examples use 4.0 MHz clocks:

```
#use delay(4000000)                  // no WDT
#use delay(4000000,restart_WDT)      // reset the WDT to
                                     // prevent timeout
```

3.10.2 THE #USE FAST_IO, #USE FIXED_IO, AND #USE STANDARD_IO DIRECTIVES

The **#use xxx_io** preprocessor directives affect the code generated when I/O ports are accessed. When I/O ports are read from or written to, the TRIS registers for those ports must be set appropriately to perform the desired reads or writes. The programmer can set these TRIS registers directly, or under CCS-PICC, the compiler can generate the required TRIS settings automatically.

Under standard I/O access (which is the default), CCS-PICC generates code to make an I/O pin either input or output every time it is used. Standard I/O access can be forced by using the **#use standard_io** preprocessor directive.

The **#use fixed_io** and **#use fast_io** preprocessor directives prevent the compiler from setting the direction of the I/O pins every time they are accessed, saving time and code space. With the **#used fast_io** directive, the user must set the TRIS pins manually some time prior to use directly, or by calling *set_tris_x()*. The **#use fixed_io** directive takes as parameters the pins to be set to outputs and sets the pins appropriately.

Any **#use xxx_io** directive stays in effect until another **#use xxx_io** directive is encountered by the compiler.

The standard forms are as follows:

```
#use standard_io(x)

#use fixed_io(port_outputs=pin[,pin,pin...])

#use fast_io(x)
```

Here *x* represents a port (A to G) and *pin* is a pin constant. Valid ports and pins are device specific. Valid pin constants can be found in the *.h* file for the selected device.

A program is shown below. This program was compiled three times. The first time, **#use xxx_io** was replaced by **#use standard_io**. In the second compile, **#use xxx_io** was replaced by **#use fixed_io** and the third time it was replaced by **#use fast_io**. The program sets Port B, pin 0 high every time, but because of the different I/O directives, the listing file is different each time. Below the program are pieces from the listing file that were created during each compile.

```
#include <16C63.H>  // #device is in this include file
#use xxx_io         // this was replaced by the appropriate
                    // #use xxx_io directive
void main()
{
        OUTPUT_HIGH(PIN_B0);
        while (TRUE)
        {
        }
}
```

Preprocessor Directive:

```
#use standard_io(B)
```

Listing File Output:

```
....................          OUTPUT_HIGH(PIN_B0);
0007:   BSF     03.5
0008:   BCF     06.0
```

```
        0009:   BCF     03.5
        000A:   BSF     06.0
```

Preprocessor Directive:

```
        #use fixed_io(b_outputs=PIN_B0)
```

Listing File Output:

```
....................        OUTPUT_HIGH(PIN_B0);
0007:   MOVLW   FE
0008:   TRIS    6
0009:   BSF     06.0
```

Preprocessor Directive:

```
        #use fast_io(B)
```

Listing File Output:

```
....................        OUTPUT_HIGH(PIN_B0);
0007:   BSF     06.0
```

3.10.3 THE #USE I2C DIRECTIVE

The **#use i2c** preprocessor directive defines an I^2C port to the microcontroller. The directive has the form:

```
#use i2c(options)
```

options may be one or more of the options listed in Table 3.9. Use commas to separate the options if more than one is used in a single statement.

I^2C Options	*Description*
MASTER	Set the master mode
SLAVE	Set the slave mode
SCL=pin	Specifies the SCL pin (pin is a bit address)
SDA=pin	Specifies the SDA pin
ADDRESS=nn	Specifies the slave address
FAST	Use the fast I^2C specification
SLOW	Use the slow I^2C specification
RESTART_WDT	Restart the WDT while waiting in *i2c_read()*
FORCE_HW	Use hardware I^2C functions

Table 3.9 *I^2C Options*

The I^2C library contains functions to implement an I^2C bus and requires the I^2C preprocessor directive for setup. The **#use i2c** remains in effect until another **#use i2c** directive is encountered. Software functions are generated unless the *FORCE_HW* is specified. The *slave* mode should only be used with the hardware SSP. The following line selects Port B, pin 0 for the data line and Port B, pin1 for the clock line and requests setup as a master:

```
#use i2c(master, sda=PIN_B0, scl=PIN_B1)
```

The preprocessor directive to use the hardware i2c on Port C and pins C3 and C4 for clock and data in slave mode is written as:

```
#use i2c(slave, sda=PIN_C4,scl=PIN_C3,address=0xa0,FORCE_HW)
```

3.10.4 THE #USE RS232 DIRECTIVE

The **#use rs232** preprocessor directive is used to define an RS232 port. Based on the transmit and receive pins defined, the implementation of the port may be either hardware or software based. If the pins specified match the hardware UART of the PIC, then a hardware port is used. If the pins specified do not match the hardware UART of the PIC, then a software UART is used. The generic form of the **#use rs232** directive is:

```
#use rs232 ([options])
```

The options allow the user to specify baud rate, transmit and receive pins, parity, data bits, stream name, and so on. The format and description of each option are listed in Table 3.10.

Option Format	***Description***
BAUD=x	Set baud rate to x.
XMIT=pin*	Set transmit pin as pin.
RCV=pin*	Set receive pin as pin.
RESTART_WDT	Flag getchar() and clear the WDT as it waits for a character.
INVERT	Invert the polarity of the serial pins (not normally needed when level converter, such as the MAX232 is being used). May not be used with the hardware UART; may only be used with a software UART.
PARITY=x	Set parity to none (x=N), even (x=E) or odd (x=O).
BITS=x	Set bits to x where x is 5-9 if the hardware UART is not being used and x is 8-9 if the hardware UART is being used.
FLOAT_HIGH	This option is used for open collector circuits where the desired operation is not to drive the output high but to allow it to float high.
ERRORS	This option flags the compiler to keep receive errors in the variable RS232_ERRORS and to reset errors when they occur.
BRGH1OK	This flags the compiler to allow bad baud rates on chips that have baud rate problems.
ENABLE=pin*	This flags the compiler to drive pin high during transmit and may be used to enable 485 transmits.
STREAM=stream_name	Associates a stream identifier, stream_name, with this RS232 port. The identifier may then be used in functions like fputc().
*pin is a pin constant . Valid ports and pins are device specific. Valid pin constants can be found in the .h file for the selected device.	

Table 3.10 *#use rs232 Options*

When selecting a baud rate, the clock rate of the PIC must be taken into consideration. Not all baud rates are available at all clock rates. If the baud rate cannot be achieved to within 3% of the desired baud rate, an error is generated. The **#use delay** directive is used to define the clock rate of the PIC and must be used prior to using the **#use rs232** directive.

The *putchar()*, *fputc()*, *getchar()*, and *fgetc()* functions all require an RS232 port to read from and write to and, thus, require a **#use rs232** directive prior to being called. *putchar()* and *getchar()* functions write to and read from the most recent port defined by a **#use rs232** directive that does not include a stream identifier. If every **#use rs232** directive includes a stream identifier, then the first stream is used as standard input and output. The *fputc()* and *fgetc()* functions write to and read from a stream name defined by a **#use rs232** directive.

When the **ERRORS** keyword is used, the **RS232_ERRORS** variable is defined according to the following rules:

With a hardware UART

- Bit 7 is 9th bit for the 9-bit data mode for receives and transmits.
- Bit 6 is set to 1 to indicate a transmit failed in float high mode.

Without a hardware UART

- RS232_ERRORS is only used during receive.
- RS232_ERRORS is a copy of the RCSTA register except bit 0 is used to indicate a parity error.

See Sections 3.3, 3.4, and 3.5 for examples and further explanation of the **#use rs232** directive.

3.11 CCS-PICC MEMORY CONTROL PREPROCESSOR DIRECTIVES

The memory control preprocessor directives give the user control over the location and size of variables in memory as well as requesting the compiler to zero the RAM and place assembler code directly into the output file.

3.11.1 THE #TYPE DIRECTIVE

The **#type** preprocessor directive allows the user to redefine variable types supported by the compiler. By default, the CCS-PICC compiler treats *short* as 1 bit, *int* as 8 bits, and *long* as 16 bits. In order to help with code compatibility, a **#type** directive may be used to allow these bytes to be changed. The generic form of the directive is:

#type standard-type=size[, standard-type = size...]

Following is an example of using the **#type** directive to redefine short variables to be 8 bits, integer variables to be 16 bits, and long variables to be 32 bits:

```
#type short=8, int=16, long=32
```

Note that the CCS-PICC compiler sample programs and include files may not work as expected if the **#type** directive is used in your program.

3.11.2 THE #BIT DIRECTIVE

The **#bit** preprocessor directive takes the form

```
#bit id = x.y
```

where *id* is a valid C language identifier, *x* is a constant or a C variable, and *y* is a constant from 0 through 7. This directive is useful to gain access to a bit in the processor's register map or to another variable. To gain access to a bit in a register, *x* is the register address. Here, the Timer0 interrupt flag is cleared by writing a 1 to it:

```
#bit TOIF = 0x0B.2  // declare TOIF as bit 2 of
                    // register 0x0B
...
TOIF = 1;           // clear TOIF by writing a one to it
```

Here is an example of checking a single bit of a variable using the **#bit** directive.

```
int result;
#bit myresult = result.0
...
if (myresult == 0)
...
```

3.11.3 THE #BYTE DIRECTIVE

The **#byte** preprocessor directive takes the form

```
#byte id = x
```

where *id* is a C identifier and x is a C variable or a constant. If *id* is already known as a C variable, then this tells the compiler to locate the variable at address *x*. If *id* is not already declared, then the compiler creates a new C variable and places it at address *x* as an 8-bit integer. In both cases, the memory at *x* is not exclusive to this variable. Other variables may be located at the same location. In fact, when *x* is a variable, *id* and *x* share the same memory location. The **#locate** and **#reserve** preprocessor directives explained later in this section can be used to reserve memory locations.

This code example creates a new variable *b_port* and locates it at address 6:

```
#byte b_port = 6
```

The next code example creates two size char variables that share an address. The first variable, *a*, is created and placed in memory by the compiler. Then the **#byte** statement causes the compiler to create a second variable, *b*. The '= *a*' portion of the **#byte** statement directs the compiler to assign the new variable *b* to the same address as the variable *a*. This is similar to the way a union would allocate the memory:

```
char a;
#byte b = a
```

3.11.4 THE #LOCATE DIRECTIVE

The **#locate** preprocessor directive takes the form

```
#locate id=x
```

where *id* is a C variable and *x* is a constant memory address. **#locate** works like **#byte**; however, in addition to creating a variable and placing it at *x*, it also reserves *x* such that no other variable can be located there. The following state declares a floating point variable, *my_float*, and places it at address 0x50:

```
float x;
#locate my_float=0x50
```

3.11.5 THE #RESERVE DIRECTIVE

The **#reserve** directive allows RAM locations to be reserved from use by the compiler. **#reserve** must appear after the **#device** directive, otherwise it has no effect. The **#reserve** directive can reserve one address or several addresses at a time by calling it as

```
#reserve address
```

or

```
#reserve address, address, address
```

where *address* is a RAM address.

Another method of calling **#reserve** reserves a block of memory addresses:

```
#reserve start:end
```

where *start* is the first address to be reserved and *end* is the last address to be reserved.

For example, to reserve the bytes at locations 0x10, 0x11, and 0x12, any one of the following methods may be used.

```
#reserve 0x10
#reserve 0x11
#reserve 0x12
```

or

```
#reserve 0x10, 0x11, 0x12
```

or

```
#reserve 0x10:0x12
```

3.11.6 THE #ZERO_RAM DIRECTIVE

The **#zero_ram** directive has no parameters and is called as

```
#zero_ram
```

This directive tells the compiler to set all of the internal registers that may be used to hold variables to zero before execution begins.

3.11.7 THE #ROM DIRECTIVE

The **#rom** preprocessor directive has the form

#rom address = {list}

where *address* is a ROM word address and *list* is a list of words separated by commas. The **#rom** directive allows the insertion of data into the *hex* file. In particular, this may be used to program the '84 data EEPROM as shown in the following example. Note that this directive does not prevent the ROM area from being used. See **#org** to reserve ROM.

```
#rom 0x2100={1,2,3,4,5,6,7,8}
```

3.11.8 THE #ORG DIRECTIVE

The **#org** statement can be formatted in several ways. They are

#org start, end

#org segment

#org start, end {}

#org start, end auto=0

where *start* is always the first ROM location (word address) to use, *end* is the last ROM location, and *segment* is the start ROM location from a previous **#org**. The **#org** statement allows the user to define to the compiler where the function following the **#org** should be placed. *end* may be omitted if a segment was previously defined and the user only wants to add another function to the segment. Each of these methods of calling **#org** is shown in the following example. The comments in the code describe what the compiler will do with the functions as a result of the **#org** statements:

```
#org 0x1E00, 0x1FFF
MyFunc()
{
        // this function located at 0x1E00
}

#org 0x1E00
        // (segment form of #org call)
Anotherfunc()
{
        // this will be somewhere 0x1E00-1F00
}

#org 0x800, 0x820 {}
// nothing will be at 800-820

#org 0x1C00, 0x1C0F
char const ID[10] = {"123456789"};
```

```
// This ID will be at 0x1C00
// Note some extra code will proceed the 12345679
```

3.11.9 THE #ASM AND #ENDASM DIRECTIVES

The **#asm** and **#endasm** directives allow for the insertion of assembly language into the C source files. **#asm** delineates the start of the assembly section, and **#endasm** delineates the end of the assembly. This appears in a C source file as follows:

```
#asm
...assembly code...
#endasm
```

or as:

```
#asm ASIS
...assembly code...
#endasm
```

The first example shows a typical call to the compiler to start and end an assembler section. If the second form is used with **ASIS**, then the compiler does not do any automatic bank switching for variables that cannot be accessed from the current bank. The assembly code is used "as-is." Without the **ASIS** option, the assembly is augmented so variables are always accessed correctly by adding bank switching where needed.

The predefined variable **_return_** may be used to assign a return value to a function from the assembly code. Be aware that any C code after the **#endasm** and before the end of the function may corrupt the value. Here is an example of a subroutine with an almost exclusively assembly body that returns a value:

```
int find_parity (int data)
{
        int count;
        #asm
        movlw  0x08
        movwf  count
        movlw  0
        loop:
        xorwf  data,w
        rrf    data,f
        decfsz count,f
        goto loop
        movwf  _return_
        #endasm
}
```

See Chapter 2 for a description of the assembly language for Microchip PIC microcontrollers.

3.12 CCS-PICC COMPILER CONTROL PREPROCESSOR DIRECTIVES

The compiler control preprocessor directives **#case** and **#opt** control the case-sensitivity and the optimization level of the compiler. The compiler control directive **#priority** also allows the user to establish the priority of specific interrupts.

3.12.1 THE #CASE DIRECTIVE

The **#case** directive causes the compiler to be case sensitive. By default, the CCS-PICC compiler is case insensitive. The following code is valid with case turned on:

```
#case
int value1;
int Value1;
void main()
{
        while(1)
                ;
}
```

This same code generates an error during compilation with case turned off citing duplicate declarations as the error.

When the compiler is case insensitive, a programmer used to working with a case sensitive compiler could make the following mistake:

```
int VALUE1 = 9;
void change_value()
{
        int value1;
        VALUE1 = 7;
        value1 = 5;
        while(1)
                ;
}
```

Here, the programmer may expect the global variable *VALUE1* to have the value of 7 after *change_value()* runs. However, because *value1* is declared as a local and is not case sensitive, the compiler only accesses the local variable *value1* while in *change_value()*. Here the local variable is assigned the value 7 then the value 5, but the global value is never changed.

NOTE: Not all of the CCS example programs, headers, or drivers have been tested with case sensitivity turned on.

3.12.2 THE #OPT DIRECTIVE

The **#opt** directive is called as

```
#opt n
```

where *n* is an optimization level from 0 to 9. The optimization level applies to the entire program and may appear anywhere in the file. The optimization level 5 sets the level to be the same as the CCS_PICC PCB, PCM, and PCH stand-alone compilers. The CCS_PICC PCW default is 9 for full optimization. This may be used to set a PCW compile to look exactly like a PCM compile. It may also be used to reduce optimization if an optimization error is suspected.

3.12.3 THE #PRIORITY DIRECTIVE

The **#priority** directive may be used to set the order in which the interrupt flags are checked and their associated handling routines are called when an interrupt occurs. As described in Section 3.7.2, some PIC microcontrollers have only one interrupt vector. When an interrupt occurs, the microcontroller jumps to the location pointed to by the interrupt vector. By default, the CCS-PICC compiler places its interrupt dispatch routine at this location. The dispatch routine checks the interrupt flags to determine which interrupts have occurred and calls the associated interrupt handling routine (see **#int_xxx** in Section 3.7.2 for defining the interrupt handling routines). The **#priority** directive gives the user the opportunity to handle some interrupt conditions before others.

The **#priority** directive has the form:

```
#priority ints
```

where *ints* is a list of one or more interrupts separated by commas. The highest priority items are first in the list. For example, if *rtcc* and *rb* are valid interrupts for a given chip, then *rtcc* could be set as the highest and *rb* as the second highest interrupts by calling **#priority** as follows:

```
#priority rtcc, rb
```

When two sources of interrupts occur simultaneously, the highest priority interrupt is always serviced first. However, once an interrupt service routine is called, it will not be interrupted even if a higher priority interrupt occurs. In this case, the higher priority interrupt will be called once the interrupt service routine in progress is completed.

In some PIC microcontrollers, there are two levels of interrupts, and prioritizing the interrupts can be performed at the hardware level. To utilize this, a flag is used with the **#int_xxx** preprocessor directive. A complete description of the **FAST** flag can be found in Section 3.7.2.

3.13 CHAPTER SUMMARY

This chapter has provided the knowledge necessary for you to use standard library functions for printing information, gathering user input, as well as debugging your programs. You have also learned how to utilize **#use rs232** to establish RS232 ports for use by the *getchar()* and *putchar()* functions such that the standard library may be used on a variety of hardware configurations.

You have learned about the use of the complier directives **#include** and **#define** and how they are used to add reusability and readability to your programs. You have also learned about conditional compilation using **#ifdef**, **#ifndef**, **#else**, and **#endif**, providing a method for having a single source code to be used for a variety of system configurations or adding temporary code for debugging purposes.

Finally, you learned about the compiler specific preprocessor directives to set up RS232 ports, I/O ports, and interrupt handling routines.

3.14 EXERCISES

EXERCISE

1. What will the contents of s be after the following executes and "123ABX" is used as the input and what is the total number of bytes required to hold the string (Section 3.5)?

```
#include <16F877.h>      // register definition file for
#fuses HS,NOWDT    // a Microchip PIC16F877
#use delay(clock=10000000)
#use rs232(baud=9600,xmit=PIN_C6,rcv=PIN_C7)

char s[12];      // declare a place to put the string
void main(void)
{
while(1)
   {
      gets(s);    // get a string from the standard input
   }
}
```

2. Fill in the missing lines below to get the *serious_code()* line compiled into the program and the *debugging_code()* line left out (Section 3.6).

```
#define DEBUGGING_CODE_OFF      1

#if_________________________
      serious_code();
#___________________________
      debugging_code();
#endif
```

3. Write a function that uses *gets()* to receive a date as three separate strings followed by the new line character '\n' in the format "MM\n," "DD\n," and "YYYY\n" and then uses *printf()* to display the date "DD-MM-YY" (Sections 3.4 and 3.5).

4. Write a function that prints its compile date and time. Use the compiler's internal tag names to get the date and time values (Section 3.8).

5. Write a function that prints its file name and compiler version. Use the compiler's internal tag names to get this information (Section 3.8).

6. Write a function that inputs a 16-bit hexadecimal value and then prints the binary equivalent. Note: there is no standard output function to print binary, so it is all up to you!

3.15 LABORATORY ACTIVITIES

1. Create a program to output the ASCII character 'G' every 50 milliseconds via the UART. Observe both the TTL level signal on the microcontroller's TXD line and the RS232 signal at the output of the media driver using an oscilloscope. On each waveform, identify the start bit, the stop bit, and the data bits of the transmitted signal.
2. Modify the program above so that the 'G' is also transmitted from the SPI bus. Use the oscilloscope to display both the transmitted signal and the clock and identify the data bits.
3. Create a program to use *getchar()* to get a serial character (from the PC) and *putchar()* to return a character that is two letters higher in the alphabet.
4. Modify the program from No. 3 so that the code for the ASCII character being returned appears on Port D (where it could be conveniently read by attaching LEDs).
5. Create a program that gets a byte of data in on Port B and uses putchar() to echo what it receives to the UART at 9600 baud. Use external interrupt 0 (set to trigger on a rising or a falling edge as is convenient to your hardware) as the strobe, which causes Port B to be read and the data transmitted via the UART.
6. Team up with a partner. One of you should create an SPI bus transmitter program (a modification of the program from No. 5) that uses the SPI bus to transmit the data from Port B when the appropriate signal is received on the external interrupt. The other partner should create an SPI bus receiver (a modification of the program from No. 4), which displays the data received over the SPI bus.
7. For a real challenge, use the programs from No. 6 and create two-way communication in which both microcontrollers read the data on Port B when triggered by their own external interrupt and send it via the SPI bus. Both microcontrollers also display data received via the SPI bus.
8. Create a program that sends "C Rules!!" out of the SPI port using printf(), passing as a parameter the name of a function to send characters out the SPI port.
9. Modify the program from No. 8 that sends "C Rules this time *ddd* !!" out of the SPI port, 100 times, where *ddd* indicates which time, from 1 to 100.

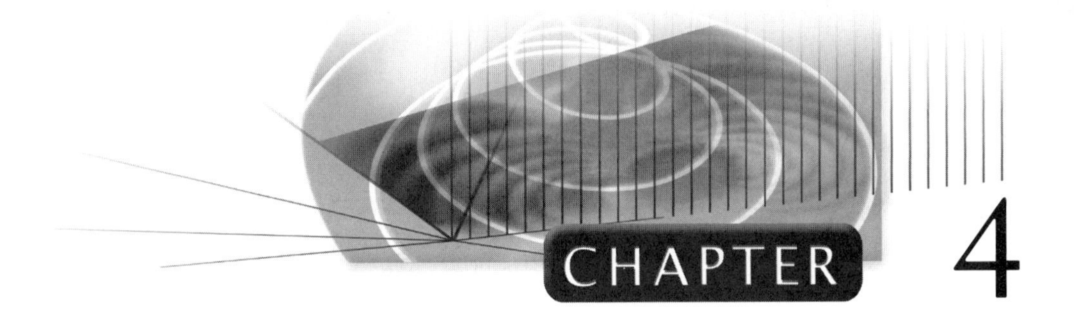

The CCS-PICC C Compiler and IDE

4.1 OBJECTIVES

At the conclusion of this chapter, you should be able to:

- Operate the CCS-PICC C compiler and integrated development environment.
- Create and compile a project.
- Apply the CCS-PICC environment to program a target device.
- Apply the CCS-PICC PIC Wizard to automatically generate shells of code.
- Apply the CCS-PICC serial port monitor to send and receive RS232 communications.
- Perform basic debugging operations using MPLAB as a software simulator.

4.2 INTRODUCTION

The purpose of this chapter is to familiarize you with the CCS-PICC C Compiler and the Microchip MPLAB development tool. The CCS-PICC C Compiler is just one of many C compilers available for the Microchip PIC microcontrollers. CCS-PICC's outstanding development environment and superb compiler determined its use in this text.

MPLAB is a development tool produced by Microchip and is available free of charge on their Web site, http://www.microchip.com. MPLAB includes a software simulator, a programming interface, in-circuit-emulator interface, a debugger interface, and a development environment capable of calling third-party compilers. A 'plug-in' is available on CCS's Web site (http://www.ccsinfo.com) to enable the CCS compiler to be called directly from within MPLAB.

Currently, CCS-PICC PCW is designed to run under Windows 95, 98, ME, NT4, 2000, or XP. The CD-ROM included with this book contains the evaluation version. It is limited in the size of the source file that can be compiled and is limited to one type of chip. There are several versions of the CCS-PICC C Compiler that handle 12 bit, 14 bit, and 16 bit opcodes. Consult the documentation on specific versions for information on which opcodes are supported. The PCW version of the CCS-PICC C Compiler is used throughout this book and includes compilation capabilities for 12-bit, 14-bit and 16-bit opcodes.

This chapter provides a look at the CCS-PICC development environment and its many features. You will learn how to create a project, set specific compiler options for the project, how to compile a project into an output file, and how to load that output file into the target device. The PIC Wizard code generator will be discussed along with the serial port monitor. Finally, you will be shown some basic features of MPLAB as a software simulator.

4.3 INTEGRATED DEVELOPMENT ENVIRONMENT

The CCS-PICC C Compiler is accessed through its Integrated Development Environment (IDE). The IDE allows the user to build projects, add source code files to the projects, set compiler options for the projects, and, of course, compile projects into executable program files. These executable files are then loaded into the target microprocessor. The goal of the following sections is to familiarize you with the usual use of the CCS-PICC IDE. There are many advanced features in the IDE that are beyond the scope of this text.

As CCS-PICC's IDE is discussed, the menu items are referenced by listing the main menu item name followed by '|' then the submenu item name. For example, a reference to the 'Save As' menu item within the 'File' main menu appears as **File|Save As**. The main menu for CCS-PICC's IDE appears below the application title bar.

4.4 PROJECTS

A project is a collection of files and compiler settings that you use to build a particular program. A few projects are provided for you on the CD included with this textbook, but it is also important to know how to create and configure a new project. All project files have a *.pjt* extension.

An open project appears in Figure 4.1. A tabbed page represents each open file. Only one source file is directly assigned to each project. Additional source files can be added to the project by using a **#include** statement in the main source file. In Figure 4.1 the project source file is *example4_1.c*. The additional source file for the project is *example4_1a.c* and is included in the project by the statement,

```
#include "C:\Program Files\PICC\Projects\Chapter4\Example4_1a.c"
```

in the project source file.

A status bar occupies the bottom of the window. The center field of the status bar across gives the full path of the selected file, which in Figure 4.1 is,

```
c:\program files\picc\projects\chapter4\example4_1.c
```

The field at the far right on the status bar shows the name of the main source file for the open project. In Figure 4.1, the main source file name is *example4_1*. The field at the far left on the status bar shows the current row and column position of the cursor in the open file.

4.4.1 OPEN EXISTING PROJECTS

Select the **Project|Open** menu command or press the **Open File** button on the toolbar to open an existing project file. When either of these is performed, an 'Open File' dialog window appears. If the **Open File** button is used, then select the project file filter in the dialog window. Browse to the appropriate directory and select the project file to open. When you open the project file, the main source file for the project is opened. Once a project has been compiled, all of the source and include files associated with a particular project can be opened by selecting the **Project|Open All Files** menu item.

4.4.2 CREATE NEW PROJECTS

New projects can be created manually with the **Project|New|Manual Create** menu item, or new projects can be generated by the PIC Wizard utility by using the **Project|New|PIC Wizard** menu item. Either method of creating a new project prompts the user for a main source file name for the project and a target device. However, when using the PIC Wizard, you are able to automatically generate code for I/O, timers, USARTs, and such.

When you select to manually create a project, the dialog window depicted in Figure 4.2 appears. The dialog prompts you to select a main source file for the project and a target device.

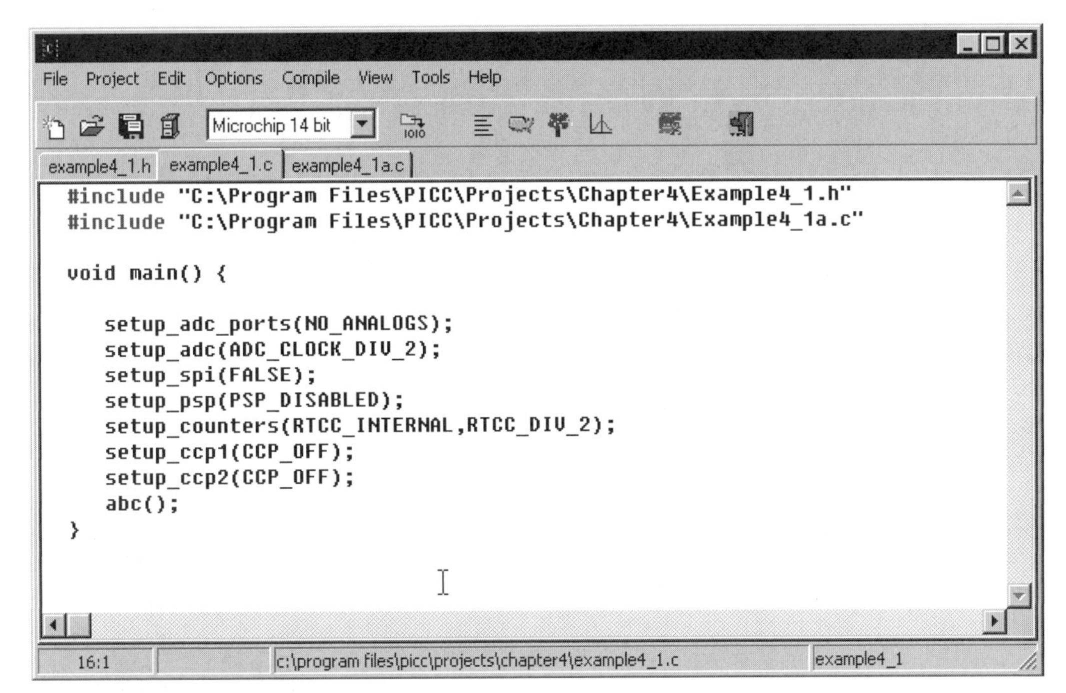

Figure 4.1 *CCS-PICC Project*

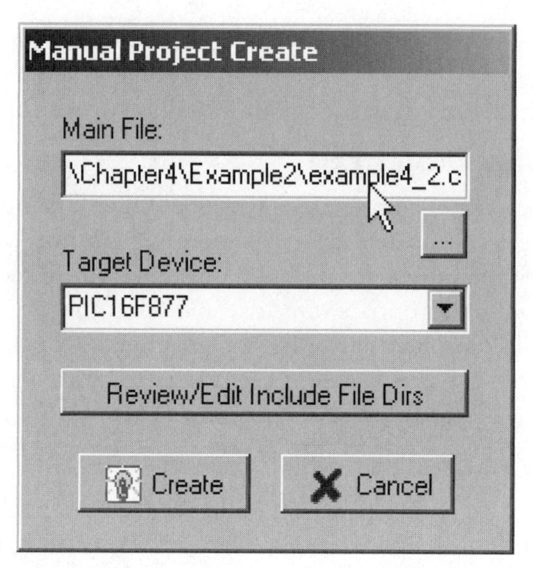

Figure 4.2 *Manual Project Create*

There is also a command button, **Review/Edit Include File Dirs**, that allows you to modify the order of and which directories are searched for included files. When the file and device fields are properly populated, click on the **Create** command button. A new project file is created with the same name as the main source file, but with a *.pjt* extension instead of the *.c* extension. Also, the appropriate include information is automatically generated in the *.c* file. Figure 4.3 shows the project created with the options used in Figure 4.2.

The PIC Wizard allows the user to select features such as timers and USARTs and their appropriate settings and then generate code for those features. The PIC Wizard is covered in more detail in Section 4.5.

4.4.3 SETTING THE INCLUDE DIRECTORIES FOR A PROJECT

The directories searched for included files for a particular project can be modified at any time. This is accomplished through either the **Project|Include Dirs…** or the **Options|Include Dirs…** menu items. Choosing either of these items opens a dialog window for selecting search directories paths and for establishing the order in which the directories will be searched during compilation. Figure 4.4 shows the Include Files dialog window.

4.4.4 COMPILE PROJECTS

Obtaining an executable program file requires *compiling* the project. The compiler is called by selecting the **Compile|Compile** menu command or the **Compile** button of the toolbar. Execution of the compiler produces several files formatted according to the formats selected in the File Formats window (accessible by the **Options|File Formats** menu). After the compilation, the 'CCS C Compiler' window opens showing the compilation results. See Figure 4.5.

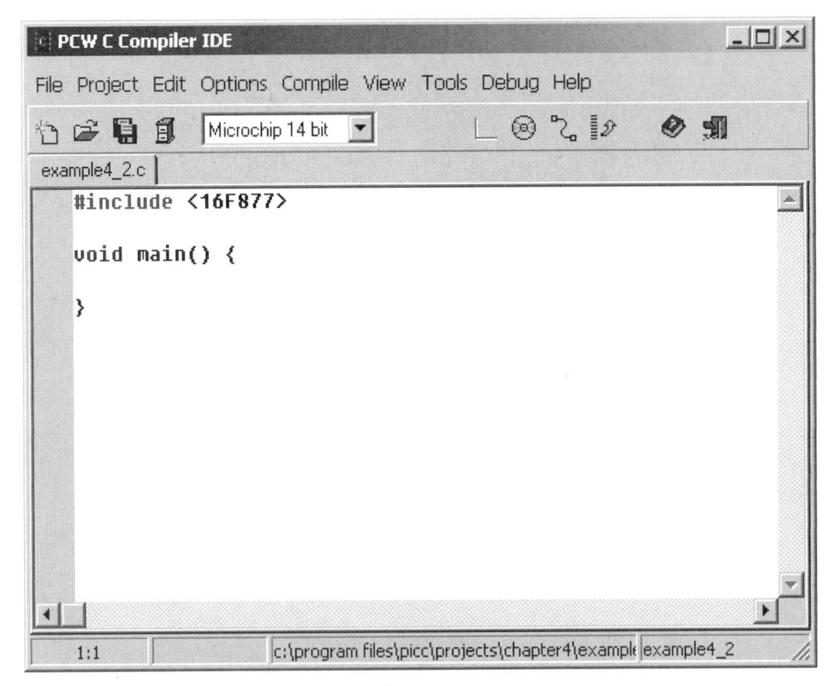

Figure 4.3 *New Project from Manual Project Create*

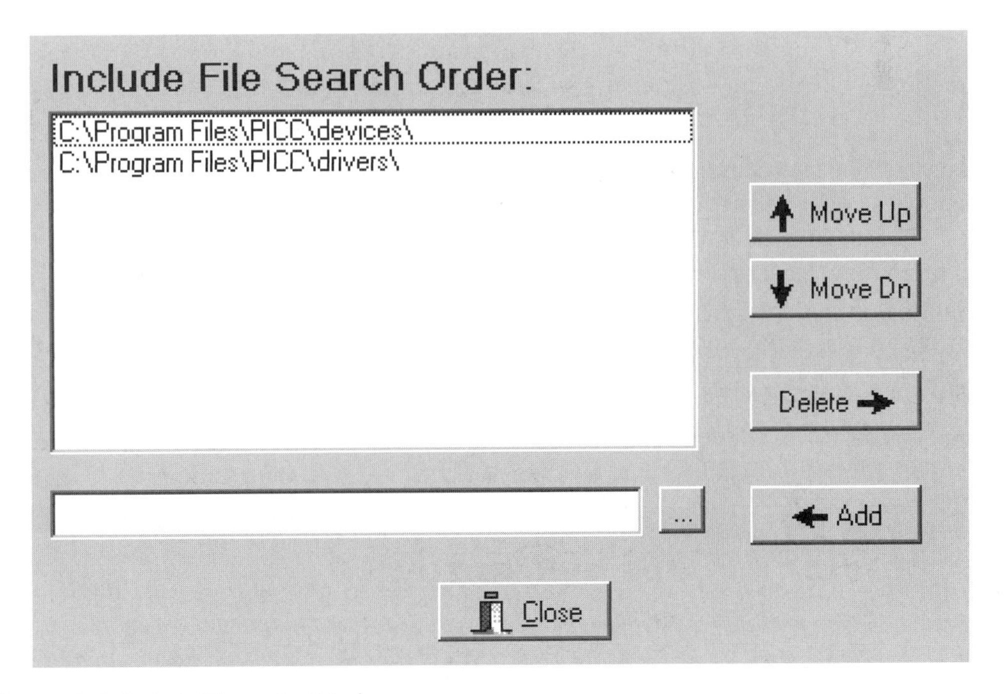

Figure 4.4 *Include Directories Window*

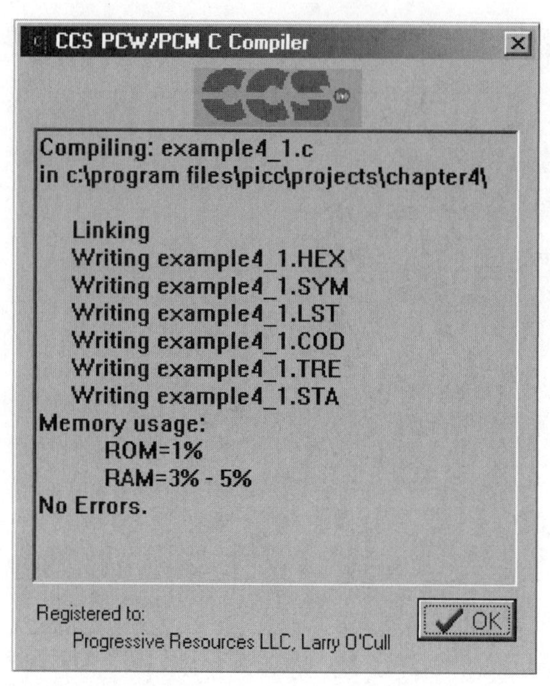

Figure 4.5 *CCS C Compiler Results Window*

The 'CCS C Compiler' window lists the main source code file at the top. Next, the statistics on the number of lines compiled and the output files are listed. The rest of the window details the ROM and RAM usage.

If errors are generated during compilation, the first error is listed in the status bar across the bottom of the IDE and the line is highlighted in the source file. In Figure 4.6, the first error is described in the status bar and the associated line of code is highlighted. The error description disappears as soon as the user begins editing the files, so to review the error the user must open the error file, *.err*, or select **View|Compiler Messages**, which opens a message window across the bottom displaying the compiler error message.

4.4.5 CLOSE PROJECTS

To quit working with the current project, use the **Project|Close Project** menu command. If the project files were modified and have not yet been saved, you are prompted to save the modified files.

4.5 PIC WIZARD CODE GENERATOR

The code wizard is constantly being updated and expanded. So, this section is not meant to be an exhaustive explanation of the features and possibilities of the code wizard, but rather an overview of the current capabilities of this tool.

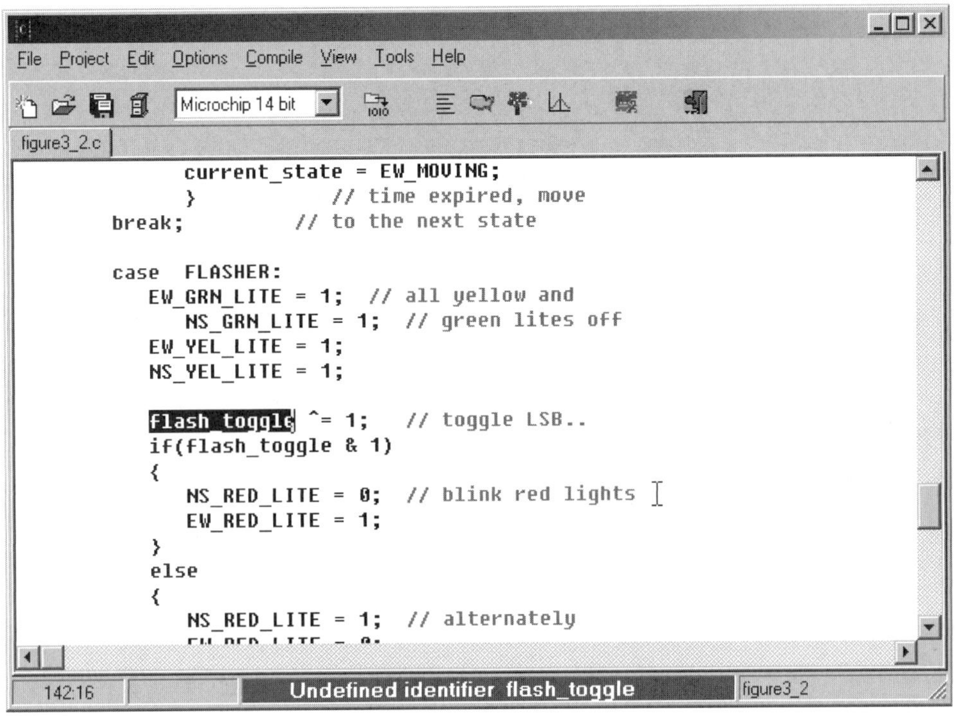

File Project Edit Options Compile View Tools Help

Microchip 14 bit

figure3_2.c

```
            current_state = EW_MOVING;
            }           // time expired, move
        break;          // to the next state

        case  FLASHER:
          EW_GRN_LITE = 1;  // all yellow and
            NS_GRN_LITE = 1;  // green lites off
          EW_YEL_LITE = 1;
          NS_YEL_LITE = 1;

          flash_toggle ^= 1;    // toggle LSB..
          if(flash_toggle & 1)
          {
            NS_RED_LITE = 0;  // blink red lights
            EW_RED_LITE = 1;
          }
          else
          {
            NS_RED_LITE = 1;  // alternately
```

142:16 Undefined identifier flash_toggle figure3_2

Figure 4.6 *Compilation Error*

When creating a new project, you are given the option of using the PIC Wizard to automatically generate a shell of code for your project. The PIC Wizard is a great tool for saving time during the start-up phase of any project. It automatically generates code to set up timers, communications ports, I/O ports, interrupt sources, and many other features. This prevents you from having to dig through the data book for control registers and their required settings.

To effectively use the PIC Wizard, you must know how the microcontroller is to be used in your project, and you must have a basic knowledge of the hardware of the processor. Chapter 2 covers many of the architectural features available in the Microchip PIC microcontrollers. Chapter 5 details good project planning and preparation for writing software. For now, carefully consider which I/O pins are used as inputs or outputs, what baud rates are required by the communications ports, what timers are required, and what interrupt sources are needed.

The PIC Wizard places the generated code directly into the source file you specify and its associated header file. This is the starting point for your program. Because you will be editing these files, any of the settings can be modified later, but you will have to look up the new settings in the data sheets for the microprocessor.

The features available through the PIC Wizard depend on the microcontroller selected. Figure 4.7 shows the tabbed frames of options available. Particular options on each frame are enabled/disabled depending on the controller selected.

This section uses the PIC16F877 as the target device and covers some of its basic options available in the PIC Wizard. The example code swatches provided in this section are specific to the PIC16F877 device. The initialization code generated by the PIC Wizard is placed at the top of *main()*. The interrupt routines are placed above *main()*.

The PIC Wizard has four command buttons: **OK, Cancel, Help,** and **View Code Generated from this tab**. Clicking **OK** will cause the PIC Wizard to generate the code according to the current settings and close. **Cancel** closes the PIC Wizard without generating a new project. The **View Code Generated from this tab** opens a window showing the code that will be generated and where the code will be placed.

4.5.1 GENERAL TAB

The 'General' tab is the first tab selected as the PIC Wizard window opens. The **Device** listbox sets the target device. Changing of the selection in this listbox updates the rest of the window. The **Oscillator Frequency** is in hertz and is entered directly into the edit field. (You will first have to clear the existing frequency by double-clicking and deleting the existing frequency or by backspacing over the existing frequency.) Selecting the correct frequency is very important for getting the other time-sensitive settings correct. The frequency along with the

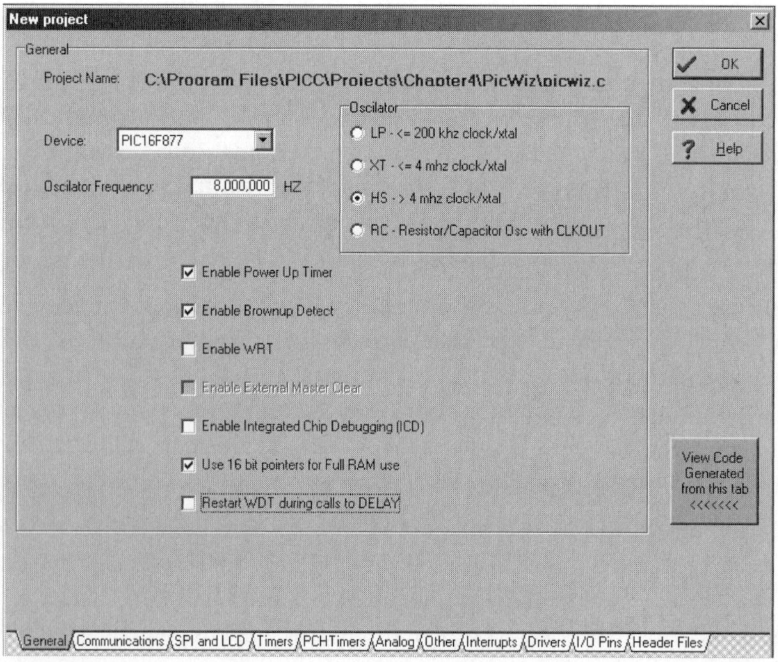

Figure 4.7 *PIC Wizard*

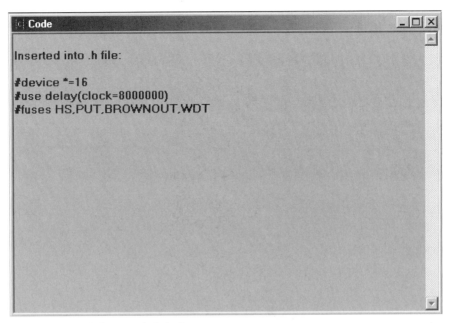

Figure 4.8 *PIC Wizard Generated Code Preview*

Restart WDT during calls to DELAY check box are used to generate the **#use delay** preprocessor directive. The watchdog timer (WDT) can be enabled on the 'Timers' tab.

The **Oscillator** frame allows you to select the type of oscillator used with the target device and is used in generating the **#fuses** preprocessor directive. The **Enable Power Up Timer, Enable Brownup Detect, Enable WRT,** and **Enable External Master Clear** check boxes also effect the generation of the **#fuses** preprocessor directive. These check boxes control the power up timer, brown-out detect, write flash, and external master clear fuses.

4.5.2 COMMUNICATIONS TAB

The 'Communications' tab allows the user to enable and set up an RS232 port, an I^2C port, and a hardware PSP port. Once a particular communication port is enabled, the options associated with that port are also enabled. Figure 4.9 shows the 'Communications' tab for the PIC16F877.

The center listbox allows you to select settings for up to four RS232 ports. The **Stream** field will be the stream identifier. Figure 4.9 shows the first RS232 port enabled for 8 bits (far right listbox **Bits**), no parity, 9600 baud, using Port C pin 6 for transmit and Port C pin 7 for receive and given the stream identifier of *pc_stream*.

In our example in Figure 4.9, the transmit and receive pins happen to be the pins for the hardware USART. If the hardware USART pins are selected, then the software generated uses

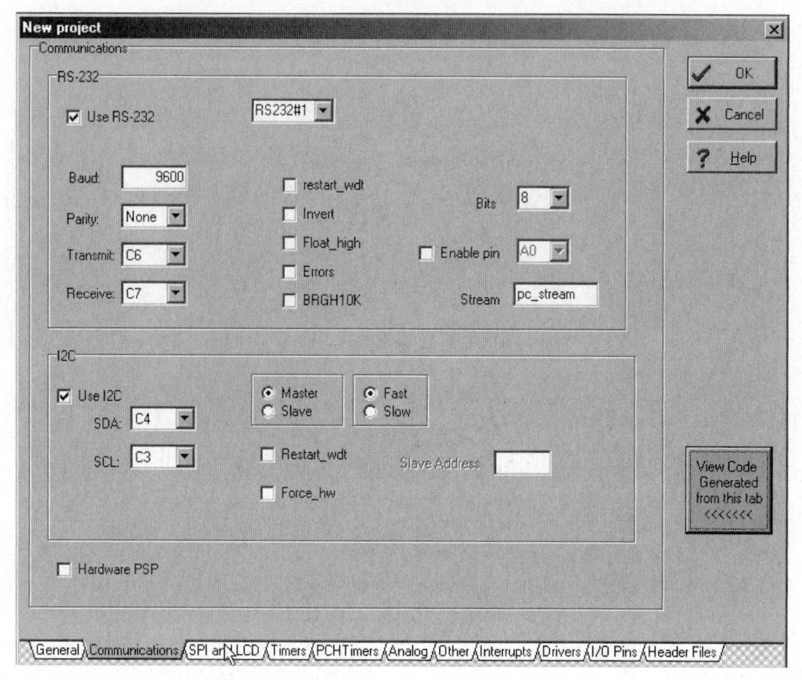

Figure 4.9 *PIC Wizard Communication Tab*

the hardware USART. However, any two general purpose input/output pins can be selected for transmit and receive and the code generated will be for a software-supported USART.

Figure 4.9 also shows that the I^2C port is enabled and configured as a master using pins 4 and 3 of Port C for data and clock, respectively. Figure 4.10 shows the code that will be generated from this tab.

4.5.3 SPI AND LCD TAB

The 'SPI and LCD' tab allows the user to select and enable the SPI and LCD support features of the chip, as they are available for the selected device. For our example, the SPI port is enabled and selected as the master with valid data on the rising edge of the clock signal.

4.5.4 TIMERS TAB

The 'Timers' tab is a very versatile tool. It covers Timer0 through Timer2 setups and the Watchdog timer setup. For this example, Timer0 is set up to use the internal clock as its clock source and to overflow at a maximum of 4 ms by setting the resolution to 16.0 µs. As each resolution is selected, the maximum time period between overflows of the timer is shown in the **Overflow** field. To reduce the time between overflows, the timer can be preloaded with a value. The interrupt associated with Timer0 can be enabled on the 'Interrupts' tab. The setup of the 'Timers' tab is shown in Figure 4.11. Figure 4.12 shows the code generated by this tab.

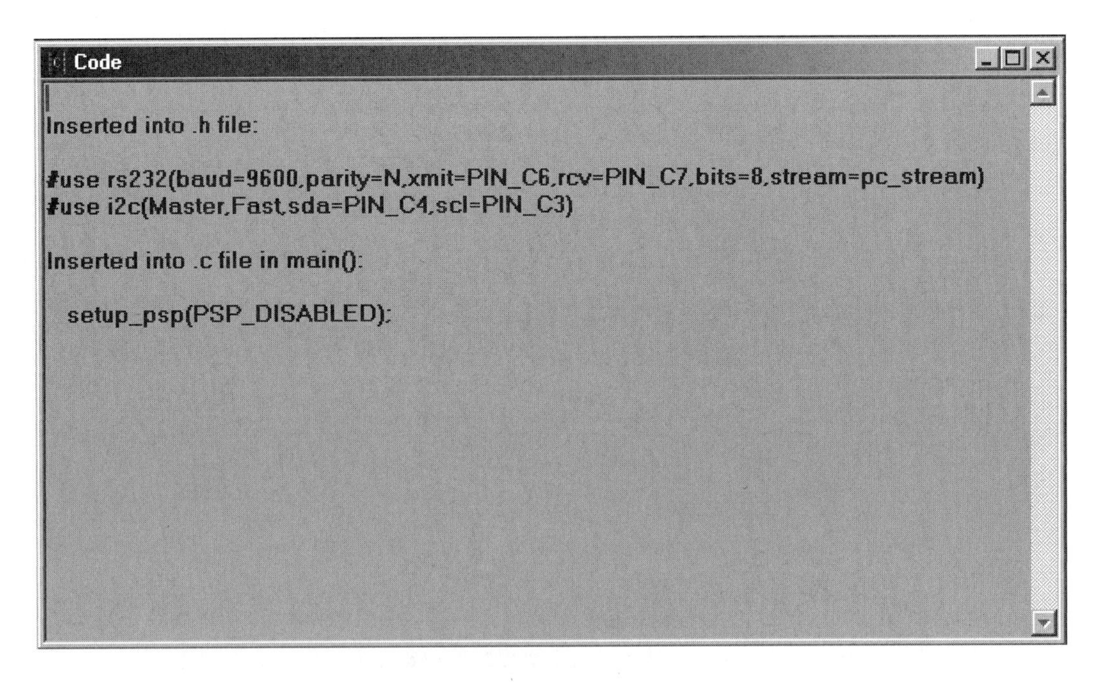

Figure 4.10 *Communication Tab Generated Code Preview*

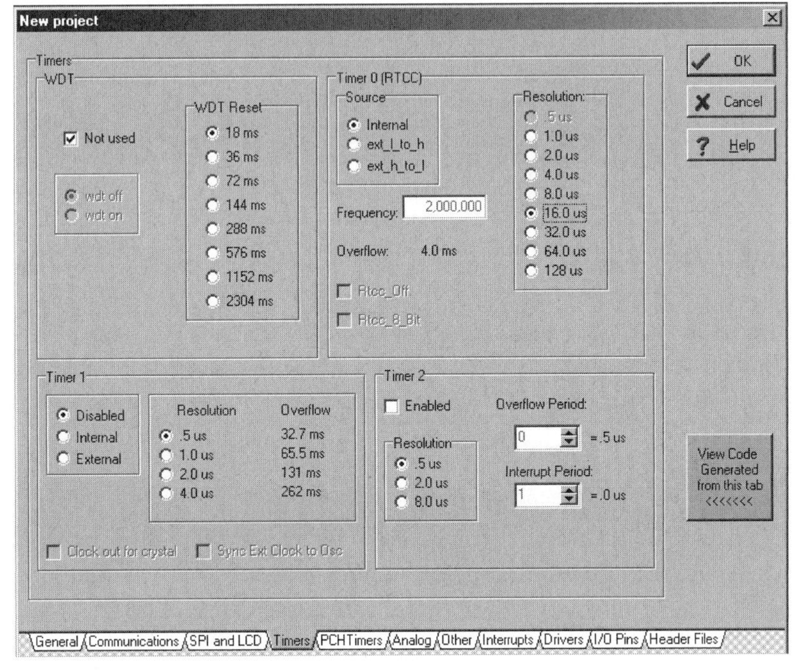

Figure 4.11 *PIC Wizard Timers Tab*

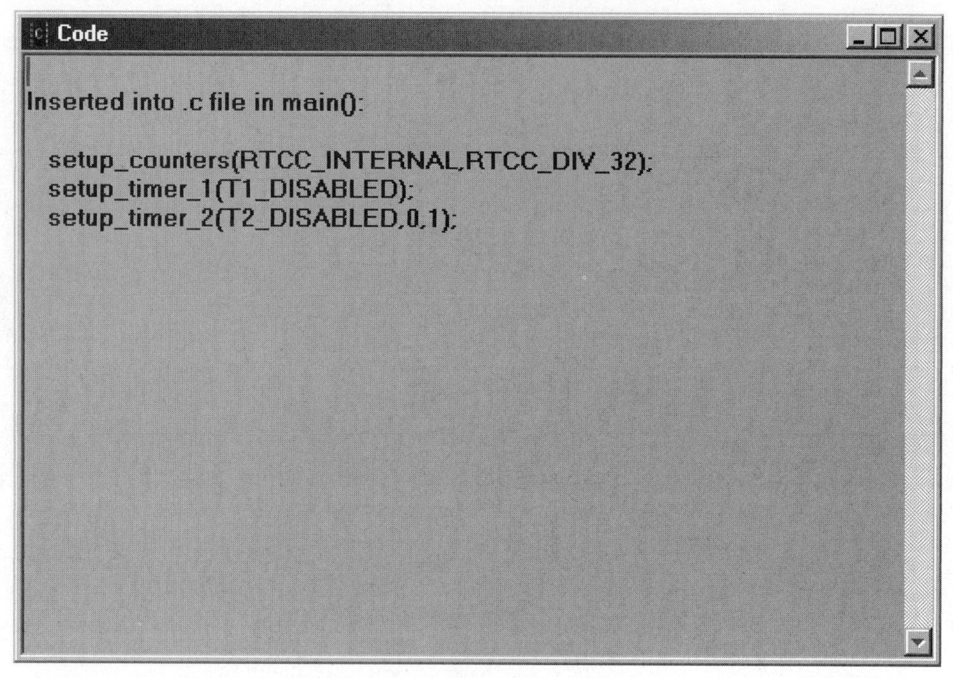

Figure 4.12 *Timers Tab Generated Code Preview*

4.5.5 ANALOG TAB

The 'Analog' tab allows you to generate code to initialize the A/D converter. The tab consists of three frames: **A/D Pins, Current, A/D Clock**. The **A/D Pins** frame allows you to select which pins should be read as analog input and which pins, if any, should be used as a reference voltage. The **A/D Clock** frame allows you to select the rate that the conversions should take place.

4.5.6 INTERRUPTS TAB

The 'Interrupts' tab consists of a series of check boxes—one for each interrupt available on the selected device. For each interrupt selected, a call is placed in *main()* to the routine *enable_interrupts()* for the interrupt, and a shell for the interrupt is placed above main. Figure 4.13 shows the interrupt tab for the PIC16F877 with the RS232 interrupts and Timer0 interrupt enabled. Figure 4.14 shows the code preview for these selections.

4.5.7 DRIVERS TAB

The 'Drivers' tab lists the software drivers available for the selected device in the CCS-PICC C compiler. Selecting a particular driver causes the associated header file to be included in the project as well as a call to the routine to initialize the driver. The I/O pins required by the driver are set up on the 'I/O Pins' tab as inputs or outputs as the driver requires. The alternate names of the pins are also displayed on the 'I/O Pins' tab as relates to the driver.

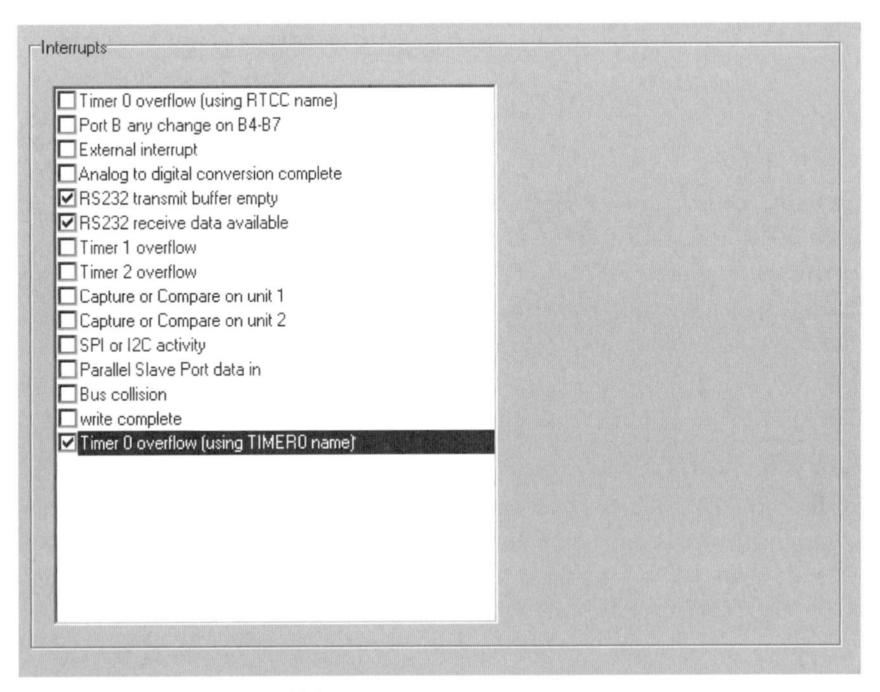

Figure 4.13 *PIC Wizard Interrupts Tab*

```
Inserted into .c file before main():

#int_TBE
TBE_isr() {

}

#int_RDA
RDA_isr() {

}

#int_TIMER0
TIMER0_isr() {

}


Inserted into .c file in main():

  enable_interrupts(INT_TBE);
  enable_interrupts(INT_RDA);
  enable_interrupts(INT_TIMER0);
  enable_interrupts(global);
```

Figure 4.14 *Interrupts Tab Generated Code Preview*

Figure 4.15 shows that 'Drivers' tab with the **3x4 Keypad driver** selected. Figure 4.16 shows the code that will be generated by this selection. Figure 4.17 shows the changes made to the 'I/O Pins' tab by selecting the keypad driver.

4.5.8 I/O PINS TAB

The 'I/O Pins' tab allows you to select the direction of each of the I/O pins on the device. Selecting the cell of the table in the 'I/O Type' column enables a drop-down list of input and output options and is represented by a down-arrow. Clicking on the down-arrow pops up the list of options. The 'I/O Pins' tab also shows which pins have been assigned special functionality based on selections from previous tabs. Figure 4.17 shows the drop-down list of input/output options for pin 2 of Port A and shows that pins 1–7 of Port B have been assigned as input/output pins by the keypad driver.

4.5.9 GENERATED PROJECT

Clicking **OK** on the PIC Wizard window causes the PIC Wizard to generate the project and close. The project consists of a source file (*.c*) and a header file (*.h*). Figure 4.18 shows the header file, and Figure 4.19 shows the source file created for the sample project using the selections made in the previous sections.

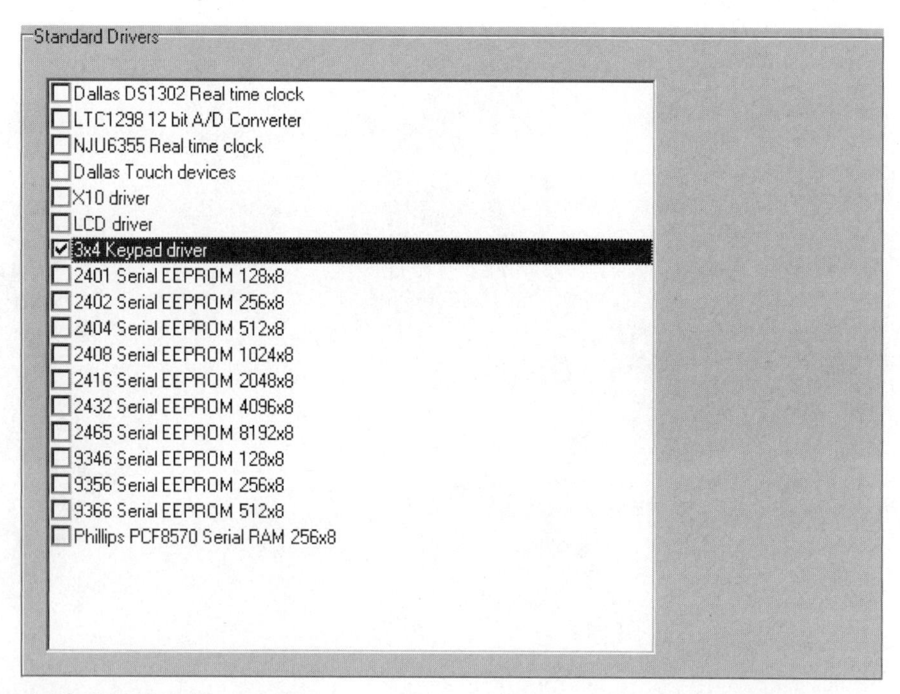

Figure 4.15 *PIC Wizard Drivers Tab*

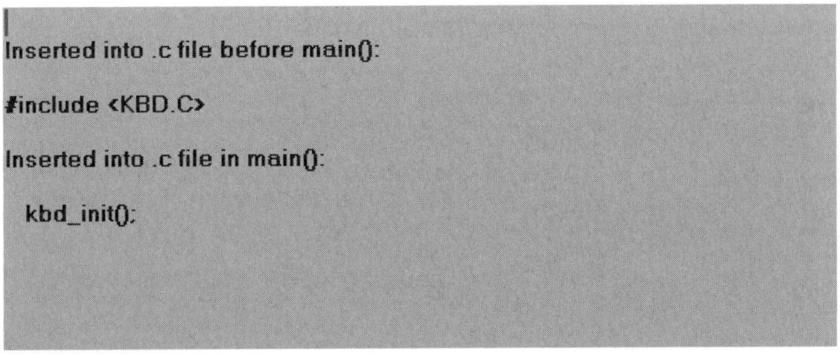

```
Inserted into .c file before main():

#include <KBD.C>

Inserted into .c file in main():

  kbd_init();
```

Figure 4.16 *Drivers Tab Generated Code Preview*

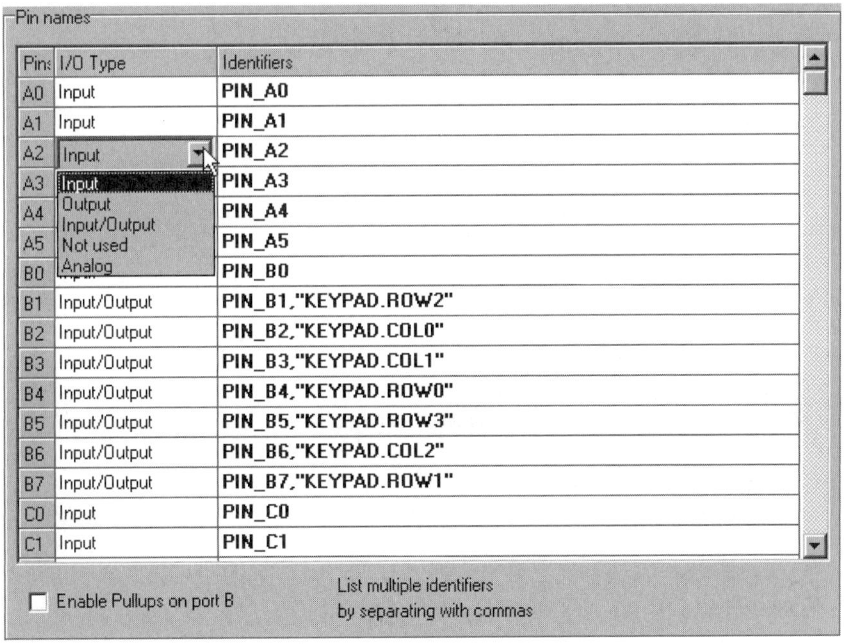

Pin names

Pins	I/O Type	Identifiers
A0	Input	PIN_A0
A1	Input	PIN_A1
A2	Input	PIN_A2
A3	Input	PIN_A3
A4	Output	PIN_A4
A5	Input/Output Not used	PIN_A5
B0	Analog	PIN_B0
B1	Input/Output	PIN_B1,"KEYPAD.ROW2"
B2	Input/Output	PIN_B2,"KEYPAD.COL0"
B3	Input/Output	PIN_B3,"KEYPAD.COL1"
B4	Input/Output	PIN_B4,"KEYPAD.ROW0"
B5	Input/Output	PIN_B5,"KEYPAD.ROW3"
B6	Input/Output	PIN_B6,"KEYPAD.COL2"
B7	Input/Output	PIN_B7,"KEYPAD.ROW1"
C0	Input	PIN_C0
C1	Input	PIN_C1

☐ Enable Pullups on port B

List multiple identifiers by separating with commas

Figure 4.17 *PIC Wizard I/O Pins Tab*

```
#include <16F877.h>
#device *=16
#use delay(clock=8000000)
#fuses HS,PUT,BROWNOUT,NOWDT
#use rs232(baud=9600,parity=N,xmit=PIN_C6,rcv=PIN_C7,bits=8,stream=pc_stream)
#use i2c(Master,Fast,sda=PIN_C4,scl=PIN_C3)
```

Figure 4.18 *PIC Wizard Generated Project Header File*

```
#include "C:\Program Files\PICC\Projects\Chapter4\PicWiz\picwiz.h"
#include <KBD.C>
#int_TBE
TBE_isr() {

}

#int_RDA
RDA_isr() {

}

#int_TIMER0
TIMER0_isr() {

}



void main() {

   setup_adc_ports(NO_ANALOGS);
   setup_adc(ADC_CLOCK_DIV_2);
   setup_psp(PSP_DISABLED);
   setup_spi(SPI_MASTER|SPI_L_TO_H|SPI_CLK_DIV_4);
   setup_counters(RTCC_INTERNAL,RTCC_DIV_32);
   setup_timer_1(T1_DISABLED);
   setup_timer_2(T2_DISABLED,0,1);
   kbd_init();
   enable_interrupts(INT_TBE);
   enable_interrupts(INT_RDA);
   enable_interrupts(INT_TIMER0);
   enable_interrupts(global);

}
```

Figure 4.19 *PIC Wizard Generated Project Source File*

4.6 SOURCE FILES

Source files are the files that contain your program source code. They are the files that you painstakingly labor over to make the microprocessor do the right thing. The IDE allows you to add and remove source files to and from the project. The IDE also has a powerful editor built into it for editing code. There are features of the IDE editor that are very specific to a code editor and are useful in developing and debugging code.

4.6.1 OPEN AN EXISTING SOURCE FILE

To open an existing file, use the **File|Open** menu command, or press the **Open File** button on the toolbar. An 'Open File' dialog window appears and allows you to browse to the directory to locate the file to open. Select the file and press **OK** to open the file. Alternately, selecting **Project|Open All Files** opens all of the files associated with a compiled project.

4.6.2 CREATE A NEW SOURCE FILE

The **File|New** menu command and the **Create New File** button on the toolbar are available to create a new source file. When either is selected, a dialog box appears, prompting you to select a filename. A new tabbed page opens for the newly created file. This file will not automatically become part of the project. You will have to either force this to be the main source file for the project or use a **#include** statement to include the new file as discussed in Section 4.4, "Projects."

4.6.3 CHANGING THE MAIN SOURCE FILE FOR A PROJECT

There may be times when you will want to select a different file to be the main source file for the project. This is accomplished by opening the desired source file and right-clicking on its tab in the editor. Right-clicking brings up the pop-up menu for the tabbed page. Select the item, 'Make file project' from the pop-up menu as shown in Figure 4.20. Now the selected file is the main source file for the project until the project is closed and reopened. When the project is reopened, it will revert back to the original main source file.

4.7 EDITOR OPERATION

The editor built into the CCS-PICC IDE supports the standard editor functions. The cursor is moved around with the Home, End, and arrow keys as well as the mouse. Clicking and dragging the mouse selects portions of text to be manipulated together. Cut, copy, and paste the selected text with the standard shortcut keys or the 'Edit' menu options or via clicking the right mouse button. Find and replace functions are supported along with the undo editing function.

4.7.1 BOOKMARKS

The editor has excellent features especially designed to facilitate editing code. Some of these features make it easier to move around in the file. Bookmarks are toggled on and off at the current cursor position by the **Edit|Toggle Bookmark** menu command or by the **Shift+Ctrl+0...9** keys. The **Edit|Goto Bookmark** menu command or the **Ctrl+0...9** keys jump the cursor to a previously set bookmark.

Bookmarks are useful for moving around within the program without paging through the code or memorizing line numbers. For example, if you are working within a function and need to reference the type and size of several global variables, press **Shift+Ctrl+0** to set a bookmark

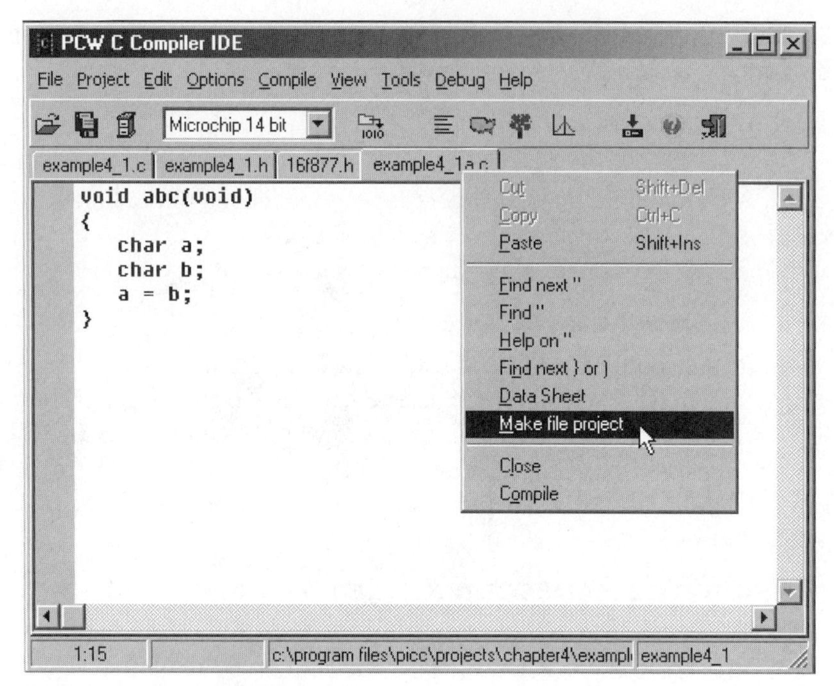

Figure 4.20 *Make file Project Pop-up Menu*

at the function. Then, go to the global declaration section and press **Shift+Ctrl+1**. Now, you can jump between the function and the global declarations by pressing **Ctrl+0** and **Ctrl+1**, respectively. This is just one example of the many uses for bookmarks.

4.7.2 INDENTATION AND TABS

The editor has two features to simplify formatting code for readability. First, the editor allows you to indent blocks of text. Highlight the text by clicking and dragging your mouse, then press the **Alt+F8** keys to indent the block of text by one tab spacing. This is useful for indenting the code in an **if** statement or a **while** statement. Second, the editor supports an auto-indenting feature. The auto-indenting feature moves the cursor directly below the start of the previous line when **Enter** is pressed instead of moving the cursor to the very beginning of the line. This feature is controlled through the **Options|Auto Indent** menu command.

Because the indentation of C code is so paramount to its readability, the user has control over the characteristics of the indentations. The indents can be comprised of spaces or of actual tab characters. The **Options|Real Tabs** menu item controls this. Tab sizes can vary from editor to editor, so sometimes it is convenient to have the editor use spaces for indentations instead of the tab character. However, when removing the indentation, it is easier to remove one tab character than all the space characters. The **Options|Tab Size** menu item allows the user to modify the width of the tab.

4.7.3 BRACE MATCHING

The editor also has a brace-matching feature that matches both curly braces and parentheses. As the cursor is positioned on an opening or the closing brace, the matching brace is momentarily highlighted. To force the matching brace to be highlighted along with the text in between the two braces, select the **Edit|Match Brace Extended** menu item or press the **Shift+Ctrl+]** keys. To jump to the matching brace, select the **Edit|Match Brace** menu item or press the **Ctrl+]** keys. This allows the programmer to check for balanced braces around functions, **if** statements, **while** statements, etc. Pressing any key or clicking the mouse hides the highlighting.

4.7.4 SYNTAX HIGHLIGHTING

The editor has a syntax highlighting features that makes the code easier to read and write. The syntax highlighting colors keywords, strings, constants, and comments special colors to make them stand out as you write and review your code. The syntax highlighting is enabled or disabled through the **Options|Syntax Highlighting** menu item. The colors used for highlighting are modified through the **Options|Editor Colors** submenu items. The list of default colors and their associated features are listed in Table 4.1.

Color	*Feature*	*Description*
White	Background	Background for the text in the project files
Dark Blue	Selection	Highlight color for the selected text
Gray	Comments	All text that is part of a comment in the source files
Bright Blue	Keywords	Keywords – *while, if, switch*, etc.
Dark Green	Preprocessor	Preprocessor directives – *#include, #use*, etc.
Black	Text	Any text that is a comment, keyword, or preprocessor directive
Bright Green	Brace Highlight	Highlight color used during brace-matching

Table 4.1 *Default Syntax Highlighting Colors*

4.7.5 OTHER EDITOR OPTIONS

The Options menu also gives the user access to the editor font and shortcut keys. The **Options|Editor Font** menu item opens a window, allowing the user to select the font used by the editor. The shortcut keys for the menu items are selectable through the **Options|Customize** menu item. The **Customize** item opens a dialog window for the user to select a menu item and assign a new shortcut key to the item.

4.8 VIEW MENU

The **View** menu gives the user quick access to information and files associated with the current project. Figure 4.21 shows the **View** menu items. Some of the files, such as binary output, symbol, listing, debug, call tree, and statistics are not available until after compilation.

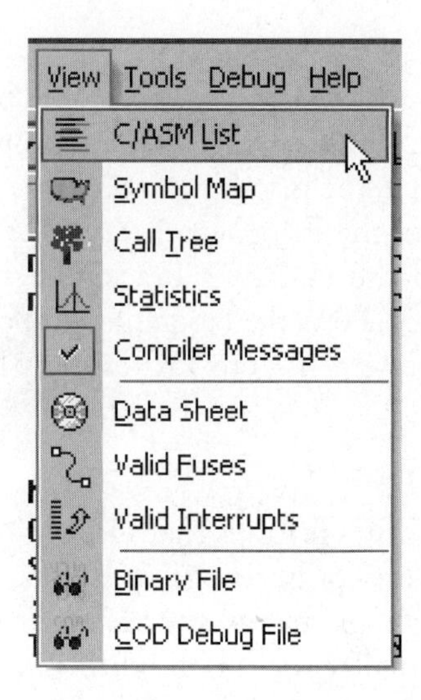

Figure 4.21 *View Menu*

4.8.1 C/ASM LIST

The **C/ASM List** option opens the listing file (*.lst*) in read-only mode. The listing file is created during project compilation. The listing file shows each C source line and the associated assembly code generated for the line. The following is part of the listing file for the interrupt-driven RS232 communication project in Figure 3.1:

```
  ....................          if(++RX_Wr_Index > RX_BUFFER_SIZE) //
wrap the pointer
  0056:   INCF    41,F
  0057:   MOVF    41,W
  0058:   SUBLW   18
  0059:   BTFSS   03.0
  ....................             RX_Wr_Index = 0;
  005A:   CLRF    41
```

4.8.2 SYMBOL MAP

The **Symbol Map** option opens the symbol file (*.sym*) in read-only mode. This file is created during compilation and contains the RAM location for each variable used in a project. The file also shows the RAM locations of other program elements such as constants and subroutines. Some locations have multiple definitions because RAM is reused depending on the current procedure being executed. The symbol map also lists the locations of the functions in the project.

Part of the symbol map for the project in Figure 3.1 is shown next and gives an example of different types of variables and their assigned locations. *TX_Rd_Index*, *TX_Wr_Index*, and *TX_Counter* are all global variables with 1 byte assigned to each variable. *RX_Buffer_Overflow* and *fPrimedIt* are single-bit variables and so their location is a RAM byte address and a bit assignment within the byte. *TX_Buffer* is an array and the map file lists the block of bytes assigned to the array. *main.k* and *bgetc.c* are local variables and are listed by *function name.variable name.*

```
044.0    RX_Buffer_Overflow
044.1    fPrimedIt
045-05D TX_Buffer
05E      TX_Rd_Index
05F      TX_Wr_Index
060      TX_Counter
061      main.k
062      bgetc.c
```

4.8.3 CALL TREE

The **Call Tree** option opens the call tree file (*.tre*) for the project. The call tree file shows each function and its subsequent function calls along with the ROM and RAM usage for the function. The ROM usage is shown after the function name in a number of the form *s/n*, where *s* is the page number of the procedure and *n* is the amount of memory required to contain the code. (Both *s* and *n* are in decimal notation.) The RAM usage is shown by the notation, *RAM=xx*, where *xx* is the total RAM required for the function. Figure 4.22 shows the call tree for the project of Figure 3.1, *rs232_interrupt.*

This shows that *main* only calls *bputc* and *bgetc* and that these functions do not call any lower level functions. (*printf* is also called by *main* but is broken down into *bputc* during compilation.) We can also see from the call tree that *main* is on page 0 and requires 68 bytes of ROM and 3 bytes of RAM. *main, serial_rx_isr*, and *serial_tx_isr* are all shown at the same level because the interrupt routines run independently from *main.*

The architecture of the PIC makes it possible to generate a call tree for the PIC, whereas it would be almost impossible to generate a call tree for other types of microcontrollers. The PIC architecture uses a hardware stack to track the returns from subroutines. This means that

```
└rs232_interrupt
  ├main  0/67  Ram=3
  |  ├??0??
  |  ├bputc  0/28  Ram=4
  |  ├@const10068  0/16  Ram=0
  |  ├bputc  0/28  Ram=4
  |  ├bputc  0/28  Ram=4
  |  ├bputc  0/28  Ram=4
  |  ├bputc  0/28  Ram=4
  |  ├bgetc  0/22  Ram=1
  |  └bputc  0/28  Ram=4
  ├serial_rx_isr  0/24  Ram=1
  └serial_tx_isr  0/25  Ram=1
```

Figure 4.22 *Call Tree*

there is a finite depth to the call tree. If the calls into subroutines are too deep, then the hardware stack overflows. When the hardware stack overflows, the program cannot return correctly from the subroutines and behaves erratically. This makes the call tree important in that it allows the programmer to verify that the hardware stack will not be overflowing during execution of the program.

The MPLAB simulator can be halted on stack overflow to trap this type of error. The option to 'Stack Overflow Break Enable' is available on the 'Break Options' tab of the 'Development Mode' window (**Options|Development Mode**). See Section 4.11 for more information on MPLAB and debugging. The stack is viewable through the menu selection **Window|Stack** and shows the return addresses for the various subroutines that have been called and not returned from.

4.8.4 STATISTICS

The **Statistics** option opens the stats file (*.sta*) for the project. This file is generated during compilation of the project. The stats file includes information such as total RAM and ROM usage, RAM and ROM usage by file and by function, and segment usage.

4.8.5 COMPILER MESSAGES

The **Compiler Messages** option opens a message window across the bottom of the IDE. This message window displays the compiler error and warning messages from the last compilation performed.

4.8.6 DATA SHEET

The **Data Sheet** option brings up Acrobat Reader with the manufacturer data sheet for the selected Microchip PIC device. If data sheets were not copied to disk, then the CCS CD-ROM or a manufacturer CD-ROM must be used to provide the data sheet.

4.8.7 VALID FUSES

The **Valid Fuses** option opens the "Fuse Review" window showing the valid fuses and the associated fuse keywords for a selected device. This is useful when determining what parameters to use with the **#fuses** preprocessor directive. Figure 4.23 shows the "Fuse Review" window for a PIC16F877.

4.8.8 VALID INTERRUPTS

The **Valid Interrupts** option opens the "Interrupt Review" window. The "Interrupt Review" window shows the list of interrupts available on the selected device. The keywords associated with these interrupts for use with the **#int_xxx** preprocessor directive are also listed. Figure 4.24 shows the "Interrupt Review" window for a PIC16F877.

4.8.9 BINARY FILE

The **Binary File** option allows the user to select and view a binary file (*.bin*). The selected file is shown in hexadecimal and ASCII notation. The CCS-PICC compiler defaults to generating a *.hex* program file. To use the **Binary File** option to view the program file, open the

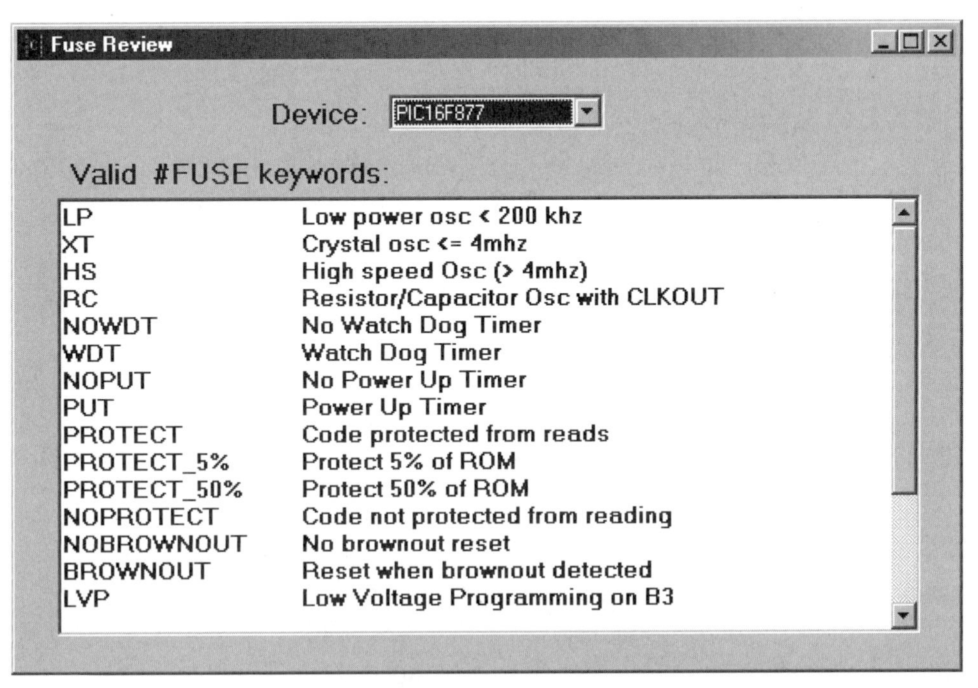

Figure 4.23 *Fuse Review Window*

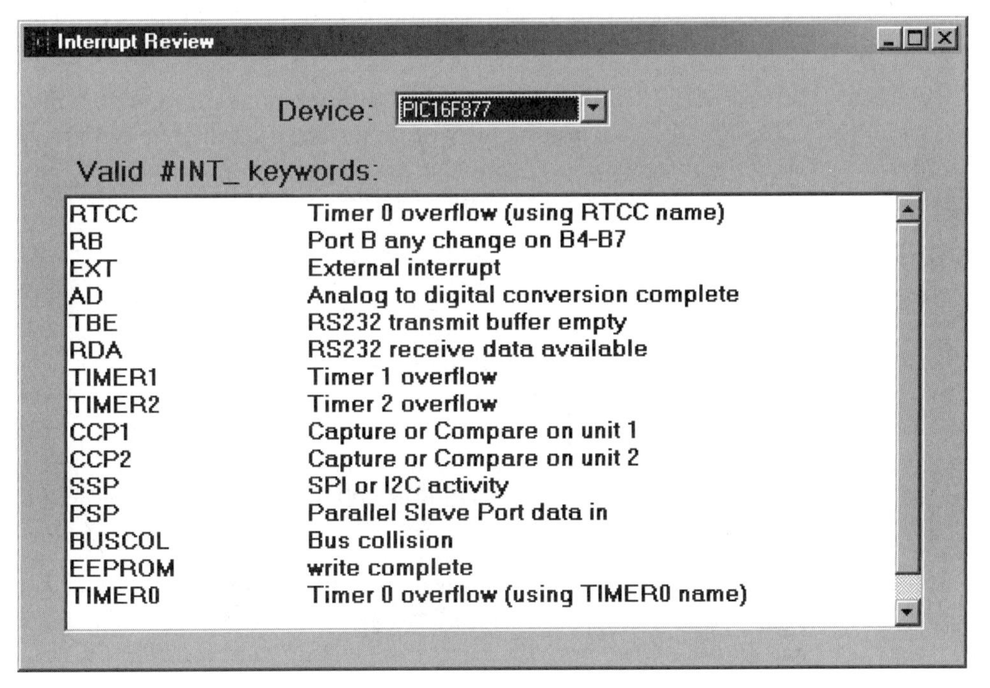

Figure 4.24 *Interrupt Review Window*

File Formats window (**Options|File Formats**) and select the **Binary** radio button in the **Object File** frame. Also, change the **Object Ext** field from *hex* to *bin*.

4.8.10 COD DEBUG FILE

Selecting the **COD Debug File** menu item opens a *.cod* debug file and displays it in an interpreted form. This file includes information about the compile date and time, the compiler version, the target device type, the program code, and the symbol table.

4.9 PROGRAM THE TARGET DEVICE

The flash-based Microchip devices support both out-of-circuit and in-circuit programming. For convenience it is often easier to program devices in-circuit, alleviating the risk of damaging the device during removal from and installation into the circuit. There are many in-circuit programmers available on the market and the CCS-PICC IDE is designed to run any number of programming tools. These tools are supported in two ways. First, they are supported through the 'Object File' frame of the 'File Formats' window (**Options|File Formats**), allowing the user to select the type of file formatting required for his or her particular programmer. Second, the programmers are supported through the ability to specify the command line of the programmer to be run.

To specify the programming tool to be used, select **Options|Debugger/Programmer….** Figure 4.25 shows the 'Device Programmer' frame with the ICD-S programmer specified. The ICD-S is an inexpensive in-circuit programming tool that interfaces to the serial port of a PC and requires an Intel 8-bit hex file. The **Tools|Program Chip** menu item runs the selected programmer. (See the Appendix C for specific information on running the ICD-S programmer.)

The **Tool** menu has a variety of helpful utilities. These utilities range from helping the user select a chip to opening MPLAB for simulating the software to comparing two files and highlighting their differences. The utilities available under this menu are briefly covered in this section.

4.10 TOOL MENU

4.10.1 DEVICE EDITOR

The CCS-PICC C compiler uses a database to track the properties of each Microchip PIC microcontroller that it supports. This database is made available to the user for updating through the **Device Editor** menu item. The Device Editor is unavailable in the Limited Edition version of the compiler included on the CD with this textbook. However, in the standard edition, the Device Editor should be used with GREAT CAUTION. If you modify this database, you are modifying the way in which the compiler will treat the selected device.

CCS maintains this database (Devices.dat); however, users may want to add new devices or change the entries for a device for a special application. Be aware that if the database is changed and then the software is updated, the changes will be lost. Save your DEVICES.DAT file before performing an update to prevent this.

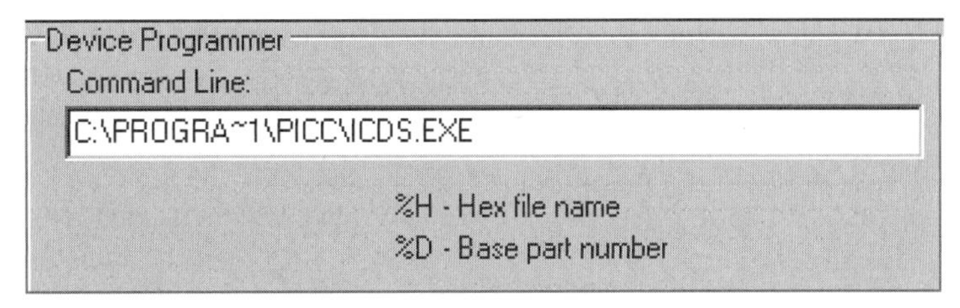

Figure 4.25 *Device Programmer Setup*

4.10.2 DEVICE SELECTOR

When available, the **Device Selector** menu item opens the 'Device Selection Tool' window. The Device Selector is not available on the Limited Edition version of the compiler included with the textbook. The 'Device Selection Tool' window aids in selecting the appropriate device for a project by allowing the user to specify the features required by the project. Figure 4.26 shows the 'Device Selection Tool' window open with the 16F87x family of devices selected by the features required in the check boxes at the bottom of the window.

4.10.3 FILE COMPARE

The **File Compare** menu item allows the user to select two files and have their differences highlighted. The file compare utility is designed for two types of files, source and list.

Device	Family	I/O	ROM	RAM	Features
PIC16F870	PCM	22	2048	128	SCI-9 ADC(5) EEPROM(1024) SPI PSP TIM0 TIM1 TIM2 CC1 I
PIC16F871	PCM	33	2048	128	SCI-9 ADC(8) EEPROM(1024) SPI PSP TIM0 TIM1 TIM2 CC1 I
PIC16F873	PCM	22	4096	192	SCI-9 ADC(5) EEPROM(256) SPI I2C-MS TIM0 TIM1 TIM2 CC
PIC16F874	PCM	33	4096	192	SCI-9 ADC(8) EEPROM(256) SPI I2C-MS PSP TIM0 TIM1 TIM
PIC16F876	PCM	22	8192	367	SCI-9 ADC(5) EEPROM(256) SPI I2C-MS TIM0 TIM1 TIM2 CC
PIC16F877	PCM	33	8192	367	SCI-9 ADC(8) EEPROM(256) SPI I2C-MS PSP TIM0 TIM1 TIM

Family
PCB 12 Bit
PCM 14 Bit
PCH PIC18
PIC17
SXC
I/O Pins
Any
1-7
8-14
15-23
24 and up
Features
UART
EEPROM
A/D
H/W SPI
Minimum RAM:
0
Minimum ROM
0
Skip Revisions
Skip Obsolete
6 Selected
Close

Figure 4.26 *Device Selection Tool*

When source file is selected, then a normal line-by-line compare is done. When list file is selected, the compare may be set to ignore RAM and/or ROM addresses to make the comparison more meaningful. For example, if an **asm** line was added at the beginning of the program, a normal compare would flag every line as different. By ignoring ROM addresses, then only the extra line is flagged as changed. Two output formats are available, one for display and one for printing.

4.10.4 NUMERIC CONVERTER

The **Numeric Converter** menu opens a window that takes as input either a signed, unsigned, hexadecimal, or floating point number and displays it as a signed, unsigned, and hexadecimal number. This can be useful when trying to determine what size of variable is required to hold a certain signed or unsigned value or when reading a value in hexadecimal notation and needing to know its value in decimal.

4.10.5 SERIAL PORT MONITOR

The serial port monitor tool is intended for debugging embedded systems that employ serial communication (RS232, RS422, RS485). The **Tools|Serial Port Monitor** menu command opens the 'Serial Input/Output Monitor' or 'Siow' window. Figure 4.27 shows the serial port monitor window.

Figure 4.27 shows the results of typing "Hello World!" on the active 'Siow' window and using the *send_hex_hello* macro (see the Macro Manager description below) to send "hello hex" to an embedded system that transmits back whatever is received on its serial port.

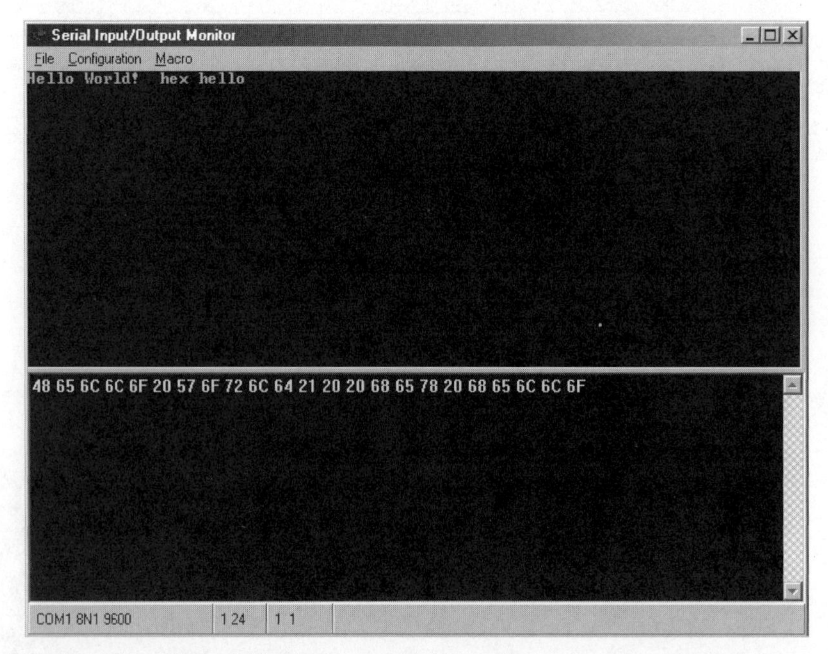

Figure 4.27 *Serial Input/Output Monitor*

The received characters are displayed in the main body of the 'Siow' window in ASCII and in the lower part of the window in hexadecimal format. Any characters typed while the 'Siow' is active are transmitted through the PC serial port in ASCII. Pressing the **Enter** key transmits a carriage return (0x0D). Receiving a carriage return sends the cursor back to the start of the current line. Pressing **Ctrl+Enter** transmits a line feed (0x0A). Receiving a line feed moves the cursor to same column of the next row. The **File|Clear Terminal** menu item clears both the ASCII and the hexadecimal displays.

To send hexadecimal data, use the **Macro** menu command to open the 'Macro Manager' window. The 'Macro Manager' allows you to assign a name to a macro that contains a series of hexadecimal data to be transmitted. To add a macro, fill in the **Name** and **Command** fields, then click the **Add** button. The **Name** field is added to the list of macros. The **Command** field should contain the hexadecimal data that you would like to transmit. Figure 4.28 shows the 'Macro Manager' window open with two macros, *send_1* and *send_hex_hello*. The *send_hex_hello* macro is selected, so its name and command data are displayed at the bottom. The command data in Figure 4.28 is the hexadecimal representation of the ASCII string, "hex hello". The **Run** command button causes the currently selected command to be transmitted out the serial port. Note that the macro must be updated to the list before modified command data will be transmitted. To modify an existing macro, select it in the list, click **Delete**, change the macro command, then click **Add**.

Figure 4.28 *Macro Manager*

The transmitted and received characters can be logged to a file by using the **Configuration|Logging** menu item to open the 'Logging Options' dialog window. Select the **Logging Enabled** check box on the 'Logging Options' window and a filename and maximum size for the log file. The transmitted and received data is time-stamped and logged. The log file is written upon closure of the 'Siow' window.

The **Configuration|Set Parameters** menu item gives the user control over the PC communication port, baud rate, data bits, stop bits, and parity of the connection. At the bottom of the 'Siow' window is a status bar that displays the

- selected PC communication port
- communication parameters
- row and column of the cursor

4.11 MICROCHIP MPLAB

Both software simulators and in-circuit emulators are useful for debugging software and are sometimes referred to as *debuggers*. Most include some method of starting, stopping, and stepping through a program. Most also allow the user to peak at and even modify registers and variable values. They are invaluable in finding out why your program is not operating as you expect it to.

CCS-PICC is designed to work in conjunction with Microchip MPLAB. MPLAB is a software simulator for Microchips's PIC parts. MPLAB also supports in-circuit emulators such as Microchips's MPLAB ICE Emulator and PICMASTER Emulator and debuggers such as MPLAB ICD Debugger. MPLAB is available free from Microchip's Web site at http://www.microchip.com/.

Although it is beyond the scope of this textbook to go into great detail about the operation of MPLAB, there are some basic operations that are useful to every programmer. These are how to load a C file for debugging; starting, stopping, and stepping through a program; setting and clearing breakpoints; and viewing and modifying variables and machine state. You are encouraged to utilize the help files included with MPLAB for more advanced operations.

Debuggers also usually require specially formatted files. These files contain information about the format of the C file, variable names, function names, and other such information. The files used by MPLAB are denoted with a *.cod* extension. CCS-PICC is designed to create this type of file and to launch MPLAB from the **Tool|MPLAB** menu item. A software 'plug-in' is required for CCS-PICC to interface to MPLAB and is available at http://www.ccsinfo.com.

4.11.1 LAUNCH MPLAB FROM CCS-PICC

CCS-PICC has two methods for launching a debugger application, a menu item and a toolbar command button. Before using either method, however, the path to the debugger must be specified. To do this, select the **Options|Debugger/Programmer** menu item and enter the appropriate command line information for MPLAB. The default path and command line parameters for MPLAB are:

C:\Program Files\MPLAB IDE\dlls\mplab.exe

CCS-PICC defaults the executable path for MPLAB; however, MPLAB may be placed differently in your system. You can perform a search on *mplab.exe*, if necessary, to locate the file. Once the file is selected, simply use the **Tools|MPLAB** menu item or the MPLAB command button in the toolbar to launch the application from CCS-PICC.

4.11.2 MPLAB WORKSPACE AND PROJECT

MPLAB requires a workspace and an MPLAB project for your program before you can begin simulating or debugging it. To create the workspace and MPLAB project, select **Project|Project Wizard.** This opens MPLAB's project wizard where the first step is to select the device type. The second step, shown in Figure 4.29, is to select the toolsuite to use with the project. Figure 4.29 shows the CCS tools selected. The check box in the lower right, **Show all installed toolsuites**, must be checked to show any installed third-party tools. Also, the software 'plug-in' for CCS must be installed for CCS tools to show in the toolsuite list. Once the CCS tool is selected as the **Active Toolsuite**, the **Toolsuite Contents** and the **Location of Selected Tool** fields are automatically populated.

The third step is to assign a project name and select the project directory. The fourth and final step is to add any existing files to the project. When CCS is selected as the compilation tool, the only file that needs to be added here is the main *.c* file of the project.

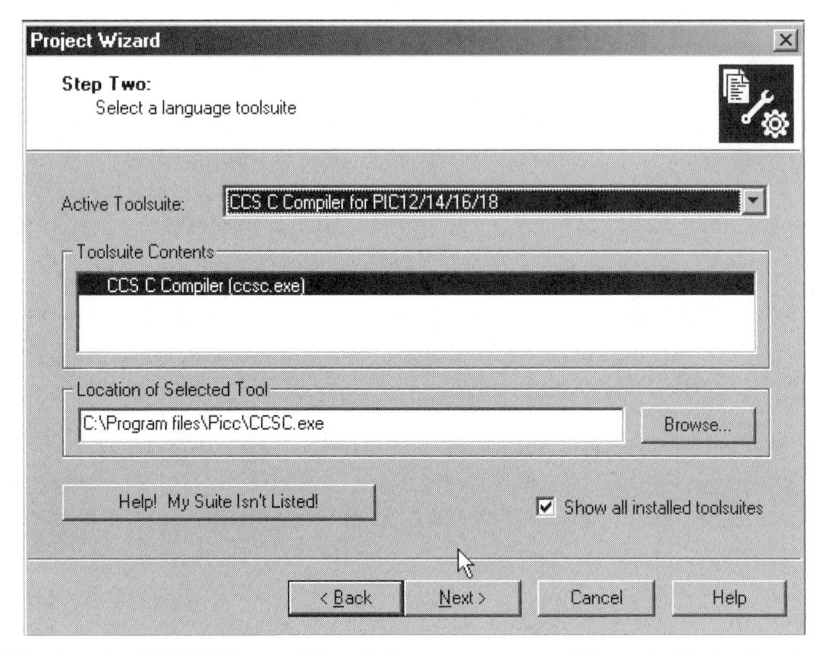

Figure 4.29 *Step 2 of MPLAB Project Wizard. (Reprint courtesy of Microchip Technology, Inc.)*

4.11.3 SIMULATOR DEVELOPMENT MODE

As the MPLAB opens, it defaults to *editor only* mode. To simulate your software, you must select the MPLAB SIM Simulator for the development mode. This is accomplished by selecting the **Debugger|Select Tool|MPLAB Sim** menu item. The status bar across the bottom of MPLAB now shows that MPLAB SIM is selected and it shows the selected device. If the device type is not correct, it can be changed under the menu option, **Configure|Select Device.**

To set the clock frequency and break point options, select the menu item, **Debugger|Settings.** On the 'Simulator Settings' window, the 'Clock' tab allows you to select the type and frequency of the oscillator to be used in your application. The 'Break Options' tab controls when the simulation is halted, including halting for breakpoints, tracing buffer full and stack overflows, and it has settings for the watchdog timer. The **Global Break Enable** on the 'Break Options' tab MUST BE checked for the simulator to halt on breakpoints! Also, if you suspect that your stack is overflowing, select the **Stack Overflow Break Enable.** This allows you to halt the simulation as the stack overflows. The stack can then be reviewed for what routines or interrupts were in process when the stack overflowed. This can account for 'mysterious' resets of your microcontroller!

4.11.4 COMPILING UNDER MPLAB

If the CCS_PICC compiler was set as the toolsuite for the project, then selecting **Project|Build All** causes MPLAB to use CCS-PICC to compile the project. The CCS-PICC compiler status window opens during compilation then disappears as MPLAB opens a 'Build Results' window. A *.COD* file or a *.COF* file is required by MPLAB to gather the necessary data to simulate, debug, or emulate a program.

The default output file for CCS-PICC when called from MPLAB is a *.COF* file. However, this file is not generated when CCS-PICC is called from MPLAB. A *.COD* file needs to be generated instead. To select the *.COD* file format, select the **Project|Build Options|Project** menu item. On the 'Build Options' window that opens, select the 'CCS C Compiler' tab. Under the 'Debug' frame, select **Expanded .COD Format.** Now, when a successful compilation is completed, MPLAB will automatically load the .COD file generated by CCS.

To change the toolsuite for the project after the project has been created, use the **Project|Select Language Toolsuite** menu item. If CCS-PICC is not the listed language tool, then you will have to add it or another compiler manually. Consult the MPLAB documentation for this.

4.11.5 SOURCE FILE AND PROGRAM MEMORY WINDOWS

As the debugger is running, the program memory window is open, showing at the assembly level what instructions are being executed. When the simulator is halted, the program memory window highlights the assembly level instruction being executed while the source file highlights the C level instruction being executed. The simulator is executing at the level of the program memory but highlights the C level instruction for you to follow along with.

4.11.6 EXECUTION SPEED

MPLAB executes the software using the PC and as such is limited by the speed and power of the PC. So the software is not running real-time as it would in the target device. Therefore, it is recommended that delays be somewhat shortened to prevent you from waiting an intolerable amount of time for delay routines to execute. For this debugging exercise, the stoplight project from Figure 3.2 is being used. The calls to *delay_ms()* in *main()* were commented out as shown below:

```
...
    while(1)
    {
          //delay_ms(250);        // 1 second delay.. this time could
          //delay_ms(250);        // be used for other needed
                                // processes
          //delay_ms(250);
          //delay_ms(250);
          Do_States();    // call the state machine, it knows
                          // where it is and what to do next
...
```

4.11.7 DEBUGGING COMMANDS

Before addressing the actual debugging commands, a couple of notes about the simulator running are in order. When the simulator or debugger is actually executing the program, the status bar across the bottom of MPLAB shows 'Running' with a progress bar filling and refilling. Also while executing, no data in any open window is updated. To see what the debugger has been up to, you must stop or halt the execution. Upon halting, the data in the open windows is updated. This includes the highlighted program memory and source code lines as well as the data that reflects the state or memory of the microcontroller.

The buttons to run, halt, step, and reset the debugger are located at the right of MPLAB's default toolbar. The function of the button is shown below the button when your mouse is held over the button for a moment. Equivalent commands plus other useful commands are also available from the **Debugger** menu. The available debug commands are run, reset, halt, step into, and step over. Many of these are intuitive, but we should review their functionality. Table 4.2 shows the shortcut keys associated with these commands and a brief description of each.

4.11.8 SET AND CLEAR BREAKPOINTS

A very useful tool of most debuggers is the ability to set and clear breakpoints. These are points in your program where you want execution to halt. Once halted, you can review variable or register values, or you can choose to start single-stepping the program to determine exactly what it is doing.

To set or clear breakpoints in MPLAB, right-click your mouse on the desired line of source code. A pop-up menu appears with (among other options) the option of either **Set**

Command	Description	Function Key
Run	Starts or resumes execution of the program. The program will be executed until the user stops it or a breakpoint is encountered.	F9
Reset	Resets the execution target. If currently executing, the program is halted.	F6
Halt	Stops or breaks execution of the program. When execution is halted, all information in all windows is updated.	F5
Step Into	Executes one line of assembler.	F7
Step Over	Executes one instruction. If the instruction is a function call, then the function is executed as well before halting.	F8

Table 4.2 *Basic Debugger Commands*

Breakpoint, if a breakpoint is not currently set, or **Clear Breakpoint**, if a breakpoint is already set for the selected line of code. When a breakpoint is enabled for a particular line of code, a red breakpoint symbol is present along the left side of the selected line.

Alternatively, breakpoints can be managed by selected **Debugger|Breakpoints** menu items or by pressing **F2** to open 'Breakpoints' window.

NOTE: The **Global Break Enable** on the 'Break Options' tab of the 'Development Mode' (**Options|Development Mode**) must be checked for the program to be halted by a breakpoint.

Once the breakpoints are set, run your program as usual. The program halts after executing the selected line. After the program is halted, all of the information in all of the windows is updated.

4.11.9 RUN TO CURSOR

A quick method of running to a particular point in your program is to use the **Run to Cursor** command. Highlight the desired line in either the source file or the program memory window and right-click it with your mouse. One option in the pop-up menu that appears is **Run to Cursor**. Selecting this command causes the program to execute until the selected line is reached or a breakpoint is encountered or you halt the execution manually.

4.11.10 WATCH

Watch is also an item in the **View** menu list and allows you to open windows for watching particular variables. The 'Watch' window allows you to select the variables and registers to be monitored. Figure 4.30 shows a watch window open with the variable *current_state* being monitored. To add variables to an existing watch window, select the variable in the right-hand list and click **Add**. Right-click a variable in the list and select **Properties** from the pop-up menu to change the format of the displayed variable. To remove a variable, select the line

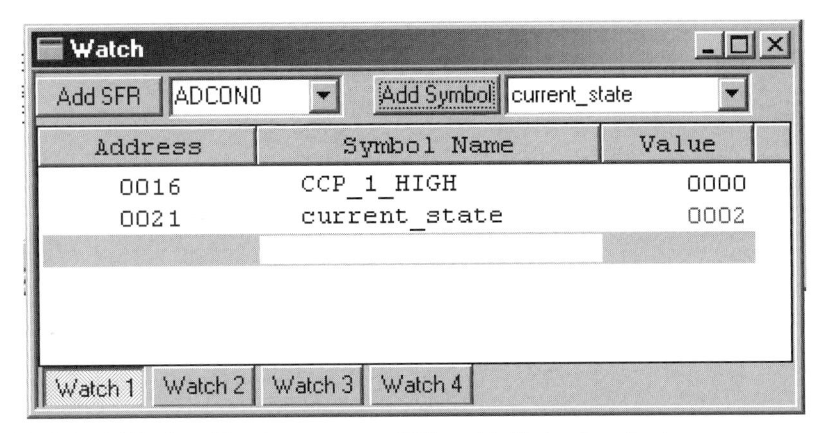

Figure 4.30 *Watch Window. (Reprint courtesy of Microchip Technology, Inc.)*

containing the variable and press **Delete** key, or right-click with the mouse to get a pop-up menu to remove the variable.

4.11.11 FILE REGISTERS (RAM) WINDOW

The 'File Registers' window (**View|File Registers**) displays the contents of RAM. This is the scratch pad for the microcontroller. To find the address of a particular variable in this area, select to view the variable in the 'Watch' window. Figure 4.30 shows a 'Watch' window for our example project. Here the symbol *current_state* has been added to the list. The address of *current_state* (left-hand column of the watch window) is 0x0021. Figure 4.31 shows the 'File Registers' window or the RAM for the simulated microcontroller. From Figure 4.31 we see that the value of *current_state* is 0x02.

A quick way to edit the value of a location in RAM is to right-click on a value in the 'File Registers' window. This brings up the **Fill Register(s)** command. Selecting this command opens the 'Modify' window with the address field populated with the address of the value you right-clicked on.

4.11.12 MODIFY MEMORY

Program, data, and EEPROM memory is modified by selecting to view the appropriate memory from the **View** menu. Once the memory is open right-click at the appropriate address and select **Fill Registers** from the pop-up menu. This opens the 'Fill Registers' window. The 'Fill Registers' window allows you to specify the starting and ending address of the memory to modify and the value to place into memory. The value can be specified in decimal or hexadecimal notation and can be auto-incrementing.

4.11.13 VIEW AND MODIFY THE MACHINE STATE

Sometimes it is very useful to get a peek at exactly what state the microprocessor is in. What are the counters doing? Where is the stack pointer? What on earth is going on with the I/O?

File Registers

Address	00	01	02	03	04	05	06	07	08	09	0A	0B	0C	0D	0E	0F	ASCII
0000	--	00	4C	0F	00	00	00	00	C6	00	00	00	00	00	00	00	-.L..... ....
0010	00	00	00	00	00	00	00	00	00	00	00	00	00	00	00	00	....
0020	11	02	00	62	00	00	00	00	00	00	00	00	00	00	00	00	...b.... ....
0030	00	00	00	00	00	00	00	00	00	00	00	00	00	00	00	00	....
0040	00	00	00	00	00	00	00	00	00	00	00	00	00	00	00	00	....
0050	00	00	00	00	00	00	00	00	00	00	00	00	00	00	00	00	....
0060	00	00	00	00	00	00	00	00	00	00	00	00	00	00	00	00	....
0070	00	00	00	00	00	00	00	02	01	00	00	00	00	00	00	00	....
0080	--	7F	4C	0F	00	3F	FF	FF	00	07	00	00	00	00	00	--	-.L..?.. ....-
0090	--	00	FF	00	00	--	--	--	02	00	--	--	--	--	00	07	-....--- ..----..
00A0	00	00	00	00	00	00	00	00	00	00	00	00	00	00	00	00	....
00B0	00	00	00	00	00	00	00	00	00	00	00	00	00	00	00	00	....
00C0	00	00	00	00	00	00	00	00	00	00	00	00	00	00	00	00	....
00D0	00	00	00	00	00	00	00	00	00	00	00	00	00	00	00	00	....
00E0	00	00	00	00	00	00	00	00	00	00	00	00	00	00	00	00	....
00F0	00	00	00	00	00	00	00	02	01	00	00	00	00	00	00	00	....
0100	--	00	4C	0F	00	--	00	--	--	--	00	00	00	00	00	00	-.L..-.- --......
0110	00	00	00	00	00	00	00	00	00	00	00	00	00	00	00	00	....
0120	00	00	00	00	00	00	00	00	00	00	00	00	00	00	00	00	....

Hex | Symbolic

Figure 4.31 *File Registers (RAM) Window. (Reprint courtesy of Microchip Technology, Inc.)*

This is accomplished by opening the 'Special Function Register' window (**View|Special Function Registers**). Part of the 'Special Function Register' window for the example project is shown in Figure 4.32.

The 'Special Function Register' window shows the name of the special function register (SFR) and the value in hexadecimal, decimal, binary, and ASCII representations. Now, you can really see what is happening. Is Port C, pin 2 low even though you are driving it high? Did you remember to set the pin as an output? Your timer is not generating interrupts—is it even running?

As with the other 'views' into the processor's state and memory, these windows are NOT updated while the debugger is executing the program. Once the debugger halts, the information in these windows is updated. Double-clicking values that are read/write accessible opens the modify window for the selected register.

4.12 CHAPTER SUMMARY

This chapter has covered many of the features of the CCS-PICC C Compiler and Integrated Development Environment (IDE). The IDE controls the projects and has a powerful code editor. The IDE gives you access to the compiler, the assembler, and their associated options. It also provides tools such as the chip programmer, the code wizard, and the serial monitor tool.

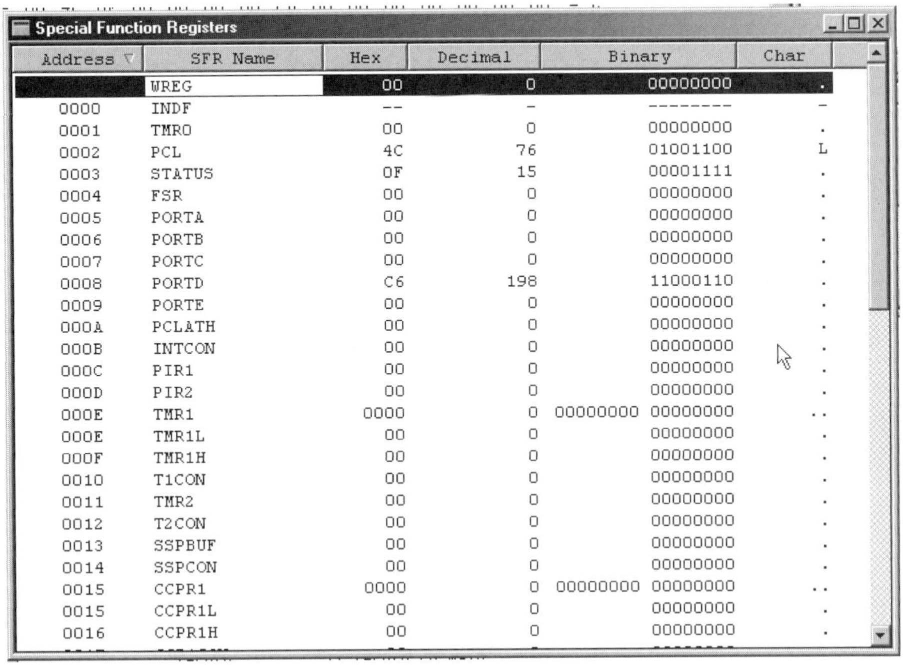

Special Function Registers

Address	SFR Name	Hex	Decimal	Binary	Char
	WREG	00	0	00000000	.
0000	INDF	--	-	--------	-
0001	TMR0	00	0	00000000	.
0002	PCL	4C	76	01001100	L
0003	STATUS	0F	15	00001111	.
0004	FSR	00	0	00000000	.
0005	PORTA	00	0	00000000	.
0006	PORTB	00	0	00000000	.
0007	PORTC	00	0	00000000	.
0008	PORTD	C6	198	11000110	.
0009	PORTE	00	0	00000000	.
000A	PCLATH	00	0	00000000	.
000B	INTCON	00	0	00000000	.
000C	PIR1	00	0	00000000	.
000D	PIR2	00	0	00000000	.
000E	TMR1	0000	0	00000000 00000000	..
000E	TMR1L	00	0	00000000	.
000F	TMR1H	00	0	00000000	.
0010	T1CON	00	0	00000000	.
0011	TMR2	00	0	00000000	.
0012	T2CON	00	0	00000000	.
0013	SSPBUF	00	0	00000000	.
0014	SSPCON	00	0	00000000	.
0015	CCPR1	0000	0	00000000 00000000	..
0015	CCPR1L	00	0	00000000	.
0016	CCPR1H	00	0	00000000	.

Figure 4.32 *Special Function Register Window. (Reprint courtesy of Microchip Technology, Inc.)*

You should now be able to create projects in CCS-PICC, add source files to them, and make them into executable files. With the proper programming cable, you should also be able to program a target device with the built-in chip programmer. You should understand the basic operation of the PIC Wizard code generator and the serial port monitor tool available in CCS-PICC.

Finally, we covered a few basics of the MPLAB debugger. You are encouraged to download this free program from Microchip and use the help files provided with it and experiment. From this section you should be familiar with loading, starting, stopping, and stepping a program in MPLAB. Also, you should be familiar with setting and clearing breakpoints and viewing and modifying variables and registers.

4.13 EXERCISES

1. What does IDE stand for (Section 4.3)?
2. For the project shown in Figure 4.33, what is the name of the main source file for the project? Also list the file name for the source file 'included' in the project (Section 4.4).

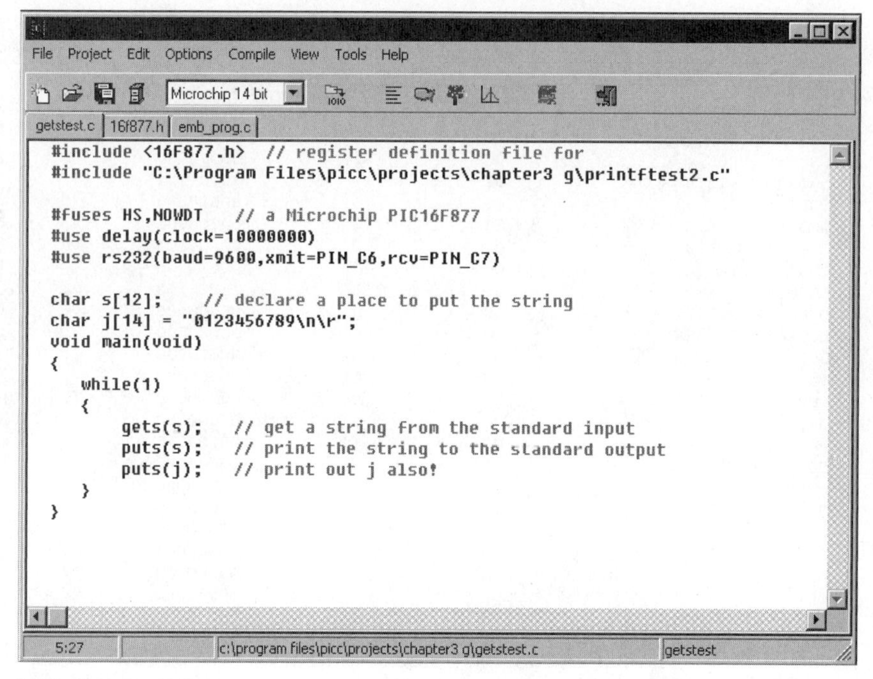

Figure 4.33 *Exercise 2 Project*

3. Name the places where error messages are displayed after compiling (Section 4.4).
4. What menu item is used to highlight all the text between a brace and its match? Also list the shortcut key (Section 4.7).
5. Figure 4.28 shows the Macro Manager window with two macros. The command line for the first macro, *send_1*, is not shown. What would be entered in the command line to send an ASCII '1'? What would be entered in the command line to send a hexadecimal value of one? How else could the ASCII '1' be sent (Section 4.10)?

4.14 LABORATORY ACTIVITIES

1. Create a new project by using the PIC Wizard to generate the source code shell. The project should be configured as follows:
   - USART – 9600 baud, 8 data bits, 1 stop bit, no parity. Transmit and receive enabled
   - Port D – All outputs
   - Port B – All inputs
   - Timer2 – Cause an interrupt to occur every 5 ms
2. Modify the generated project above to send the value present at Port B to the USART once per second. Do not send the value from an interrupt routine.

3. Use the serial port monitor available in CCS-PICC to read the data read from Port B in the above program in hexadecimal notation.
4. Send data, in hexadecimal notation, from the serial port monitor to the target device. Place the value received by the USART on the target device on Port D. Verify that the correct value is present at Port D.
5. Use the PIC Wizard to generate a shell to turn on an LED when a falling edge occurs on a Port B pin 4 and turns it off when a rising edge occurs on a Port B pin 5 interrupt. Complete and test the program

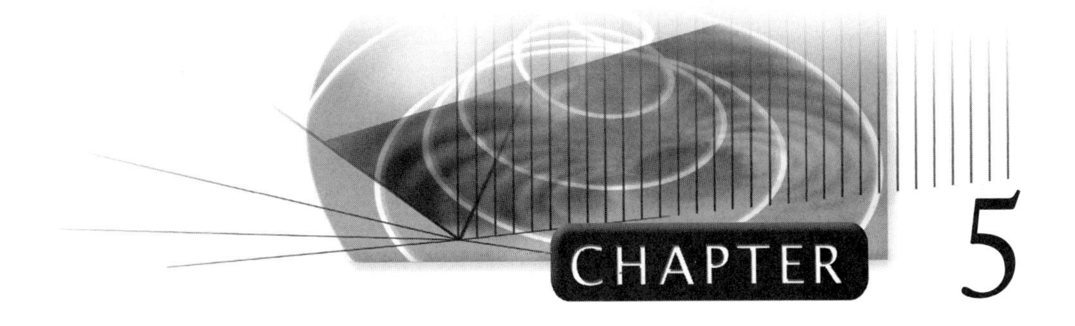

Project Development

5.1 OBJECTIVES

The sole objective of this chapter is for the student (you) to be able to develop electronic projects that involve microcontrollers for either commercial or personal purposes.

5.2 INTRODUCTION

Electronic products involving microcontrollers are most efficiently developed using an orderly approach to the *process*, using a progression of steps, from conception through accomplishment, that virtually always results in success.

In this chapter, the process will be described and then applied to a real project to further demonstrate how the process works.

5.3 CONCEPT DEVELOPMENT PHASE

Every project is based on an idea or concept, which comes from any need—somebody "wants one." The need may be to fill a gap in a product market, to improve a production process, to meet a course requirement, or simply to create something that has not been created before. Because projects are often started to satisfy a need, the original description of the project is sometimes called a problem statement or a need statement.

5.4 PROJECT DEVELOPMENT PROCESS STEPS

The steps you should follow in the process of developing a project are as follows:

1. Definition phase
2. Design phase
3. Test definition phase
4. Build and test the prototype hardware phase
5. System integration and software development phase

6. System test phase
7. Celebration phase

5.4.1 DEFINITION PHASE

The objective of defining the project is to clearly state what the project is to accomplish. This step involves specifying what the device is to do, researching to ascertain that the project is, in fact, feasible, developing a list of specifications that fully describe the function of the project, and, in a commercial environment, providing a formal proposal to go ahead with the project. This step is sometimes called the *feasibility study*.

The purpose of the research conducted during the definition phase is to ensure that the project can, in fact, be accomplished. The early portion of this research will result in a coarse or macro-level block diagram; an example is shown in Figure 5.1. This block diagram will show, in somewhat general terms, the circuits that will make up the final project. If the research shows that certain portions of the project are very sophisticated or are a new use of an existing technology, some of the circuitry can be simulated or built and tested as a 'proof of concept' that the circuitry can be used in this project. The goal of the research is for the designer to be reasonably certain that the project will work.

At the end of the definition phase, a complete set of specifications for the project is developed. These include electrical specifications as well as operating specifications that detail the operation and human interface to the project. Although spending time on the human factors aspect of the project is usually against the engineer's instincts and inclinations, it is important to ensure that the finished project not only works but also fulfills its purpose in terms of operation. In many systems, there is a lot more effort in creating an operations specification than there is in creating an electrical or functional specification.

In a commercial environment, a Project Proposal is written to summarize the definition phase of the project. In addition to summarizing the research and feasibility testing and providing a written list of the project specifications, the proposal will include an anticipated budget for the project and a schedule for completion of the project. The purpose of this proposal is to give the customer (or upper management) confidence that investing money in this project will result in a successful outcome. Figure 5.2 presents one possible outline for a project proposal.

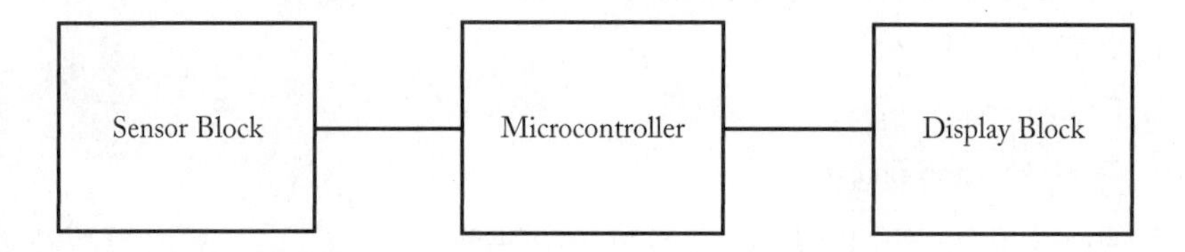

Figure 5.1 *Basic Block Diagram*

Project Title

A. Description of project.

This section fully describes the project, its overall function, the reason or rationale to complete the project, and an indication of the benefits expected to arise from completion of the project. The purpose of this section is for the reader to understand what the project is, and why it is important.

B. Summary of Research.

The project research completed during this phase of the project is summarized including research done on the market and/or competing products, on similar or related projects or products, as well as the results of any proof-of-concept testing completed. The purpose of this section is to show that the project has been thoroughly researched to assure that the relationship of this project to other, existing, projects is clear.

C. Block Diagram.

The overall, macro-level, block diagram is presented in this section. Following the block diagram is a discussion of the *function* of each block *as it relates to the project's overall function*. The purpose of this section is to give the reader confidence that the project can be completed successfully.

D. Project Budget.

A detailed budget is presented here to show all the expected expenditures, including labor and materials, associated with this project.

E. Project Schedule.

The project schedule shows the milestones for starting and completing each block, for starting and completing each phase of the development process, for review or reporting points in the project, and for the final completion and presentation of the project. The purpose of this section is to show, realistically, the time it will take to complete the project and to give the reader confidence that the project has been thoroughly planned out.

F. Appendix.

The appendix includes the bibliography developed during research and any other information that is too large to be included in the text.

Figure 5.2 *Proposal Outline*

The definition phase is a relatively short but very critical step that consumes approximately 10% to 15% of the total project time. This step is crucial in ensuring that the completed project will do what it is designed to do.

5.4.2 DESIGN PHASE

The major goal of the design phase of the project is to fill in the macro blocks developed during the definition phase with actual circuitry and to plan out, in flowchart form, the project software. Because the hardware and software of the project are intimately related, the hardware design and the software plan must be accomplished concurrently. When thoroughly done, this step will consume 40% to 50% of the total project time investment. Unfortunately, the following two characteristics seem to reflect the approach of many engineers to the definition phase of the project:

- Designing the project is the step that, more than any other step, determines the success of the project both in terms of function and efficient development—for *efficient*, read "on time and under budget."
- Designing the project is the step that is most neglected, disliked, and even hated by engineers who vastly prefer hooking up hardware or writing programs to doing research and paperwork.

Designing the project involves going from the somewhat hazy notion of how the project might work, which was developed during the definition phase of the project, all the way to the completed schematics and software flowcharts. The design phase of the project ***does not*** involve any prototyping of hardware or writing of any software. The process of designing the project is shown separately for hardware and software below:

Hardware development steps:

1. Start with the basic block diagram developed during the definition phase. This block diagram will very likely have blocks that encompass more than one function. For example, if you were developing a device to record automobile parameters such as speed of the vehicle, gasoline consumption, and the deflection of the springs, you would probably have one macro block that says "sensors" with a direct line to the microcontroller. Or you might have a single block for each sensor with a line to the microcontroller. And you might have one more block for a display, as shown in Figure 5.1. The objective of this step is to provide a starting point in the form of blocks that you know will provide the desired results, if you can fill in the blocks with appropriate circuitry.
2. Thoroughly research both relevant components and circuits to determine what sort of circuitry and components could be used to fill each block. As you determine what components and circuits you are going to use, you will naturally be breaking up the project blocks into smaller blocks. The objective is to break each block into the smallest pieces that you plan to *individually test*. When complete, each block should be labeled with its input and output voltage levels and/or signals. These signal definitions are then used to develop tests for the individual blocks. Figure 5.3 shows only the speed sensor (one-third of the sensor block in Figure 5.1) broken up into reasonable, testable blocks.

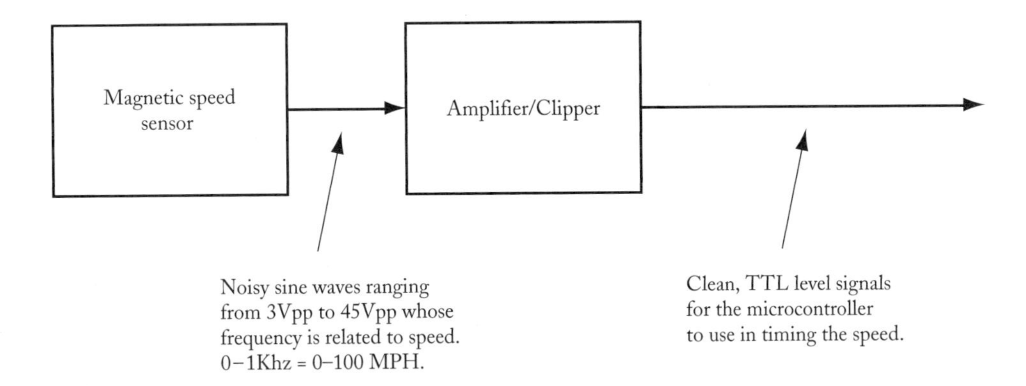

Figure 5.3 *Expanded Sensor Block*

3. Using your research as a basis, develop a tentative circuit schematic for each of the blocks that contain electronic circuitry. Use a circuit simulator, such as Spice or Electronic Workbench®, to simulate the operation of your circuitry. Modify your circuitry as required until the simulator says it works correctly to fulfill the purpose of the block.
4. Use the individual block circuits to create an overall schematic for the project. Use 'cut and paste' techniques or, alternatively, combine the circuit schematics so that errors are not introduced by redrawing the block circuits into a single schematic.

Software development steps:

1. List the tasks to be completed by the software.
2. Prioritize the tasks and determine which ones are critical and will need to be handled on an interrupt basis, and which ones are less critical. Using our example, timing the speed pulses would probably be considered critical and, hence, a candidate for an interrupt, whereas updating the display would probably be considered a low priority task that could be completed whenever the processor is not doing more critical tasks.
3. Create a software flowchart or outline for each interrupt function.
4. Create a software flowchart or outline of the program code to complete the project's entire task.

When the project has been thoroughly designed, you will have a detailed block diagram (showing all blocks and signals), a complete set of schematics (as yet untried in hardware), and a set of software flowcharts or outlines.

5.4.3 TEST DEFINITION PHASE

The test definition phase of the project has the goal of ensuring that the project meets all of its specifications and goals determined in the project definition phase through the development of a test specification for the project. The test specification is divided into two parts: intermediate tests and final, or system, tests.

The test specification for intermediate tests is a list of functional tests to be performed on each block or group of blocks in the project. The tests are designed to ensure that the section of the project being tested will fulfill its function. For instance, if you are testing a Butterworth filter, you will need to specify the frequency range, the number of data points, and the required results for the block test.

The final test specification is a document that includes all of the test procedures that will be used to verify that the project prototype meets all of its intended specifications. In the commercial environment, the final test specification would very likely need to be approved by the customer and the results of the final testing presented to the customer upon completion.

5.4.4 BUILD AND TEST THE PROTOTYPE HARDWARE PHASE

Building and testing the hardware should be relatively self-explanatory. In this step you should construct and thoroughly test the hardware for the prototype project. Use the actual input and output devices for test wherever possible. If using the real hardware is impractical (such as the speed sensor on a vehicle), then use electronic test equipment and other circuitry as necessary so that you will be able to supply simulated inputs from each of the sensors to test the blocks or sections of the project as specified in intermediate test specifications. Keep any signal simulators developed handy so they can be used during system integration and software development.

At the completion of this step, you should be absolutely sure that the correct input signals are being applied to the microcontroller from each sensor and that the correct output signals from the microcontroller will drive the output circuitry in the expected manner. The purpose of this step is to remove the quandary that occurs when some portion of the program does not seem to be interacting with the hardware correctly. Often this leads to uncertainty on the developer's part as to whether the problem lies in the circuitry or in the software, but in this case, because the circuitry has all been tested, the developer will be relatively sure that the problem lies with the software.

In some cases, you will find that it is easier to test those portions of the hardware that are driven by the microcontroller through the use of the microcontroller itself. If you elect to do this, combine the effort with the next step, System Integration and Software Development Phase (also known as Write and Test Software), as you develop the functions to drive the output devices. In this way, you will be spending your time developing useful functions rather than simply developing test code that must be discarded later.

5.4.5 SYSTEM INTEGRATION AND SOFTWARE DEVELOPMENT PHASE

This step is the software corollary to the preceding hardware step. Writing and testing the software means to develop and *individually* test each of the functions that you flowcharted or outlined when you designed the project. Use a simple *main()* function to exercise each of the individual input device functions and output device functions so that, as with the preceding hardware steps, you are absolutely sure that each of the individual input and output functions work correctly. Again, this step is to remove the uncertainty that will occur later when things do not work as you expect them to.

When the individual functions are working correctly, write the overall software in a stepwise fashion—adding one additional function to the code and debugging it before adding more—until the entire project is functioning correctly using the real or simulated inputs.

It is often useful to develop the code to run the output devices first so that you can use output devices such as displays to show results when testing the input devices. It is also useful to use the serial terminal program in CCS PICC to display intermediate results as a means of debugging and testing your software

The goal of this step in the process is to have a fully working project using the real or simulated inputs and outputs.

5.4.6 SYSTEM TEST PHASE

The system test is the activity that actually puts the project into use. During the system test phase, the project should be tested in accordance with the final test specification developed earlier, to ensure that all of the specifications defined in the definition phase of the process are met.

For a commercial project, the system test phase will also include a demonstration for the customer, and it will often include transfer of the *intellectual property* (IP) relating to the project. The IP includes complete documentation and records relating to the project and is the property of the customer.

5.4.7 CELEBRATION PHASE

Successful completion of a project is always a good reason to celebrate. It is beyond the scope of this textbook to suggest the ways, means, and extent of your celebration. Just enjoy it.

5.5 PROJECT DEVELOPMENT PROCESS SUMMARY

The project development process is summarized in Figure 5.4. This figure shows each of the steps, the expected intermediate results, and the final results or deliverables from each step.

5.6 EXAMPLE PROJECT: AN ELECTRONIC SCOOTER

5.6.1 CONCEPT PHASE

As an example, suppose we are engineers at the Phly Bynite Corporation, which has recently begun looking for ways to expand its product offerings. At lunch one day with our fellow employees, someone comments on how he is wearing out a pair of new sneakers, every month, walking to work. This leads to a conversation on the impracticality of driving such a short distance on busy roads, the high prices of automobiles, insurance, and gasoline, and the availability of the new, wide, bicycling and walking paths. Scooters suddenly become a "hot topic" of conversation, with the possibilities of an electronically driven scooter rising to the top.

Process Step	Intermediate Expectations/Results	Deliverables
Definition Phase	Research Basic block diagram	Complete project specifications Schedule Budget Proposal
Design Phase	More research Part Selection Circuit simulation results	Final, detailed block diagram Block schematics Flowcharts for each interrupt function Flowchart for the project
Test Definition Phase		Test specifications for each block or section Final test specifications
Build and Test Prototype Hardware Phase	Individual block test results	Working hardware
System Integration and Software Development Phase	Individual function test results	Working software
System Test Phase		A completed, fully functional project Project documentation
Celebration Phase	Up to you	Up to you

Figure 5.4 *Project Development Process Summary*

Further discussion leads to some speculation on what features would be useful to make the scooter as safe and useful as possible, and eventually a napkin is used (projects often start with a cocktail napkin and a pen) to start sketching out the attributes of such a system.

In the design of the electronic scooter, our list might look like this:

- Easy installation
- Collection of as many types of operational data as possible
- Low cost
- Simple construction

Expanding on our list of attributes, we speculate that the electronic scooter would consist of two units: a drive unit and a display unit.

5.6.2 DEFINITION PHASE

Our list of wants from the concept phase could then be filled in with more details, embellishing or defining each concept that has been put on paper. Notice that no attempt has been made yet to determine what is feasible or possible at this point. Remember that the goal of this phase of the project is to develop a very basic block diagram of the system and a complete set of specifications for the project.

We start by expanding our list of concepts:

1. Easy installation
   a. Simple setup (we would like to be able to just add the electronics to a readily available, inexpensive scooter)—something in the form of a "saddle bag" that can just be bolted on
   b. Simple, unintelligent LCD on display unit to save cost and battery life
   c. Buttons that will allow the user to view different parameters and switch the display contents easily while cruising along on the scooter
2. Collection of as many types of operational data as possible
   a. Speed of scooter
   b. Battery health
   c. Motor current
   d. Braking input
   e. Trip odometer
   f. Auxiliary analog input for future feature expansion
3. Low cost
   a. Evaluate the costs of each type of parameter (i.e., speed, distance, etc.) to be monitored
   b. Keep the overall construction inexpensive (use as few parts as possible)
4. Simple construction
   a. Parts should be available from convenient sources
   b. Use simple feedback systems—nothing too exotic

This expanded list provides a better "mental picture" of the envelope that we are trying to work within. As we expanded the list, the wants and wishes formed. At the same time, apprehension of complexity and cost was also noted.

This is the same process whether you are designing something for yourself as a hobby or you are sitting in an engineering/marketing meeting discussing the development of the next product at your company (as we are doing in this example).

5.6.2.1 Preliminary Product Specification

Now that we have a wish list, we can expand the definition to build a performance specification in which a set of parameters is assigned to each function or feature, indicating its range of performance and tolerance. Performance specifications do not always need to be created. They can be found, in many cases, looking at similar or competitive products that already exist. The preliminary functional specifications for an electronic scooter may look something like those shown in Table 5.1.

Parameter	*Min*	*Max*	*Measurement Tolerance*
Speed	0 mph	15 mph	+/- 0.5 mph
Trip Odometer	0.0 mile	99.9 miles	+/- 0.1 mile
Capacity		200 lb	
Range		10 miles	
Charge Time	6 hours	24 hours	
Battery Voltage	9.5 V DC	13.8 V DC	+/- 0.2 V DC
Motor Current	0	100 amperes	+/- 0.2 ampere
Braking Input	0	100%	+/- 1%

Table 5.1 *Electronic Scooter Preliminary Specifications*

Now may be a good time to comment on safety. Based on the specifications we have come up with and the nature of the device itself, we should be aware of the dangers of working with high current sources, motors, pinch points such as pulleys and belts or sprockets and chains, and the need to always wear the appropriate protective gear when working with or using a device such as this electronic scooter.

5.6.2.2 Operational Specification

Because there is a human-to-product interface involved in our electronic scooter, an operations specification should also be developed.

A good way to begin is to make some sketches of what the display may indicate and create a list or even an "Operator's Manual" describing how the button presses affect the operation of the device. Taking these steps early will greatly reduce the software development time in many cases simply because of the psychological effects of having a list. As items on the list are completed, check them off. The operational specifications for our electronic scooter may look like this:

Drive Unit Operational Specification:

1. Keep units basic, A/D and timer counts are okay.
2. Measure the velocity (speed) of the scooter.
3. Measure the operating motor current.
4. Measure battery voltage.
5. Measure braking requirement from a basic hand control.
6. Transmit the information to the display unit, once or twice per second.
7. PWM drive will be used for the motor.
8. PWM electronic brake will be used, using the motor to assist in stopping the scooter.

9. CAN bus interface is used to communicate with the display (in case we want to add antilock brakes or some automotive type diagnostics later).
10. A "cruise control" type of algorithm will be used to establish the speed of the scooter, eliminating a wrist–controlled "throttle" and increasing the battery life by reducing the motor current demands from a standing start.

Display Unit Operational Specification:

1. Gather packets of information from the drive unit as they come in.
2. Using simple, unintelligent LCD, represent as much information as possible with this display, indicating speed and battery health in their relative units.
3. A "Select" button will allow the user to select the desired data to be read such as the vehicle speed, battery voltage, and motor current.
4. A "Reset" button will allow the user to set an onboard trip odometer or other similar feature.
5. A "low battery" indicator, LED or LCD, will flash when the drive unit battery is too low for reliable operation.
6. A "status" LED will flash when the drive unit is experiencing a condition such as an error or low battery. It should be on steady when the scooter is enabled and ready to move.

5.6.2.3 Basic Block Diagrams

Using the data that we have put together so far—the wish list, the expanded definition, and preliminary performance and operational specifications—we can draw some basic block diagrams of the system. The diagrams shown in Figure 5.5 and Figure 5.6 show the basis for the hardware architecture, such as the power supply, which parameters are measured, and the method of outputting the processed data whether it is by actions taken on the motor or information sent to the display.

The block diagrams show a higher level of detail than that described in the "list writing stage" of the definition phase. On the block diagram, the inputs to the microcontroller are named by the hardware function of the input. The following section describes how each of the measurements are taken and what kind of input to the microcontroller is required. This is an important step in this phase of the project because the quantity of special inputs (such as analog input or timer/counter input) is limited on the microcontroller.

Using the block diagram as a guide, we can choose an appropriate microcontroller with sufficient I/O and special features. As we go on through the design process, we can then check our chosen microcontroller against the project needs to ensure that we have made a good choice with the microcontroller.

The PIC18F458 was chosen for the prototype development. The PIC18F458 is pin-for-pin compatible with the smaller memory model PIC18F448. This allows some freedom during the design phase. At the end of the project, and after the features have ceased to creep, if the

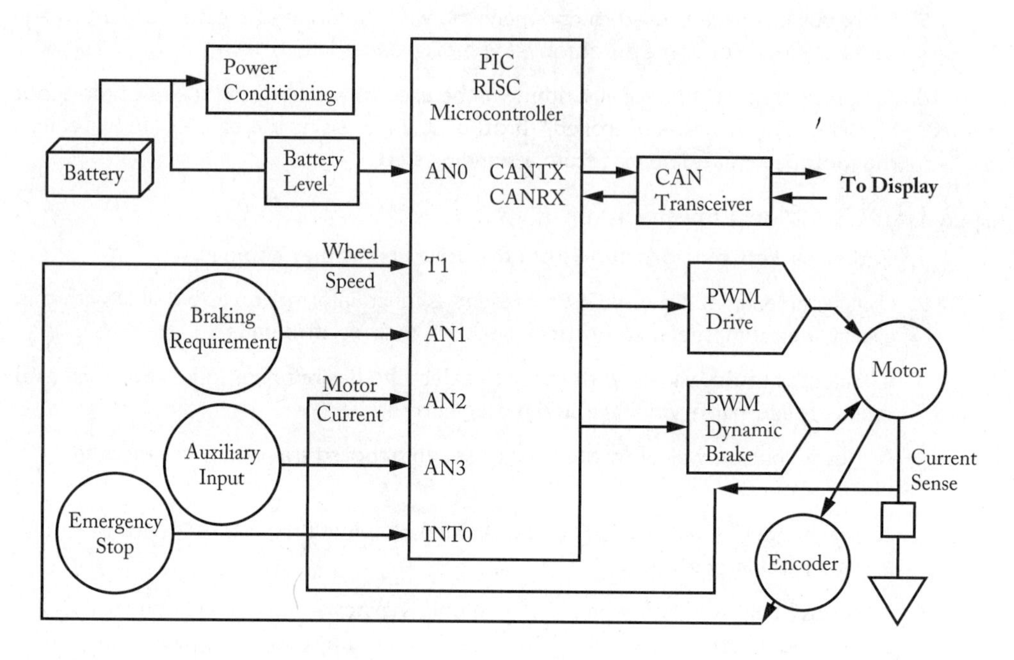

Figure 5.5 *Block Diagram, Drive Unit*

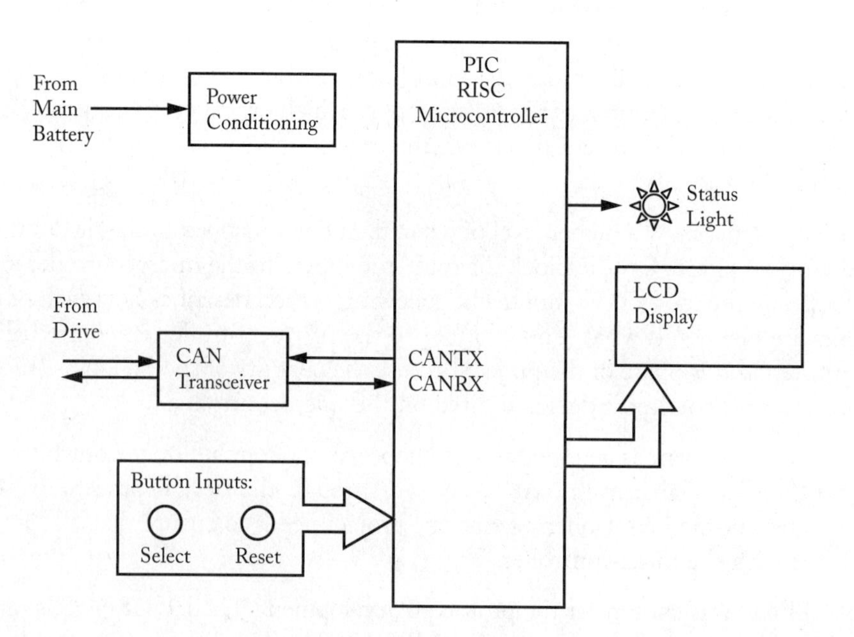

Figure 5.6 *Block Diagram, Display Unit*

code could fit into a lesser memory size and potentially less expensive component, that choice is available as well.

The other, perhaps more obvious points of selection are the availability of A/D converters, two PWMs, a UART, or CAN interface, as well as enough discrete I/O for switches, LEDs, and other digital signal requirements. Both the PIC18F448 and the PIC18F458 have an eight-channel 10-bit A/D, two PWMs, a UART, and a CAN interface.

5.6.3 SYSTEM CONSIDERATIONS FOR THE DESIGN

The final step in the definition phase is a feasibility study or 'sanity' check of the system. Now that we have specified what we *want* the system to do, can we meet those expectations or do they need to be revised based on physical or electrical limitations? This section describes the steps taken to ensure that our project is feasible before we move on to the design phase.

It is hard, however, to completely separate this final step of the definition phase from the design phase. In many cases, it is necessary to develop very specific details about the design to ensure we can measure to our specifications. So, this section crosses into the design phase in many areas as we explore the physical and electrical requirements of electronically driving a scooter.

Each parameter in a system may require a different method of determining its value. In some cases, the parameter may be measured in several ways and it becomes more of a matter of selecting what is appropriate based on performance, cost, or both. The design considerations for each parameter being measured are discussed next.

Once the method of measurement is selected, it is checked for total range and resolution. When dealing with voltages and A/D converters, the resolution of the A/D is the pacing factor. When bringing pulses into a counter, the size of the counter and the rate in which the counts are checked determine the range and resolution of the readings.

In this electronic scooter project, braking, battery voltage, and motor current are all converted to voltages that are processed by the A/D converters. These are calculated out to verify that everything is within range and that the specifications can be met as the sections are designed. The speed of the scooter will be derived from a pulse-counting method described later in this chapter. The requested speed is a different matter. It could be measured as a voltage from a potentiometer in a handlebar control, but for very specific reasons the requested speed will be taken from the same input that gives us the scooter's running velocity.

5.6.3.1 Drive Requirements (According to Newton)

The size of the motor and the battery to properly drive the scooter are based on some of Newton's laws as they are applied to inclined planes. Because the scooter has mass, it also has weight. The gravitational force (F_G) is pulling the scooter straight down. The surface of the incline pushes up on the scooter, providing the normal force (F_N), which is always perpendicular to the surface. As the scooter moves, the force of friction opposes it (Figure 5.7).

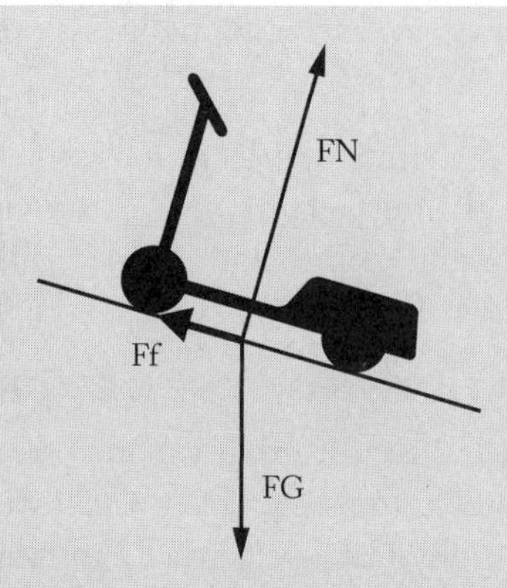

Figure 5.7 *Forces Acting on a Scooter at Rest*

Every vector has two perpendicular components: one horizontal and one vertical. In this case, horizontal and vertical are in reference to the plane (Figure 5.8).

It is the x component of the gravitational force that causes the scooter to want to move down the incline. It is important to note that the scooter does not actually have to be moving for us to look at the sum of the forces. Even if the scooter is resting, gravity is still acting on it and it is the balance of the x component of the gravitational force and the force of friction that holds it in place. They are also balanced any time the scooter is moving with a constant velocity. The sum of the forces on the x axis is:

$$\sum F = F_{Gy} - F_N + F_{Gx} - F_f$$

and because we know that F_N will always be equal to F_{Gy} (unless we are sinking into the ground!) then:

$$F_{Gy} - F_N = 0$$

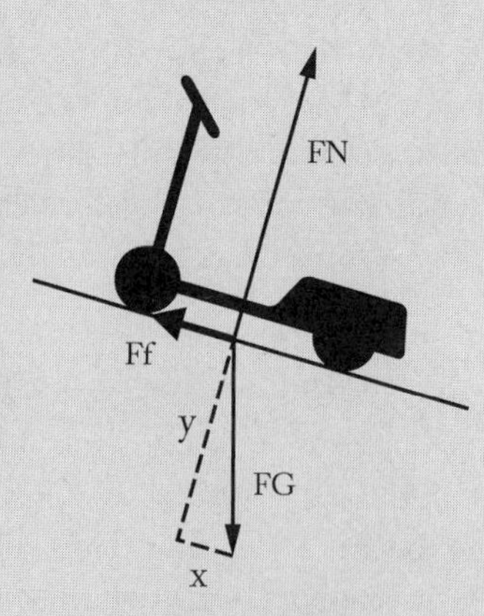

Figure 5.8 *Perpendicular Components of Acting Forces*

so the total force is,

$$\sum F = F_{Gx} - F_f$$

If the scooter is moving down a hill, the applied force is F_{Gx}. Using trigonometry, we can solve for the *x* component. F_{Gx} is the opposite leg of the right triangle; therefore,

$$F_{Gx} = F_G \sin \theta$$

where θ is the angle of the incline, with 0° being level ground. The sum of the forces is now

$$\sum F = F_G \sin \theta - F_f$$

and because force is equal to mass times acceleration,

$$F = ma$$

We can now solve for the acceleration of the scooter along with anything else we need:

$$F_G \sin \theta - F_f = ma$$

$$a = (F_G \sin \theta - F_f)/m$$

Because $F_G = mg$, we can substitute

$$a = (mg \sin \theta - F_f)/m$$

$$a = ((mg \sin \theta)/m) - (F_f/m)$$

$$a = g \sin \theta - (F_f/m)$$

In the absence of friction (and with the ball-bearing wheels in this new two-wheel scooter, we are almost completely absent of friction), we find that the acceleration of the scooter is independent of the scooter's mass and dependent only to the angle of the incline.

$$a = g \sin \theta$$

This is reasonable because we already know that in the absence of friction everything accelerates at the same rate in a gravitational field regardless of its mass. Because we are concerned with what it will take to drive the scooter up a slope, let us calculate the required force to accelerate uphill. In this case, the only difference is that the applied force is independent and not the *x* component of the force of gravity. It also will be in the opposite direction of gravity, meaning the force of friction is now in the same direction as the *x* component of the gravitational force. Remember, if you stop driving the scooter, it will roll down the hill or, at the very least, stop moving (Figure 5.9).

Looking at Figure 5.7, we see that the force of friction *and* the *x* component of the gravitational force are opposite the applied force. The sum of the forces up the hill is

$$\sum F = F_A - F_f - F_G \sin \theta$$

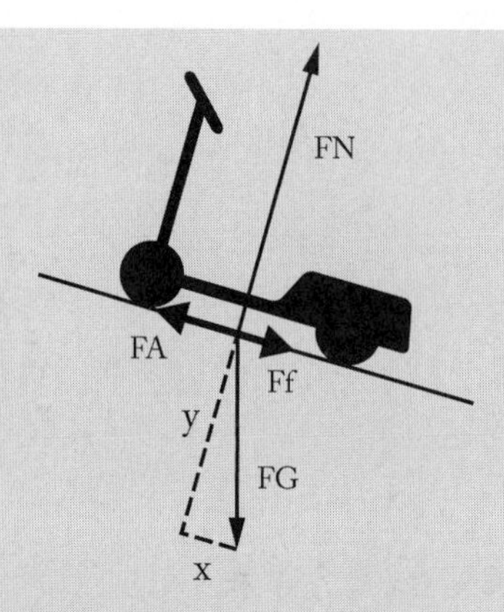

Figure 5.9 *The Applied Force Component Total*

If we assume that a 200-pound individual will operate the scooter and add an extra 25 pounds to account for the weight of the scooter and all of the additional parts, such as the motor and battery, we can compute the amount of force to accelerate the scooter on a flat, level surface. Because at 0° of incline, $F_G \sin \theta = 0$ then,

$$\sum F = F_A = ma$$

The weight (225 pounds of force) must be converted to a mass by removing the acceleration of gravity, which is 32 feet per second^2.

$$m = F_G / g$$

$$m = 225 \text{ lb} / 32 \text{ ft/s}^2$$

$$m = 7.031 \text{ lb}_{\text{mass}}$$

If we then set a design goal of accelerating this total load to 15 mph in a period of 5 seconds, then the acceleration value would be

$$a = \frac{15 \text{ miles}}{1 \text{ hour}} * \frac{1}{5 \text{ seconds}} * \frac{5280 \text{ feet}}{1 \text{ mile}} * \frac{1 \text{ hour}}{3600 \text{ seconds}} =$$

$$a = 4.4 \text{ ft/s}^2$$

Applying this acceleration, we then find that the required force applied to the load is

$$F_A = 7.031 \text{ lb}_{\text{mass}} * 4.4 \text{ ft/s}^2 = 30.93 \text{ lb}$$

Where applying Newton's law begins to indicate the issue of reality is when we want to go up a hill on the scooter. For example, if you wanted to accelerate at the same rate up a 10° slope, an increase in load places an increase on demand on the motor by the additional contribution of gravity to the force equation. As stated earlier, the contribution of gravity on acceleration is:

$$a_g = g \sin\theta$$

$$a_g = 32\ \text{ft/s}^2 * \sin(10°)$$

$$a_g = 5.55\ \text{ft/s}^2$$

So now the total required force is:

$$F_T = F_A + ma_g$$

$$F_T = 30.93\ \text{lb} + (7.031\ \text{lb}_{mass} * 5.55\ \text{ft/s}^2) = 69.95\ \text{lb}$$

10° does not seem like much incline, but it rapidly increases the amount of force required to move the load.

5.6.3.2 MOTOR SELECTION

From the physics exercise in Section 5.6.3.1, we found that the amount of force required to accelerate a loaded scooter (225 pounds) on a flat surface to 15 mph over 5 seconds is:

$$F_A = 7.031\ \text{lb}_{mass} * 4.4\ \text{ft/s}^2 = 30.93\ \text{lb}$$

and the force required to do the same on a 10° slope is:

$$F_T = 30.93\ \text{lb} + (7.031\ \text{lb}_{mass} * 5.55\ \text{ft/s}^2) = 69.95\ \text{lb}$$

Now that we have computed the forces required to move the scooter, we have to somehow translate this into a motor specification. In our electronic scooter preliminary design specification (see Table 5.1), we set a design goal of 15 mph as a maximum speed. Combined with the required torque, this speed will establish the motor size and type.

The maximum speed will be directly related to maximum motor revolutions per minute (rpm) and any gear or pulley ratio that we may apply in order to get the desired torque.

Torque is a measure of the ability of a force to produce a rotation of a body about a particular axis. Torque is a vector quantity, having both magnitude and direction.

The magnitude of the torque is given by the equation:

$$\tau = F r \sin\theta$$

where F is the force and r is the distance from the axis to the point where the force is applied.

In converting the force required, which is tangent to the wheel, θ is 90°, so $\sin\theta = 1$. Therefore, to convert from force to torque, the equation is simply

$$\tau = F r$$

Using the flat surface (0°) and 10° surface force values from our previous work, we can compute the torque required to turn the 4-inch wheels on the scooter with the maximum load. At 0°, the torque would be computed as:

$$\tau = 30.93\ \text{lb} * 2\ \text{in.} = 61.87\ \text{lb-in.}$$

Because most motors are specified in ounce-inches of torque, we simply scale the result.

$$\tau = 61.87 \text{ lb-in.} * 16 \text{ oz/lb} = 989.96 \text{ oz-in.}$$

At 10°, the torque requirement would be computed as:

$$\tau = (69.95 \text{ lb} * 2 \text{ in.}) * 16 \text{ oz/lb} = 2238.4 \text{ oz-in.}$$

That is quite a load—too much for a small DC motor!! This is where a small gear ratio can make a big difference. If the motor output was reduced to the wheel with a 3:1 gear ratio, the required outputs for the two conditions, 0 and 10°, would be reduced accordingly:

$$\tau_{0^\circ} \text{ @ 3:1} = 989.96 \text{ oz-in.} \div 3 = 329.98 \text{ oz-in.}$$

$$\tau_{10^\circ} \text{ @ 3:1} = 2238.4 \text{ oz-in.} \div 3 = 756.13 \text{ oz-in.}$$

At a 4:1 ratio, the results get even better:

$$\tau_{0^\circ} \text{ @ 4:1} = 989.96 \text{ oz-in.} \div 4 = 247.49 \text{ oz-in.}$$

$$\tau_{10^\circ} \text{ @ 4:1} = 2238.4 \text{ oz-in.} \div 4 = 559.60 \text{ oz-in.}$$

There is an all-powerful law of physics that keeps balance in the universe. It is, "you can't get something for nothing." In this case, by gear reduction we get a more desirable torque at the wheel, but we are trading off the motor speed to get it.

Referring back to our electronic scooter preliminary design specification (see Table 5.1) we set a design goal of 15 mph as a maximum speed. With a 4-inch wheel, the scooter moves 1.047 feet per revolution of the wheel. At 15 mph, the wheel must turn at

$$\text{rpm @ wheel} = ((15 \text{ miles/hour} * 5280 \text{ ft/mile}) \div 60 \text{ min/hour}) \div 1.047 \text{ ft/rev}$$

$$\text{rpm @ wheel} = 1260.50 \text{ rpm}$$

With that in mind we must now locate a motor that can provide the proper amount of torque and have a top speed of at least 3781 rpm when used with a 3:1 gear ratio, or 5042 rpm when used with a 4:1 gear ratio.

Some searching on the Internet yielded these inexpensive motor choices:

- Sullivan Products S98600: high-torque motor, 12 V DC, maximum 80 amps, 210 in.-oz (148 N-cm) of stall torque, 5500 rpm (no load)
- Sullivan Products S98603: "Dynatron" high-torque motor, 12 or 24 V DC operation, maximum 80 amps, 340 in.-oz (240 N-cm) of stall torque at 12 V, 680 in.-oz (480 N-cm) at 24 V, 4800 rpm no load at 12 V, 9600 rpm no load at 24 V
- Sullivan Products S980511650: "Megatron" high-torque motor, 12 or 24 V DC operation, maximum 100 amps, 600 in.-oz (424 N-cm) of stall torque at 12 V, 1200 (848 N-cm) at 24 V, 2800 rpm no load rpm at 12 V, 5600 rpm no load rpm at 24 V

These motors are all quite good in terms of size and performance. But looking at the torque and rpm available, more design decisions must be made before we make a final motor selection.

In our drive system we will be implementing a cruise control–type speed control algorithm. This eliminates the need for a wrist-controlled "throttle," which will greatly reduce the demands on the drive. The cruise control speed control allows the rider to manually accelerate the scooter to a speed, and then the control simply maintains that speed. This system conserves the tremendous amount of power required to get the loaded scooter up to speed from a standing start by having the rider supply it. This will increase not only the range of the scooter (battery life) but will improve the life of the motor by reducing the stress of stall currents in the motor windings. Super high currents can potentially demagnetize the motor's stator, reducing its efficiency, which would result in increased currents and yet more damage.

5.6.3.3 Vehicle Speed Measurement

The speed of the scooter is directly proportional to the speed of the motor turning the drive wheel, so by measuring the motor rotational speed we can derive the vehicle's linear speed. A device that measures the rate at which something is turning is called a tachometer. Several possible tachometer types are used for rotational speed measurement.

One often-used form of a tachometer is that of a small electric motor or generator. As it turns, it generates a DC voltage that is related to the rate at which it is spinning. In this case, the resulting DC voltage could be measured using the microcontroller's A/D converter and the rotational speed interpreted from the DC voltage. Some motor speed controls utilize the back EMF of the motor (a voltage that is generated by the motor itself), but this voltage is usually tricky to separate from drive voltage and is usually only an approximation of the motor's speed at best.

Another version of tachometer is the *digital tachometer*. A digital tachometer typically generates a pulse train, associated with the rate of spin, which can be measured as a frequency and converted to *revolutions per minute* (rpm). A digital tachometer is, in most cases, a less expensive alternative to an analog tachometer. It can be easily constructed from a slotted disk and a photo interrupter or optocoupler. It can also be an off-the-shelf solution in the form of an optical encoder. Because there are no magnets and much less inertia due to the weight of the armature, the digital tachometer offers an additional advantage over the analog tachometer in that it usually imposes less drag on the system.

If we place an encoder on the system that delivers four pulses for each revolution of the motor, geared 3:1 with the 4-inch wheel, at 1 mile per hour, it would yield 16.8 pulses per second (pps or hertz). At 15 mph, a pulse train of 252 pps would be generated. If this signal were sampled at 1-second intervals using an 8-bit counter, the maximum speed that could be measured would be one that generates 255 pps. Working this backward, we can find the maximum speed that can be measured at a 1-second interval to be:

255 pps ÷ 16.8 pps per mph = 15.178 mph

What this cross-check shows us is that at a 1-second sample rate, 8 bits is barely enough. If the motor were geared at a 4:1 ratio or our maximum speed were greater than 15.2 mph, we will need to sample the counter two times per second, or every 500 ms, to prevent the counter from rolling over during our measurement. It also shows that an 8-bit counter would still be adequate if the sampling rate is simply increased.

Another method for improving the measurement capacity is to simply get a bigger counter. Timer1 is a 16-bit counter and can hold a value 256 times larger. This means that we can effectively sample at one-fiftieth the rate, or once every 10 seconds! If the sampling rate is even once per second, the collected number in Timer1 would effectively be five times larger than what could be measured within the 8-bit counter.

5.6.3.4 Battery Health Measurement

Battery health can be as exotic as one chooses it to be. It can be as simple as measuring the voltage, or as complicated as monitoring the charging currents and determining the battery's internal impedance changes. In most cases, an embedded system is concerned with the simple requirement of "is there enough voltage to operate correctly?" This can be accomplished in most cases with a resistive divider, to scale the voltage being measured, and an A/D converter. Or, we could use the analog comparator peripheral in the PIC device, if we are interested in simply a "go / no-go" reading.

In the case of the scooter, we will want to use the A/D and maintain a reasonable resolution because this input will be used to issue warnings to the operator, make predictions about remaining operating time, and eventually shut down the drive when the battery voltage is too low to safely operate the motor driver.

5.6.3.5 Motor Current Measurement

Monitoring of motor current during the operation of the scooter provides for the intelligent protection of the drive system. The current can indicate to the drive control when an overcurrent situation is occurring and allow the software to decide whether or not that the system should be shut down. The measured current, when combined with the battery voltage, additionally allows for the prediction of battery usage and remaining range of the scooter.

Measuring current in a DC drive like this one is typically done with some type of sense resistance in series with the motor. The motor specifications listed in Section 5.6.3.2, "Motor Selection," show stall currents of 80 to 100 amperes. Therefore, to sense the current we will need a shunt or sense resistor capable of carrying 100 amperes of current. Also, because some voltage drop must exist for a resulting sense voltage, power is being dissipated across the device.

For example, a sense resistor of 0.01 ohm will dissipate 100 watts of heat at 100 amperes of current. A 100-watt device is physically very large. Fortunately, in this system the durations of the 80- to 100-ampere periods should be very short. This allows us to utilize a lesser power rating for the current sensing.

Once the large current is converted to small voltage, some amplification and filtering will need to be done and the result sent to the A/D converter. With the previously mentioned sense resistor and a small gain stage, a 5 V = 100 A current monitor can be constructed. The PIC18F458 A/D has 10 bits of resolution, which means that the current can be monitored at a resolution of 97 mA per bit.

5.6.3.6 Brake Control Measurement

The user will apply the braking control input to the scooter by using a common handbrake system. Because the drive controller will perform the actual braking electronically, we will be converting the handbrake position into a voltage by using a potentiometer.

The handbrake control shown in Figure 5.10 moves the cable linearly as the operator squeezes the control. Using a bell crank and a return spring, a potentiometer can be rotated approximately 90°. This allows for about a 30% of the total range of a single-turn potentiometer to be used. That means that a potentiometer connected between 5 V DC and ground could deliver a voltage range of 1.5 V DC. This small voltage range could appear anywhere in the total 5 V DC range available, depending on where the wiper of potentiometer is placed with respect to the control range.

Placing the potentiometer in the center of its range for the center of the control range would provide voltages back to the PIC18F458's A/D of 1.75 to 3.25 V DC. Because the A/D has 10 bits of resolution, this would be converted into a range of counts from about 358 to 665.

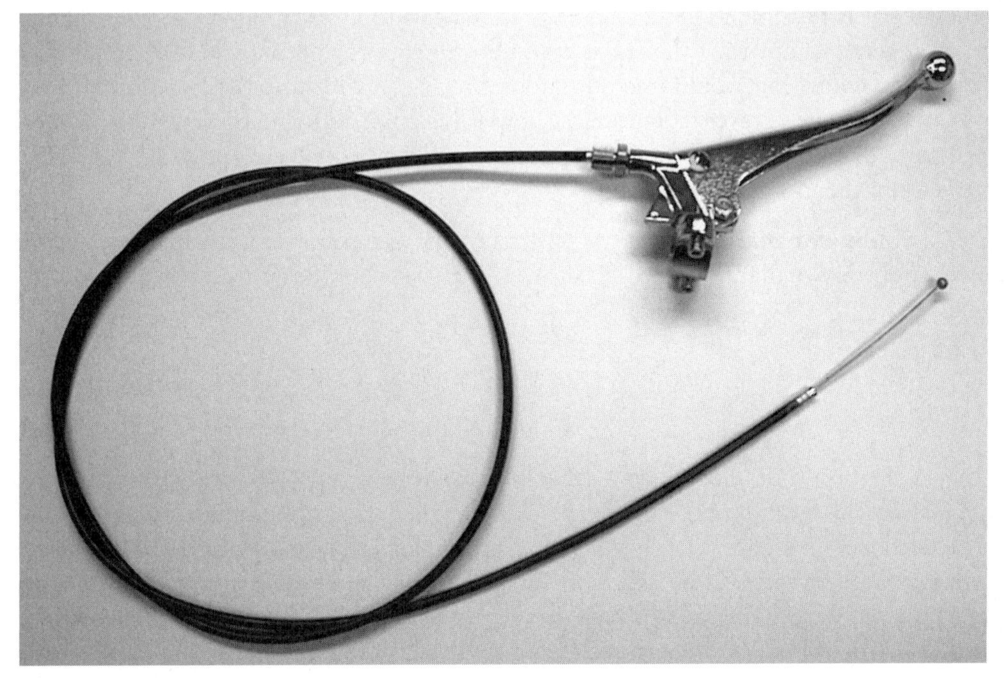

Figure 5.10 *Typical Handbrake Control Assembly*

A little over 300 counts! This is more than adequate for a percentage of "squeeze input." Additionally, there is built-in error detection. By having the voltage centered with a limited range, out of range values can be interpreted. This allows the system to know if there is a problem with the braking control, or if the potentiometer has become disconnected for some reason.

Because we are implementing a cruise control system, the braking input serves multiple purposes. It provides for the ability to determine if the user wants to coast, keeping the scooter from going to a higher cruising speed, or to reduce his or her current speed by braking. If the user wants to actually brake, the system unlocks the cruise control and applies a percentage of braking resistance based on his or her braking input.

5.6.3.7 Electronic Braking

Electronic braking is simply a method of utilizing the motor that is typically used to drive a system to resist the motion it creates. This can be an exotic process depending on the level of braking required. In an AC motor system, a DC current is applied to the motor to stop and to even hold the motor to prevent it from rotating. The scooter uses a DC motor, so applying a DC current in the same way would cause the motor to turn—not the effect we are looking for. Applying an AC current, on the other hand, would stop and hold the motor and prevent it from spinning, but it would be difficult to do because we are running on batteries instead of an AC wall outlet.

Our goal is to keep this system simple and low cost. Because a DC motor also functions as a generator, it is possible to get a stopping resistance simply by applying a load to the output power generated by the spinning motor. Shorting out the output would yield the quickest possible braking but would also put undue stresses on the motor. In our system, we will deposit the generated power to an external, low-value, high-wattage resistor. By using a value of resistance that keeps the motor braking current inside of its stall current rating, we are helping to ensure the motor's longevity.

For a 12-volt motor at 80 amperes (the stall rating of our examples), the braking resistance should not be lower than:

$$R = V \div I$$

$$R = 12\text{ V DC} \div 80\text{ A}$$

$$R = 0.15\text{ ohm}$$

In theory the power developed across this load could be nearly 1000 watts!! But in actuality, the available power and the braking duty cycle keep this from happening. A lot of heat can be developed though, especially when braking and going down a long, long, hill. This consideration could be factored into the software, allowing safe braking durations for the given rpm of the motor. Braking of this type usually ends up being a bit of an empirical process, and that is why we build prototypes. If we assume that the power issued to the brake load resistor is 5% of maximum due to losses and braking duty cycle, we can start off with a 50-watt resistor at 0.15 ohm and do some testing.

Another feature is to be able to apply a variable amount of braking. The harder you squeeze the control, the more brake is applied. This can be handled in the same way as the drive itself. A PWM can be used to generate a percentage of applied load to motor by switching the load resistance on and off across the motor.

5.6.4 HARDWARE DESIGN - DRIVE UNIT

Now that we have proven to ourselves that the project really is possible, we can start designing the final hardware. By using the block diagrams, specifications, and feasibility research from the definition phase, we can quickly pull together the schematics for the hardware. First, each block of the drive unit is developed, and then the blocks are combined together into one schematic.

The drive unit is battery powered, with the idea that a fully charged battery will power the unit for 10 miles over average terrain. Because the battery is capable of delivering huge amounts of current, and the motor is capable of using it, fusing the unit is not operationally very practical. To improve the safe operation of the device, a relay is designed into the system to disconnect the battery from the drive circuits. A relay with high current contacts is placed in the final drive output. This relay requires a two-way "agreement" in order to operate. First, the unit controller must be switched on and the controller software must be cycling the input to the power control watchdog circuit. Second, the operator must actuate a foot switch, indicating that he or she is onboard the scooter. In the event that neither of these conditions is met, the motor drive is physically disconnected from the battery by the high-current relay.

Conservative power utilization was considered in the design of the monitoring circuits. Even though there is a large amount of power available, there is no sense in wasting it and potentially reducing the range of the vehicle. The speed of the scooter is monitored using a photo interrupter and processed as a strictly digital signal using a counter. The braking input and motor current sensing are monitored by the A/D of the microcontroller. A "motor stalled" detector was designed in to provide an immediate indication of a maximum motor current condition to the microcontroller. This allows for a quick decision to be made without waiting for data to be available from the A/D converter.

The battery voltage is also monitored by the microprocessor's A/D such that the "health" of the power system can be reported to the display unit. The schematic developed for the scooter drive unit is shown in Figure 5.11. Specific portions are discussed in the following paragraphs.

Speed Input

The speed of the vehicle is monitored as a pulse train from an optical encoder. A photo interrupter (in this case a disk with some holes in it) is used to block a beam of infrared (IR) light

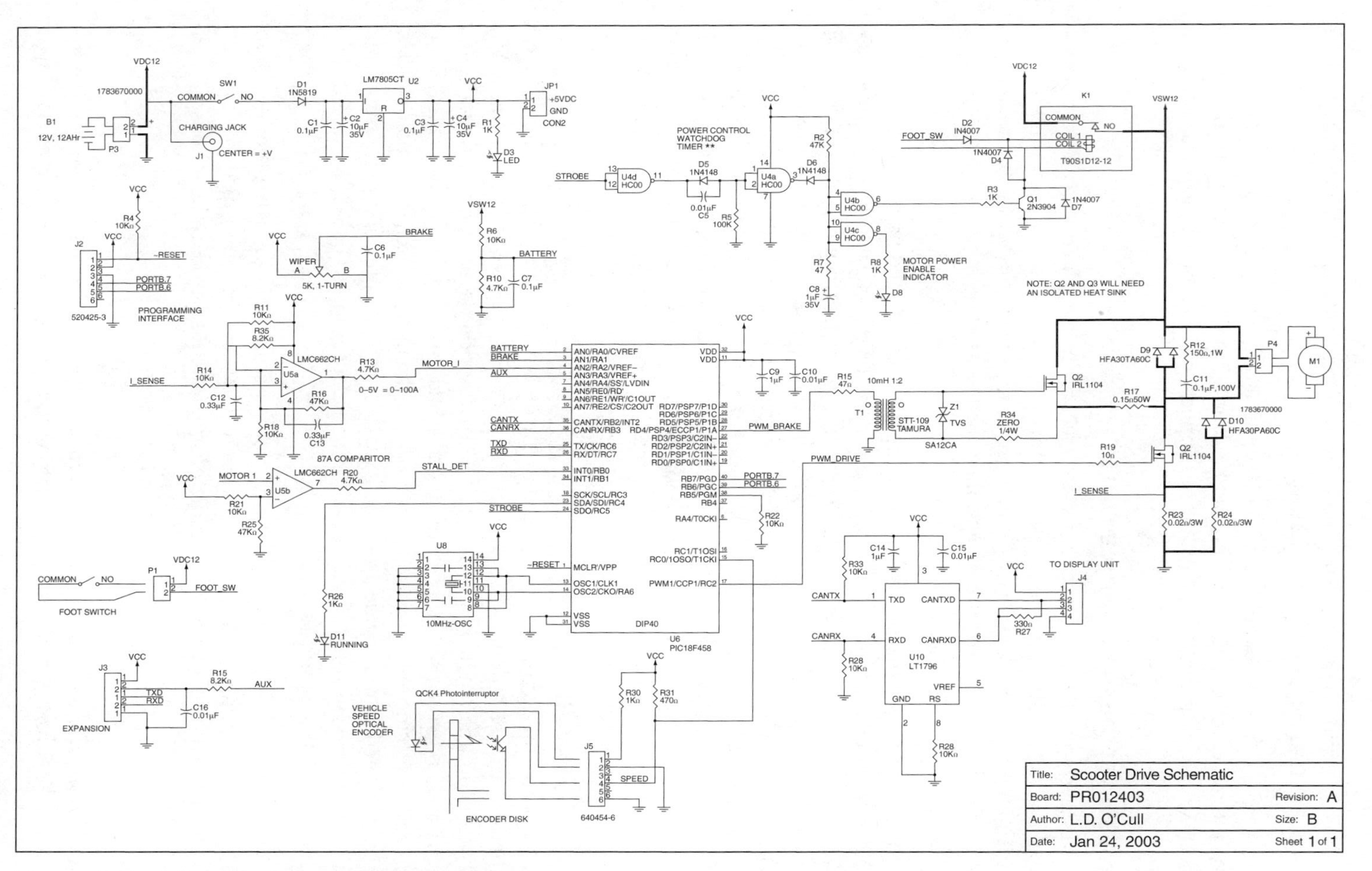

Figure 5.11 *Schematic, Electronic Scooter Drive Unit*

from reaching its destination, a phototransistor. When the IR light strikes the surface of the phototransistor, the transistor conducts, causing the signal SPEED to be pulled low. When the disk breaks the IR beam, the SPEED signal returns high. The result is a waveform that is directly associated with the vehicle speed.

As discussed in Section 5.6.3, "System Considerations for the Design," if we place an encoder on the system that delivers four pulses for each revolution of the motor, geared 3:1 with the 4-inch wheel at 1 mile per hour, it would yield 16.8 pulses per second (pps or hertz). At 15 miles per hour, a pulse train of 252 pps would be generated. The SPEED signal is being counted by the Timer1 input (T1CKI). Timer1 on the PIC18F458 can be programmed to function as a 16-bit counter, so there is plenty of resolution for our speed-challenged scooter.

Braking Input

The braking input is one of the simplest signals in the system. The brake hand lever pulls a cable that goes to a potentiometer on the drive unit. A return spring is used to push the cable back out as the operator releases the hand lever. The potentiometer then converts the hand lever position into a voltage. The potentiometer is connected to VCC and GND so that a 0–5 V DC is possible. The hand lever has a limited amount of motion, so only about 30% of the potentiometer can be used. This allows for some fault detection, such as an open or short in the braking potentiometer, or even a loose cable from the hand lever, allowing the spring return on the cable to pull the potentiometer to one end of its travel causing a 0-volt or 5-volt reading.

Motor Current Monitoring

To monitor the large motor currents in this drive system, a small voltage is measured across a sense resistance. The sense resistance is 0.01 ohm. This will produce a 1-volt drop at 100 amperes.

The drop in voltage in a sensing circuit like this one must remain low. If too much voltage drop is allowed, then several undesirable outcomes could result. First, the sense resistor itself could heat up to the point of burning. Second, the increased drop could decrease the ability of the system to adequately switch on the Metal Oxide Semiconductor Field Effect Transistor (MOSFET) switching transistors, used to power the motor, causing the MOSFET to heat and eventually fail.

Two 0.02-ohm resistors are used in parallel to increase the distribution of heat and power. The current across the sense resistors will show up as pulses, because the motor is driven from a pulse-width modulator (PWM). The sense voltage needs to be converted into something simpler to measure. An operational amplifier (op-amp) is used to filter, integrate, and amplify the sense voltage into a smooth, large signal that can be measured by the PIC's A/D converter.

In addition to the A/D reading of the current, a second op-amp is used as a comparator. When the current reaches a motor stall level, in this case greater than 80 amperes, the comparator output goes to logic high and issues a digital signal to the microcontroller. This signal can be used as a polled digital input or it can generate an interrupt. Stall level currents are okay for short durations, but too much time in stall can damage the motor, the drive output, and, potentially, even the battery and wiring. It is for this reason that redundancy in measuring methods is in order.

Motor Power Control

The motor drive is a PWM direct-switching system. Because the motor windings effectively make a huge inductor, currents switched into the motor are integrated or averaged over time. Therefore, the duty cycle, which is the percentage of on time of the switch compared to its off time, is almost directly proportional to the amount of current being fed to the motor.

This configuration is very efficient because it reduces the amount of power lost in the transistors used to switch these currents to the motor, which, in turn, reduces the heat and the physical size of the drive. The selected MOSFET for this output is the IRL1104 made by International Rectifier. This device is capable of switching more than 100 amperes with a TTL signal at its gate, which is really rather amazing! In addition to its "brute force," it has a very low turn-on resistance. This means that when the transistor is on, there is very little voltage drop across the output of the device, drain to source; therefore, there is very little power dissipation.

When the motor has current deposited into it by the MOSFET switch, it develops the big magnetic field that it needs internally in order to move. When this field collapses from the switch opening, a large fly back voltage is generated. This voltage is trapped in the drive unit by high-speed, high-current, fast-recovery diodes. These diodes not only protect the MOSFET from the high voltages but also create the additional feature of keeping the energy circulating in the motor to squeeze out as much performance as possible for this simple configuration. Another benefit of these diodes is that when the motor is coasting and working more as a generator, they provide a path for the current back to the battery, thus working as a battery charger.

The battery voltage is connected to the final output of the drive through a high-current relay. This relay functions as a safety switch. If the MOSFET were to fail in a shorted condition, the motor would continue to run. Because the relay has to be enabled by the software properly pulsing the STROBE signal of the power control watchdog timer circuit as the operator is holding down a foot switch, the user only has to release the foot switch to overcome a failure of this type.

Electronic Brake

The electronic braking of the scooter is somewhat similar to the method of driving the scooter. A PWM type of switching drive is used to load the freewheel motor-now-generator. In the example drive system schematic, the same type of MOSFET is used to load the motor. One very noticeable difference is that a pulse transformer is used to couple the PWM signal from the microcontroller to the MOSFET gate. This is because the gate voltage must be higher than its source voltage by at least 4 volts. Using a pulse transformer allows for a simple translation of the signal, ground referenced on the primary of the transformer to an indeterminate reference on the secondary.

The only shortfall of this type of coupling is that a 100% duty cycle is not allowed. Only alternating current (AC) can pass through a magnetic transformer, so the signal must be moving or completely off. We thought that a 97% duty cycle was "close enough" to allow for this simple, straightforward connection of the MOSFET.

A software cautionary note is, therefore, in order. The braking and the drive can never happen together. If they were to coexist, the motor would be bypassed by the braking MOSFET, causing a direct connection from the battery to the supply relay to the drive MOSFET through the sense resistors to ground. In other words, "BANG!!" This is the kind of scenario where we want to test the software thoroughly with a current limited supply before connecting up the battery and "going for it." In order to get this condition, both PWM outputs would need to be enabled (drive and braking) and be on at the same time. Sounds a little complex, but it is easy enough to do. Operating under software control allows for so much power and flexibility!

CAN Interface

Controller area networking (CAN) is a commonly used method of communication for automotive, commercial, and industrial applications. The most common usage is in motor vehicles. There are several reasons for its popularity, such as the CAN transceiver itself. The transceiver in our design, a Linear Technology LT1796, is capable of being shorted out, abruptly open circuited, or even subjected to an overvoltage of up to +/- 60 V without damage. It has high input impedance (about 100 K ohms), which allows for more than 200 devices to be connected together in the local area network. These devices maintain this high impedance even when powered off, such that they do not interfere with other devices on the network. The structure of the drivers within the CAN transceiver allows for data rates of up to 1 million bits per second. Other features include built-in slew rate limiting to lower electromagnetic interference (EMI) when using inexpensive wire to connect devices, as well as thermal and electrostatic discharge (ESD) protection.

The PIC18F448 and PIC18F458 contain an on-chip CAN interface module. This module has the capability of preprocessing messages going in and out of the CAN transceiver. Messages can be addressed and prioritized on the way out as well as filtered and sorted on the way in. This hardware handling of the data traffic greatly reduces the amount of CPU cycles required to process the communications going on between micros. For this scooter project, CAN is a bit of a "big hammer for a small nail" but it gives us a chance to do a basic CAN setup and implementation that can be expanded at a later time.

In our system, the scooter driver will stream data to the network continuously, and the display will be configured to listen continuously, extracting and displaying the pertinent information as required by the operator.

Power Supply

The power supply for the entire system is, of course, the battery. In actuality, not much of the system can operate at battery voltage, so a regulator is used on the drive unit to down convert the battery to 5 V DC (VCC). The power supply, as well as the battery switching, is diode isolated to help prevent damage from an accidentally reversed battery connection. Because the display unit requires very little power and is fairly close to the drive unit (within a few feet by wire), the VCC is provided to the display unit through the CAN communication cable. There is an ON/OFF switch to control power to the scooter. A charging jack is also provided to allow the unit to be easily plugged into an appropriate battery charging system.

Battery Selection

The initial battery selection is made on a couple of assumptions. In the preliminary specification, the scooter is to have 10 miles of range. If the scooter is traveling at a maximum speed of 15 miles per hour, it will take about 40 minutes to cover the 10 miles (which is much faster than the walking method). Making another assumption that we are on smooth, level ground the whole time, we can make a guess at the approximate average motor current. Using a "fish scale" to pull the scooter across a parking lot at about 4 mph, we measured approximately 6 lb of frictional forces. So converting that into a torque requirement (as shown in Section 5.6.3.2, "Motor Selection"), we get:

$$\tau = (6 \text{ lb} * 2 \text{ in.}) * 16 \text{ oz-lb} = 192 \text{ oz-in.}$$

and with a 4:1 gear ratio, the required motor torque is:

$$192 \text{ oz-in.} \div 4 = 48 \text{ oz-in.}$$

The Sullivan Dynatron motor is specified at 80 amperes at 340 oz-in. (240 N-cm) of torque at 12 V DC. If we assume that current and torque are linearly related, then the average flat-surface running current would be:

$$\frac{340 \text{ oz-in.}}{80 \text{ amperes}} = \frac{48 \text{ oz-in.}}{x}$$

Therefore,

$$x = (80 \text{ amperes} \div 340 \text{ oz-in.}) * 48 \text{ oz-in.}$$

$$x = 11.29 \text{ amperes}$$

This shows that the average flat-surface running current is 11.29 amperes. A battery can be selected from this assessment.

A Panasonic LC-RA1212P maintenance-free battery will discharge to 9.6 V DC in 40 minutes at a 12-ampere continuous load, which meets our requirement of 40 minutes or 10 miles. This is a good starting point during the prototyping phase. This battery capacity may be found inadequate during the system test phase, but you have to start somewhere.

This battery, as well as many batteries of this size, can also produce more than 200 cold cranking amperes (CCA). This is important as well because there will be times when the motor will need short-duration high current to pull the scooter up small slopes, negotiate bumps, or accelerate after suffering a speed loss due to soft surface conditions.

5.6.5 SOFTWARE DESIGN - DRIVE UNIT

Designing the software is really just planning out how the software will function. When the design of the software is complete, we have a list of tasks and their priorities and a set of flowcharts describing how the tasks are tied together.

In our scooter project, the drive unit and the display unit operate independently. The drive unit performs all the control, braking, and management functions, as well as sends a status stream out on the CAN bus. The display unit is simply a listener on the CAN bus and makes the appropriate information available to the rider based on the rider's input.

From a system perspective, the scooter drive remains inert, unless the user requests the drive system to contribute. It is up to the user to manually accelerate, request cruise control, or request braking. The user applies a small squeeze to the handbrake control to indicate to the drive system that the cruise control should be engaged. Squeezing the handbrake control more than 5% but less than 15% is an indication of "cruise request" or, in some cases, "coast." If the user pushes the scooter up to the desired speed, lightly presses and releases the brake control lever, the current speed of the scooter is the speed the drive will try to maintain. If the user goes down a hill to gain speed, the drive will try to hold speed or simply return to the previously memorized speed once level ground is reached. If during the descent on the hill the handbrake was lightly squeezed indicating "coast," then while the handbrake is held, the controller will perform no speed control. When the handbrake is released, the speed the scooter is traveling will be the new target cruising speed. This allows the user to get the scooter up to speeds that he or she may not be able to achieve by pushing the scooter. There are other interlocks, which would include but not be limited to the following: the foot switch, which must be enabled, indicating that the passenger is onboard, a minimum of 3 mph for the cruise control to take effect, and a maximum of 15 mph for a memorized speed.

Any handbrake applied beyond 15% indicates that the cruise function is to be disabled and braking is to be applied. The range of 15 to 100% of handbrake would correspond to 0 to 100% applied electronic braking.

The purpose of the cruise control is to maintain the scooter at a selected speed. Conditions outside the control system of the scooter can contribute to or detract from the speed of the scooter. This means that we will not always get the same speed with the same amount of current applied to the motor. If we are going uphill, we need more current. If we are going downhill, we need less current or maybe even some braking applied. We need an intelligent method of combining the input from the user with the information from our feedback system to control the speed of the scooter.

This intelligent control of the scooter speed comes from a simple Proportional Integral Derivative (PID) loop. The purpose of the PID loop is to compensate for outside influences on the system by using the requested output in conjunction with the error (or difference) between the actual output and the requested output to form a new input to the system. This input is also referred to as the *command* of the system, which in our project is the current going to the motor. In this system, the motor current is being controlled by the PWM drive applied to the motor. So *command* is analogous to PWM drive.

The PID loop actually has four components: proportional command (P), integral of error (I), derivative of error (D), and an offset. Because the actual dynamics of the system are somewhat of an unknown, the PID equation components will require scaling and "tuning" to make the scooter respond quickly but smoothly. This scaling is accomplished by multiplying each component by a constant to determine how each component contributes to the final *command* of the system. The equations are:

cmd = (Kp* P) + (Ki * I) + (Kd * D) + Ko

P = target_speed

I = I + (measured_speed – target_speed)

D = (measured_speed – target_speed)

In the equations, *Kp*, *Ki*, and *Kd* are the gain constants for the proportional, integral, and derivative components, respectively, and *Ko* is the constant offset of the system. *measured_speed* is the scooter's actual speed. *target_speed* is the memorized speed. *cmd* is the output of the system and the current being applied to the motor.

The proportional command is the amount of command, or, in this case, motor current, directly applied by the drive as a result of user input. This can be thought of as the direct input of the system. If you were turning a wrist throttle, the proportional output would be the portion directly applied to the drive based on the amount you turned the wrist control. In our system, the proportional command is based on the requested speed. The gain for this component is based on the amount of command required to reach the desired output in a perfect system.

The derivative of error is the instantaneous amount of error that exists between the actual drive output and the requested drive output. The amount of error is computed, indicating how far the drive is ahead or behind the user's requirement. This computed error is inversely applied to the command. If the drive is behind, more command is applied. If the drive is ahead and we are going faster than we want to, then some command is removed.

Because this component of the command is based on the instantaneous amount of error, a gain too high for this component can cause the output to oscillate around the target value, bouncing from too low to too high. A gain too low for this component may cause the output to move too slowly toward the target value.

The integral of error is the accumulation of the error over time. The error is the difference between the actual drive output (actual speed) and the requested drive output (requested speed). The integral portion adds the error back to itself each time the PID loop runs, then inversely applies itself to the command. If the actual system output is consecutively lower than the target output, then the integral portion continuously adds the negative error to itself, growing more negative. If the output is constantly higher than the target output, then the integral portion continuously adds positive error to itself, becoming a larger positive number. (Conversely, the derivative component would remain constant with constant error.) If the output is oscillating around the target, then the integral portion starts getting smaller and smaller as the negative and positive errors are added together.

The integral portion slowly makes up for gains and losses that the other two components are unable to compensate for; in other words, the integral portion is the contributor that supplies the accuracy to the final output of the drive. As a result, if the gain is too high for the integral component, then the system may never 'rest' as it continues to try to drive the output to the exact requested value. If the gain is too low for the integral component, then the system may not get as close to the target value as the user may want.

The PID loop runs on a fixed time interval, for example, every 100 ms. Even though during a scooter ride things may appear to be happening very quickly, compared to computer time these events are in slow motion. With testing we may find 100 ms to be too coarse, at which point we will increase the frequency the loop is executed to increase the resolution of the output to the drive. Another item to consider is the actual measurement of speed. The resolution of the measurements and input fed to the PID loop dictate the resolution available at the output. So to keep the resolution high, and the mathematics simple, all the calculations will be based on wheel encoder counts and a resulting percentage of output to the PWM drive, from 0 to 100%. Remember that the display and drive are independent. This allows us to work in whatever units we please at the drive, where it invisible to the user, and then convert those units to something meaningful at the display for the user's sake. These types of divisions can make system designs move quicker and easier through the prototyping stages, from concept to reality.

In addition to the primary functions of cruise control and braking, there are a couple of other functions that the drive software must perform. One is fault detection. There are several conditions that can be dangerous to the health of the drive system. These conditions include but are not limited to: low battery voltage, sustained high motor current, braking input malfunction, and, perhaps, excessive braking time. The second software function is communication. The drive unit should continuously share its status with other systems on the CAN bus. In this simple scooter system, there is only one other device on the bus: the display. Nonetheless, the drive controller will need to package data that is considered necessary or useful and place it out on the CAN bus. This information should include items such as the wheel speed, battery voltage, motor current, and drive status or error conditions.

So, based on the previous description, the drive unit software has four major tasks to perform:

1. Cruise control
2. Braking
3. Fault monitoring
4. Communications to the controller network area (CAN)

Figure 5.12 depicts the fundamental structure used for the Drive Unit software. The main loop monitors the hand control and set values that apply to cruise or braking. Timer interrupt service routines are used to update PWMs, compute the PID values used to control the motion, and periodically send messages on the CAN bus. In the actual code, faults are monitored in both the interrupt service routines and the main loop. A more detailed flow chart might indicate the types of faults monitored and where the monitoring occurs.

5.6.6 HARDWARE DESIGN - DISPLAY UNIT

Now we repeat the design process for the display unit. The display unit has a CAN receiver module and transceiver for gathering information from the drive unit. The display unit uses the microprocessor as the LCD controller, eliminating the need for an expensive intelligent display. The scooter-to-human interface consists of buttons, an LED, a beeper, and an LCD. The power for the display is provided by the drive unit, simplifying the electronics even more (Figure 5.13).

LCD Interface

The liquid crystal display (LCD) selected for the display unit is a very basic, off-the-shelf Varitronix VI302DP display. These displays can be commonly found in digital clocks, panel meters, bench-top instrumentation, and in many other consumer and commercial applications.

Many designers have been led to believe that an LCD requires special hardware in order to create the display because the signals seem complex. This is not true! The biggest difference between an LCD and an LED, for example, is that an LCD requires an AC signal to make a segment visible, whereas an LED is visible with the presence of a DC signal.

An LCD is structured in layers. There are segments and back planes. The VI302DP has a single back plane and many segments. Other varieties may have multiple back planes to support more segments, providing more information with fewer electrical connections. What causes a segment to be visible is a moving difference in potential between the segment and the back plane. This is usually referred to as the *phase* relationship. When the LCD is being driven, typically between 30 and 120 hertz, the segments that are out of phase with the back plane are visible, whereas the segments that are in phase with the back plane are not. Figure 5.14 gives an example of this phase relationship.

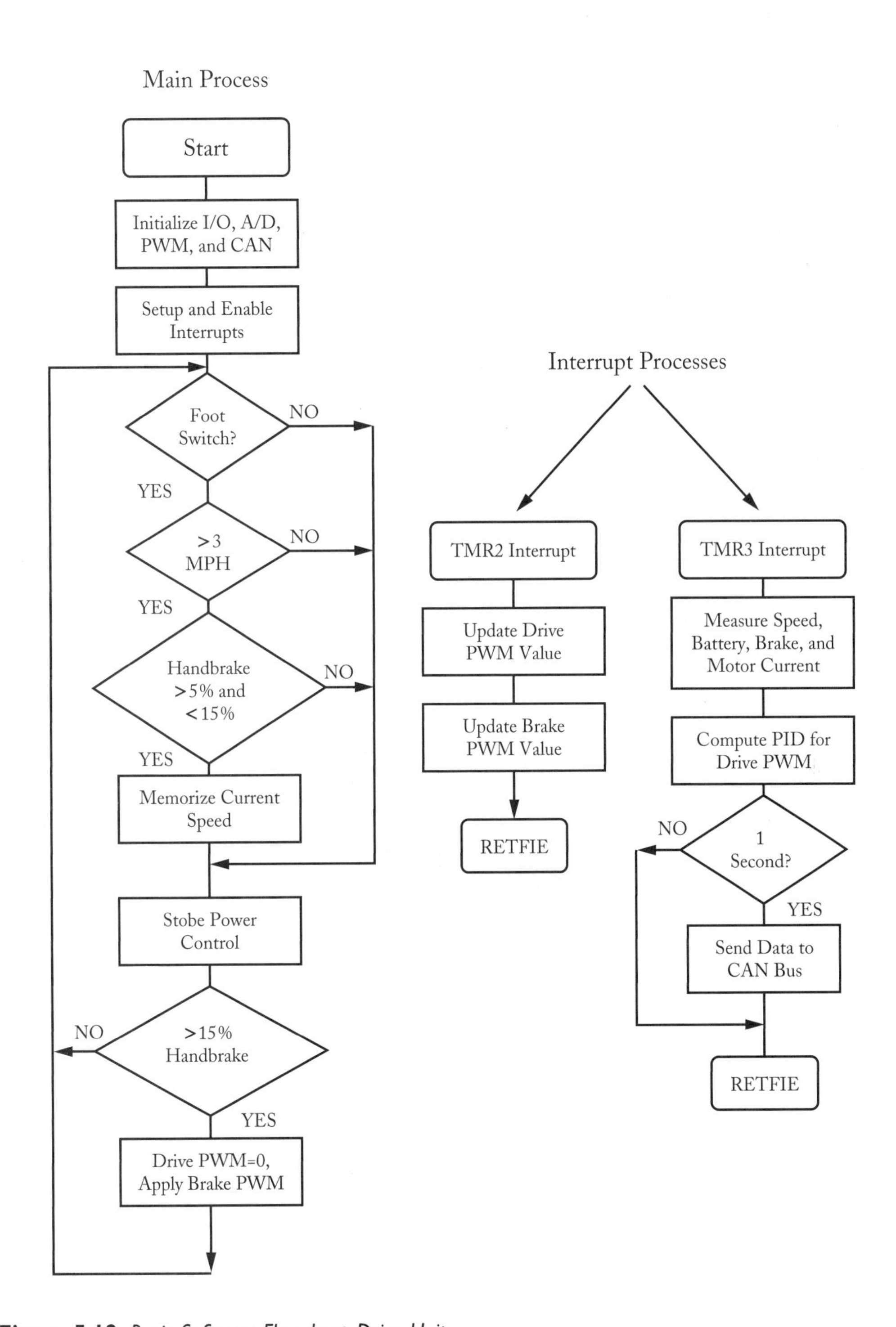

Figure 5.12 *Basic Software Flowchart, Drive Unit*

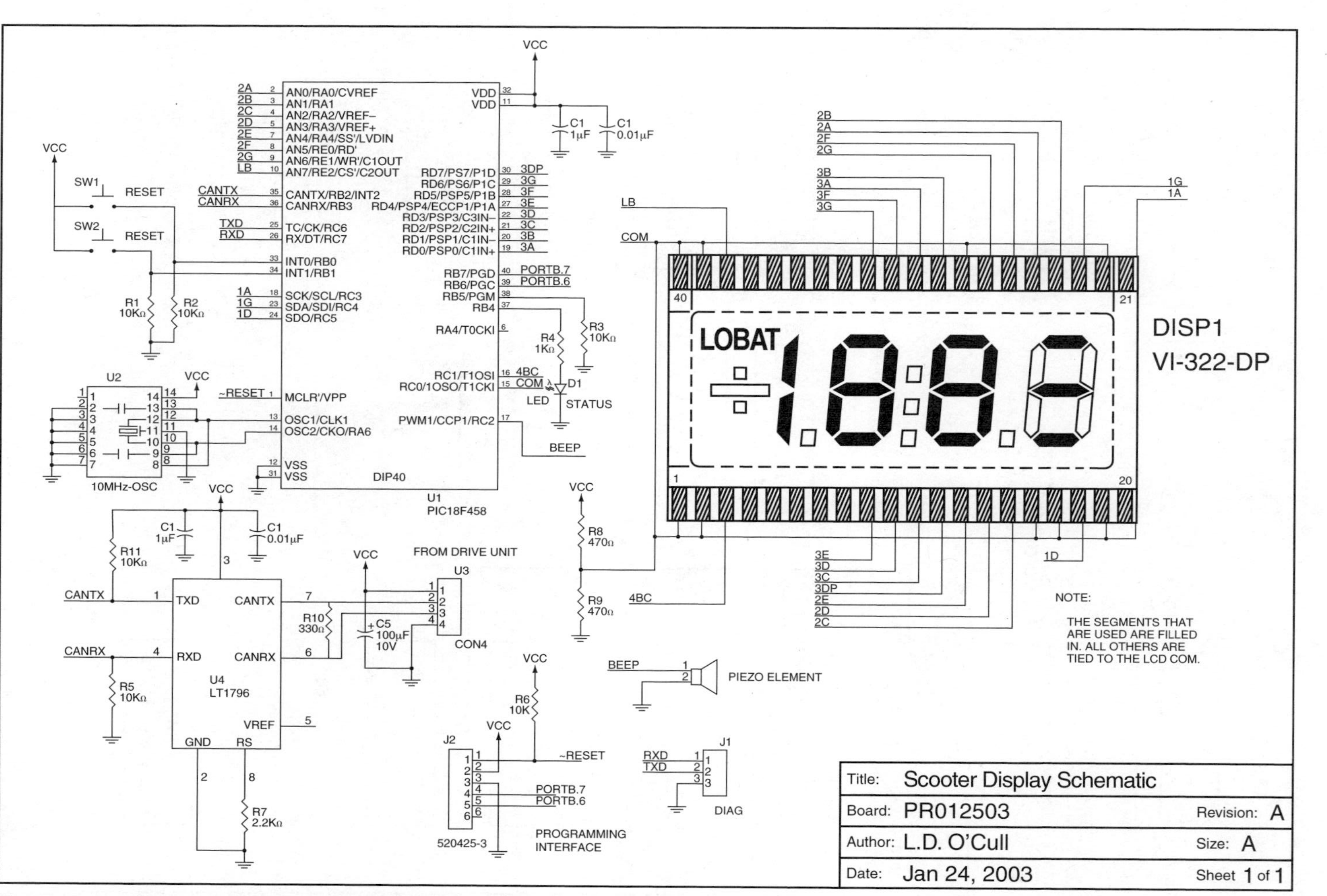

Figure 5.13 *Schematic, Electronic Scooter, Display Unit*

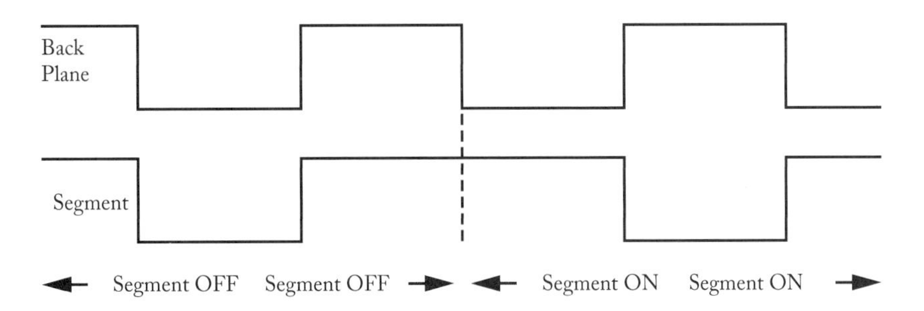

Figure 5.14 *LCD Segment-to-Back Plane Phase Relationship*

We are using a very basic display for the scooter display unit. It only has one back plane, and we do not even need all of the segments. This keeps the number of connections down and keeps the controlling software simple. Table 5.2 shows the pin/segment definitions for the VI302DP display. Note that pins 1 and 40 are called COM. These pins are both connected to the single back plane. The display is organized as a group of 3-1/2 seven-segment digits forming "1.9.9.9" with digit 4 on the far left and digit 1 on the far right. To better understand the organization of the segments within the digits refer to Figure 5.15.

Buttons, Lights, and Sound

The display unit's purpose is to provide information to the rider but should not be a point of distraction, so the more basic the display is, the better it is for the rider. At the same time,

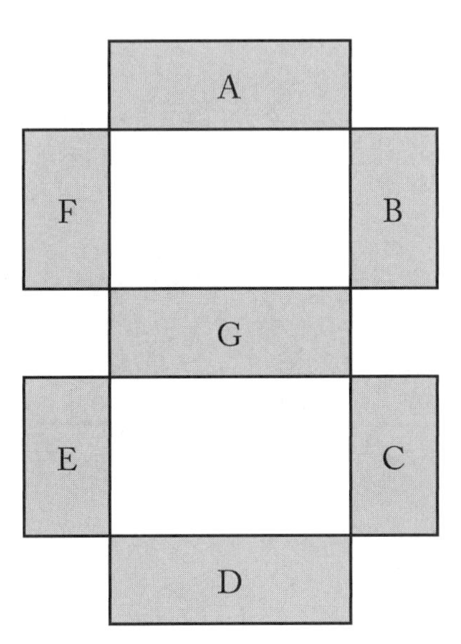

Figure 5.15 *Seven-Segment Display Organization*

Pin No.	Segment	Pin No.	Segment	Pin No.	Segment	Pin No.	Segment
1	COM	11	3C	21	1A	31	3F
2	-	12	3DP	22	1F	32	3G
3	4B, 4C	13	2E	23	1G	33	
4		14	2D	24	2B	34	
5		15	2C	25	2A	35	
6		16	2DP	26	2F	36	
7		17	1E	27	2G	37	
8	4DP	18	1D	28	Colon	38	LO BAT
9	3E	19	1C	29	3B	39	: (+)
10	3D	20	1B	30	3A	40	COM

Table 5.2 *Pin/Segment Definitions, VI302DP LCD*

we want the user to have the ability to get other useful information than just vehicle speed. The user might be interested in knowing about a possible fault condition, or he may just be more interested in how far he has gone than he is in how fast he is getting there.

The display is equipped with two buttons: one is for selecting the item to be displayed (mph, kph, volts, amperes, etc.), and the other is for resetting error codes, the trip odometer, or whatever other future features may need an extra input. A small beeper is used to indicate to the user a button press, but it can also be used to warn of a fault or low battery.

The beeper is a small piezoelectric diaphragm device that requires a 2 KHz to 4 KHz input in order to make sound. The microcontroller's PWM1 (Timer2) signal is used because it can be configured in software for an optimum frequency as well as volume by controlling the duty cycle.

An LED was also included as an additional visual. The LCD is not "eye-catching" in that it is black-on-gray in color. But the LED is bright red and, more importantly, different in visual texture than anything else on the display unit, making it stand out. The LED is intended as a status light. When the user enables the system by turning on the power and stepping on the foot switch, the LED should light to indicate a ready condition. In the event of a fault or warning, the LED can flash, helping to draw the user's eyes to the display to see what is going wrong.

5.6.7 SOFTWARE DESIGN - DISPLAY UNIT

The display unit has two basic functions: collecting and converting data and interface with a human being. The collection process is fairly simple in that the drive unit packages information on a regular basis and places it on the CAN bus. The CAN messages are in fixed lengths of up to 8 bytes per packet. Because all the work of setting up and tearing down these packets is performed by the hardware, we will simply need to break the data out and convert the information into displayable units.

There are several parameters that we can display and a couple of choices as to the units those parameters could be displayed in. So, from our Preliminary Product Specification we can concoct a list of what the rider may wish to see (Table 5.3).

Speed in mph
Speed in kph
Trip Odometer in Miles
Trip Odometer in Kilometers
Battery Voltage
Motor Current
Future Feature

Table 5.3 *Display Parameter List*

Because the scooter display is inexpensive and unintelligent, some thought must be put into how a user will be able to quickly interpret what it is that he or she is looking at. This is a prototype device, so we utilized some segments down the right-hand side of the display as unit markers. In a production situation, we would have actual units made into the display like mph and kph. For this prototype we will use the markers to point to some labeling on the display enclosure to help the rider understand what it is we are displaying (Figure 5.16 and Figure 5.17).

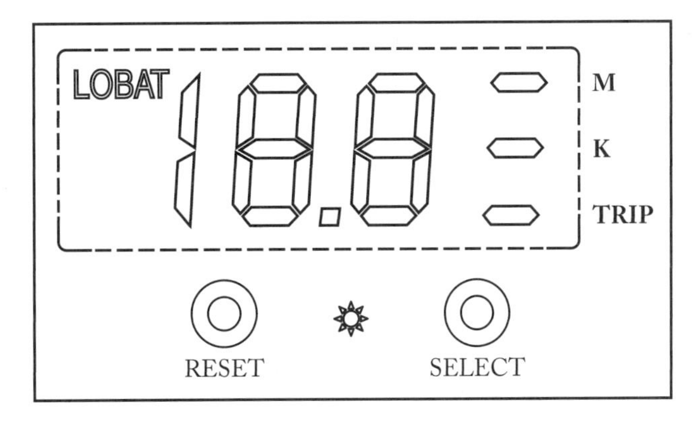

Figure 5.16 *Display Unit LCD and Button Layout*

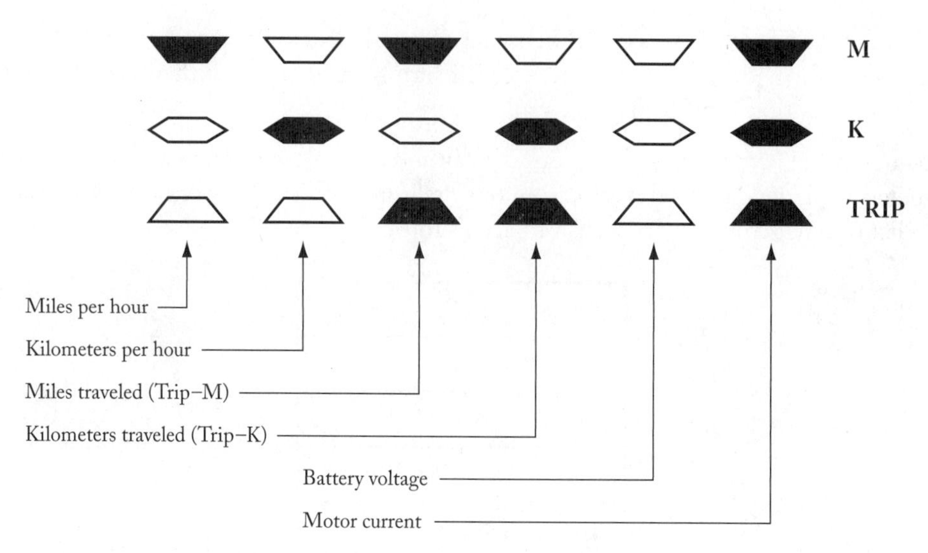

Figure 5.17 *Display Information Unit Key*

Overall operation of the display unit is simple. Pressing the SELECT button "rolls" through the different parameters in order, with each item's units being notated by the key shown in Figure 5.17. Each time the button is pressed, the beeper sounds to give the user confirmation on the button press. In the case of the "Trip" odometers, the RESET button allows the rider to reset the trip accumulation back to zero.

The low battery indicator on the LCD will be visible whenever the scooter is enabled and the battery voltage is below 10 volts. The STATUS LED will also flash whenever any type of error occurs, including low battery. If an error occurs, the error indication will also be displayed on the LCD. Pressing the RESET button could clear the condition until the condition presents itself again; however, in this system, an error of this nature will require the drive unit to have its power reset, thus resetting the display. Because the display is made of seven-segment digits, an error can be displayed as the letter "E," followed by the error number. Some possible error messages are shown in Table 5.4.

E1	Motor Overcurrent
E2	Max Speed Exceeded
E3	Brake Input Fault
E4	Brake Duration Exceeded
E5	Motor Enable Fault

Table 5.4 *Example Scooter Error Messages*

5.6.8 TEST DEFINITION PHASE

The test definition phase of the project outlines specifically how we intend to test the individual pieces of the project and the complete project. Some measurements are very easy to test, whereas others require much more work and creativity. Some may also require writing software for the microcontroller.

The test definitions for various parts of the project are shown next.

Braking Input

Input – 0 to 5 V DC via a 10 K ohm potentiometer

Expected Results – 0 to 5 V input to microcontroller

Method – Apply different voltages to the input through a potentiometer to simulate a handbrake input, and measure the voltage input to the microcontroller with a voltmeter.

Vehicle Speed

Input – 15 mph (assuming a 4:1 gear ratio from wheel to motor)

Expected Results – 336 Hz signal to the microprocessor

Method – Use a variable power supply to spin the motor at a known rate, and measure the pulse train to the microcontroller using a frequency counter, an oscilloscope, or the microcontroller itself. You may need to write software to count the pulses and transmit them to the display unit for display, if a portable method of counting the pulses is required.

Battery Health

Input – 8 to 15 V DC via a variable power supply

Expected Results – 2.5 to 4.8 V input to microcontroller

Method – Apply different voltages to the drive power input using a variable power supply to simulate a battery, and measure the voltage input to the microcontroller with a voltmeter.

Motor Current

Input – 0 to 1 V DC via a 10 K ohm potentiometer

Expected Results – 0 to 5 V input to microcontroller

Method – Apply different voltages to the I_SENSE input using a potentiometer to simulate the sense resistors, and measure the voltage input to the microcontroller with a voltmeter. NOTE: The actual sense resistors will need to be out of circuit in order to perform this test.

System Test for the Complete Project

The extent of the system test definition will vary according to the end user of the project. If the electronic scooter project were to be destined for hobby use only, the test might be no more extensive than comparing the scooter to one that your friend uses to see if the results are "about right." On the other hand, if the project is for commercial usage or for a customer, more extensive testing is required. In this case, it would be appropriate to use temperature- and humidity-controlled chambers to run the drive and display units through the entire range of operating conditions and record the results. Whatever the extent of the tests imposed for the system, the intent is the same: to give the end user confidence that the unit is performing up to specifications as set down during the definition phase.

5.6.9 BUILD AND TEST PROTOTYPE HARDWARE PHASE

Now that the project hardware is completely designed and the expected results of the sections of the hardware are specified, it is time to build the hardware. After assembling the hardware, the tests from the previous phase are executed. This section briefly discusses some real-life methods of executing the tests and measuring the results. The software required to execute the tests is also discussed.

Fundamental tests of the hardware are best performed with voltmeters and oscilloscopes. When a PIC microcontroller is completely erased, all of its pins are inputs and it has no impact on the surrounding circuitry. This allows for some basic checking of the hardware functionality. No matter how good the software is it can never make up for hardware that just does not work. Following are some step-by-step checks for both the drive unit and display unit hardware. This example method of validation applies to many types of designs. Knowing for certain that the hardware performs as expected greatly reduces the software development and integration time, as well as the amount of time you will spend "chasing your tail" when things do not seem to make sense.

Drive Unit Checkout

1. Set the lab supply for 12 V DC with a current limit of approximately 1 ampere. At this point, no battery should be connected.
2. Make sure that VCC is at 5 V DC. If the power supply is not correct, nothing will be quite right.
3. Using a voltmeter, verify that the BATT signal (pin 2 of the PIC18F458) is approximately 3.8 V DC.
4. Connect the wheel sensor to the appropriate inputs.
5. Using an oscilloscope, verify that the SPEED signal (pin 15 of the PIC18F458) has pulses of adequate level (TTL level, which is greater than 3.3 V for a "high" and less than 1.5 V for a "low"). These pulses will only exist while spinning the wheel, so give it a whirl!
6. Using the test program listed next, program the PIC18F458 with the following fuse bit settings. These settings guarantee the proper operation of the oscillator and brownout detector used to reset the microcontroller:

```
/*

   Drive Unit Test Software.

*/
#include <18F458.h>
#use delay(clock=10000000,RESTART_WDT)
#fuses HS,PUT,BROWNOUT,NOWDT,WDT128
#use fast_io(D)

/* definitions for expanded comparator module */
#byte ECCP1CON = 0xFBA
#byte ECCPR1L  = 0xFBB
#byte ECCPR1H  = 0xFBC
#byte TMR1L    = 0xFCE
#byte TMR1H    = 0xFCF

int temp;

#int_TIMER3
TIMER3_isr()          // 26ms tick at 10MHz
{
   output_bit(PIN_B2, temp++ & 1);
                  // toggle CANTX pin to check driver
   output_bit(PIN_C4, temp & 1);
                  // toggle WORKING LED
}

#int_TIMER2
TIMER2_isr()
{
   output_d(TMR1L & 0x0F);
                  // put counts out on port D lower bits
}

void set_pwm1E_duty(int16 duty)
{
   int t;

   ECCPR1L = (duty >> 2) & 0xFF;    // top 8 bits
   t = duty << 4;                   // bottom 2 bits
   t &= 0x30;
   ECCP1CON = (ECCP1CON & ~0x30) | t;
}
```

```
void main()
{
  int16 x;

  setup_adc_ports(A_ANALOG);        // setup analog inputs
  setup_adc(ADC_CLOCK_INTERNAL);
  setup_psp(PSP_DISABLED);
  setup_spi(FALSE);
  setup_wdt(WDT_ON);

  setup_timer_1(T1_EXTERNAL|T1_DIV_BY_1);
                // setup timer 1 as counter
  setup_timer_3(T3_INTERNAL|T3_DIV_BY_1);
                // setup timer 3 as real-time interval

  setup_timer_2(T2_DIV_BY_4,255,1);
                // setup timer 2 to source PWMs

  setup_ccp1(CCP_PWM);
                // setup CCP1 as Drive PWM
  setup_comparator(FALSE);

  set_tris_d(~0x1F); // setup ECCP1 as Brake PWM
  ECCP1CON = 0x0C;   // setup ECCP1 as single, simple PWM

                // enable timer and external interrupts
  enable_interrupts(INT_TIMER2);
  enable_interrupts(INT_TIMER3);
  enable_interrupts(global);


  while(1)
  {
    for(x=0; x < 1000; x++)     // ramp up drive..
    {
       delay_ms(5);
       set_pwm1_duty(x); // put x% duty on Drive Signal
    }
    set_pwm1_duty(0);     // put 0% duty on Drive Signal
    delay_ms(100);

    for(x=0; x < 1000; x++)     // ramp up brake..
    {
       delay_ms(5);
       set_pwm1E_duty(x);
```

```
                        // put x% duty on the Brake Signal
        }
        set_pwm1E_duty(0);
                        // put 0% duty on the Brake Signal
        delay_ms(100);

        for(x=0; x < 1000; x++)     // enable power relay..
        {
            output_bit(PIN_C5,1);
            delay_ms(2);
            output_bit(PIN_C5,0);
            delay_ms(2);
        }
        delay_ms(250);   // let the power relay time out
    }
}
```

7. With no motor connected, and using an oscilloscope, verify that the drive and brake MOSFET gates are alternately receiving PWM input from the processor. The above-listed program ramps up the drive and then the brake.
8. Spin the wheel encoder, and note the counts from Timer1 appearing on the lower 4 bits of Port D. This can be done with an oscilloscope or with some LEDs.
9. Press the foot switch. The power relay should come on as Port C bit 5 (pin 24 of the PIC18F458) is toggled and should go off when the signal goes steady. The relay should only function when the foot switch is pressed.
10. Using an oscilloscope, verify that pin 7 of the LT1796 CAN transceiver has a square wave, generated from pin 35 of the PIC18F458.
11. Verify that the "Running" LED is toggling.

Display Unit Checkout

1. Set the lab supply for 12 V DC with a current limit of approximately 1 ampere. At this point, no batteries should be connected.
2. Make sure that VCC is at 5 V DC. If the power supply is not correct, nothing will be quite right. (Does this sound familiar? Well, it is still true.)
3. With the drive unit running its test program as a CAN signal source, verify that data is getting to the processor by looking at the CANRX input (pin 36) of the PIC18F458 with an oscilloscope. (The waveform should look something like the CANTX output [pin 35] of the drive unit.)
4. Using the test program listed next, program the PIC18F458 with the following fuse bit settings. These settings guarantee the proper operation of the oscillator and brownout detector used to reset the microcontroller:

```
/*

    Display Unit Test Software.
*/

#include <18F458.h>
#use delay(clock=10000000,RESTART_WDT)
#fuses HS,PUT,BROWNOUT,NOWDT,WDT128

int temp;

#int_TIMER3
TIMER3_isr()        // 26ms tick at 10MHz
{
   output_bit(PIN_B2, temp++ & 1);
                    // toggle CANTX pin to check driver

   if(temp & 1)
   {
   output_d(0xFF); // "light" some of the LCD segments
   output_a(0xFF);
   output_bit(PIN_C3,1);
   output_bit(PIN_C4,1);
   output_bit(PIN_C5,1);
   output_bit(PIN_C0,0);    // COM pin opposite..
   }
   else
   {
   output_d(0x00);
   output_a(0x00);
   output_bit(PIN_C3,0);
   output_bit(PIN_C4,0);
   output_bit(PIN_C5,0);
   output_bit(PIN_C0,1);    // COM pin opposite..
   }
}

#int_TIMER2
TIMER2_isr()
{
}

void main()
{
   int16 x;
```

```
    setup_wdt(WDT_ON);

    setup_timer_3(T3_INTERNAL|T3_DIV_BY_1);
            // setup timer 3 as real-time interval
    setup_timer_2(T2_DIV_BY_4,255,1);
            // setup timer 2 to source PWMs

    setup_ccp1(CCP_PWM);   // setup CCP1 as Beeper PWM
    setup_comparator(FALSE);

    enable_interrupts(INT_TIMER2);
            // enable timer and external interrupts
    enable_interrupts(INT_TIMER3);
    enable_interrupts(global);


    while(1)
    {
      delay_ms(10);

      if(input(PIN_B0))
         set_pwm1_duty(512);
                   // set beeper to 50% at ~2500Hz
      else
         set_pwm1_duty(0);          // set beeper to off

         if(input(PIN_B1))
            output_bit(PIN_B4,1);    // status light on
         else
            output_bit(PIN_B4,0);    // status light off
    }
}
```

5. Verify that the "wired" segments on the LCD become visible ("18.8").
6. Press the SELECT button. The beeper should sound at approximately 2500 Hz as long as the button is pressed.
7. Press the RESET button. The STATUS LED should light as long as the button is pressed.
8. Using an oscilloscope, verify that pin 7 of the LT1796 CAN transceiver has a square wave, generated from pin 35 of the PIC18F458.

These tests should create confidence in the hardware and allow you to proceed on to integration and software development.

5.6.10 SYSTEM INTEGRATION AND SOFTWARE DEVELOPMENT PHASE, DRIVE UNIT

The first step in putting together the software for the drive unit is to define the unit operation. In this case, a system decision can be made as to where the conversion of the displayed data is to occur, at the display unit or the drive unit. If the data were processed at the drive unit, then each of the parameters would have to be calibrated, and common units, like mph or kph for speed, would need to be defined for the data that is to be transmitted. This thought process leads us to leaving the data collected in the most basic form possible and letting the display unit do the conversions at the end. Besides, the display unit's sole purpose is to interface with the human being, so it has plenty of time to do the conversions (human beings are, after all, much slower than microcontrollers).

So, in this system the most basic units would be ADC counts, Timer1 counts, and the state or operating condition of the drive unit. This data can be combined and transmitted as a packet of values on a periodic basis. The transmitted period of the packets is not really critical, but specifying a period and keeping it consistent is important, because mph and kph are measured as counts over time. The parameters are not particularly fast moving, but human beings like to be updated fairly frequently. For instance, a typical digital voltmeter offers at least a 1-second sampling rate. The only critical sampling that is taking place is the vehicle speed. Even though it will not be sampled quickly or with much resolution, an average of the readings is all that is necessary for an adequate speed representation.

The next step in getting the software started is to define the inputs and outputs of the microprocessor that will be used. This is a great way to get started in that another round of sanity checking gets performed on the schematic, and once the definitions are made, the need for constantly referring back to the schematic is reduced. The definitions for the drive unit may appear as those shown in Figure 5.18. A reminder: comments following the #define statements need '/*' and '*/' as delimiters.

Once the pins have been labeled, the actual inputs, outputs, and peripherals need to be configured. This is usually performed at or called from the top of *main()*. In Figure 5.19, note that Timer3 is set to create an overflow interrupt. Based on a 10.0 MHz oscillator, the inter-

```
#define WORKING_LED  PIN_C4
#define POWER_RELAY  PIN_C5
#define OVERCURRENT  PIN_B0

enum { BATTERY,BRAKE,MOTOR_I,AUX_INPUT }; // analog data channels
#define LAST_CHANNEL AUX_INPUT
int a2d_chan;
int16 analog_data[LAST_CHANNEL+1];        // analog data queue
```

Figure 5.18 *Example Definitions, Drive Unit I/O*

rupt will be issued every 26 milliseconds. This is a base time interval that will be subdivided in the software to provide 100 milliseconds and 1-second intervals. This real-time tick will be used to sample the wheel speed sensor as well as to determine when it is time to transmit the data out on the CAN bus to be picked up by the display unit. Timer1 is configured as a counter to capture the actual pulses produced by the wheel speed sensor. Timer2 is configured as the PWM timing source for CCP1, driving the motor output MOSFET, as well as ECCP1, driving the regenerative braking MOSFET. The Timer2 interrupt is used to allow the PWM values to be updated at the rollover point of the counter, eliminating any glitches from the PWM output. The external interrupt, INT_EXT, is enabled to allow detected motor overcurrent or stalled conditions to be immediately captured. The UART is configured for 19,200 baud, 8 data bits, and 1 stop bit for any debugging messages or information we want to send to the "Expansion" connector. The ADC is set up to convert at the slowest rate, using the PICs internal A/D RC oscillator, which generally provides the most noise-free data on completion of each conversion. The CAN interface is also initialized by the CCS library function call *can_init()*, described later in this section (Figure 5.19).

Vehicle Speed

Timer1 tracks the vehicle speed. Each time a transition is generated on the optical encoder by the photo interrupter disk, coupled to the wheel, a count is generated in Timer1. There are two pieces of information that can be gathered from this single source. One is the speed of the scooter, based on the number of counts during an interval. Another is the distance traveled, which is the total number of counts collected. In Figure 5.20, note that the timer is read two or three times to guarantee that good data is being sampled. This is done because there is a possibility that during the few microseconds that it takes to read both halves of the timer, the wheel sensor may issue another pulse to the timer, generating a new value. To prevent a "half-baked" answer from the timer, the timer is read twice to see if the same reading occurred. If so, then the timer was not altered mid-sample. Otherwise, if the timer was altered between or during the first and second readings, then it is nearly a certainty that a change will not occur during the third reading, making it valid.

The variable *delta_counts* is used to get the number of counts in the last 100 ms interval. This value is used to compute the amount of drive required by the Proportional Integral Derivative (PID) control loop to keep the scooter moving at a smooth, consistent speed. The *out_data.total_pulses* variable is part of the CAN communication packet structure and represents the amount of pulses collected over a 1-second interval. The display unit uses this to compute the scooter's speed as well as track the total distance traveled.

When the scooter is under cruise control, the *delta_counts* are also the speed of the scooter and are used by the PID equations to compute the amount of required drive at the PWM. The PID parameters are computed each 100 ms, and the resulting duty cycle is updated in the variable *Drive_Command. Drive_Command* updates the physical PWM hardware in the Timer2 overflow interrupt routine.

```
#include "18F458.h"
#use delay(clock=10000000,RESTART_WDT)
#fuses HS,PUT,BROWNOUT,WDT128
#use fast_io(D)
#use rs232(baud=19200,parity=N,xmit=PIN_C6,rcv=PIN_C7)
#zero_ram

      . . .

main()
{

      . . .

   setup_adc_ports(A_ANALOG);         // setup analog inputs
   setup_adc(ADC_CLOCK_INTERNAL);
   setup_wdt(WDT_ON);                 // enable Watchdog Timer

   setup_timer_1(T1_EXTERNAL|T1_DIV_BY_1);
                                      // setup timer 1 as counter
   setup_timer_3(T3_INTERNAL|T3_DIV_BY_1);
                                  // setup timer 3 as real-time interval

   setup_timer_2(T2_DIV_BY_4,255,1); // setup timer 2 to source PWMs

   setup_ccp1(CCP_PWM);               // setup CCP1 as Drive PWM

   set_tris_d(~0x1F);                 // setup ECCP1 as Brake PWM
   ECCP1CON = 0x0C;                   // setup ECCP1 as single,
                                      //simple PWM

   can_init();                        // initialize the CAN interface

   enable_interrupts(INT_TIMER2);     // enable timer and external
   enable_interrupts(INT_TIMER3);    //interrupts
   enable_interrupts(INT_EXT);

   enable_interrupts(global);

      . . .

}
```

Figure 5.19 *Drive Unit, I/O Initialization*

```
#int_TIMER3
TIMER3_isr()            // 26ms tick at 10MHz
{
      . . .

   pass++;
   if(pass > 3)         // ~100ms
   {
      pass = 0;
      wheel_counts = TMR1H << 8; // read the timer..
      wheel_counts+= TMR1L;
      wheel_test = TMR1H << 8;    // read it again.. are they the same?
      wheel_test += TMR1L;
      if(wheel_test != wheel_counts)
      {                    // caught a transition, so read again..
         wheel_counts = TMR1H << 8;
         wheel_counts += TMR1L;
      }
      delta_counts = (int16)(wheel_counts - last_counts);
                           // this delta is the speed..

      out_data.total_pulses += delta_counts;

      last_counts = wheel_counts;
   }
      . . .

}
```

Figure 5.20 *Wheel Sensor Readings Using Timer1*

As each parameter of the PID equation (Figure 5-21) is calculated, the values are scaled to the number of counts available to the wheel sensor as well as the total available PWM duty. In this case, 100% of the PWM is equivalent to 1023 counts. Using the #defines *MAX_PWM* and *MAX_PULSES*, the equations can be easily rescaled if a change in the PWM or wheel sensor ranges should be needed at a later time. Each of the PID parameter contributions is also scaled by the defined values *P_GAIN*, *I_GAIN*, and *D_GAIN*. This allows the cruise control's responsiveness to be tuned by simply adjusting a few constants. Once everything has been computed, and before the PWM is updated, limits are checked to make certain that no bogus values make it to the PWM. In a speed control system like this one, bad values can lead to erratic operation as well as damage to the drive output MOSFETs. So, exercising a little caution, and doing some range checking, is a good thing.

```
signed int32 P_val,I_val,D_val,T_val;    // P I D amounts
#define P_GAIN 100       /* percentages that each parameter's */
#define I_GAIN 50        /* value that can contribute to the result */
#define D_GAIN 20        /* A measure of "aggressiveness" */

unsigned int16 wheel_counts, wheel_test, last_counts;
signed int16 delta_counts, target_counts;

int16 Drive_Command,Brake_Command;  // PWM values

#define MIN_PULSES 6     /* 3 MPH minimum for cruise engagement */
#define MAX_PULSES 34    /* maximum pulses per sample at 15MPH @ 4:1 */
#define MAX_PWM  1023    /* Maximum Drive PWM (100%) */
#define MAXB_PWM 1000    /* Maximum Braking PWM (97%) */
#define MIN_PWM   0      /* Minimum PWM.. applies to bo

#int_TIMER3
TIMER3_isr()          // 26ms tick at 10MHz
{
      . . .

   if(flags.cruising)      // this part executed at 100ms intervals
   {
         // compute the Drive Command direct proportion
      P_val = (((target_counts * P_GAIN) * MAX_PWM) / MAX_PULSES) / 100;

         // compute the additional Drive Command as a result of Diff.
         //error
      D_val = ((((target_counts-delta_counts)*D_GAIN)*MAX_PWM)
      /MAX_PULSES)/100;

          // formulate an integration of error over time, and
          // apply it to the Command
      if((target_counts-delta_counts) > 0)
         if(I_val < MAX_PULSES) I_val++;

      if((target_counts-delta_counts) < 0)
         if(I_val > (-MAX_PULSES)) I_val--;

      T_val = (((I_val * I_GAIN) * MAX_PWM) / MAX_PULSES) / 100;
      T_val += P_val + D_val;

      if(T_val < MIN_PWM)       // limit the output to prevent PWM
                                // rollover
```

Figure 5.21 *The PWM and PID Control Loop (continues)*

```
            T_val = MIN_PWM;
        if(T_val > MAX_PWM)     // limit the output to prevent PWM rollover
            T_val = MAX_PWM;

        Drive_Command =  (int16)T_val;    // apply command to DRIVE PWM
    }
    else
    {                           // if not cruising..
        I_val = 0;              // reset integrator
        Drive_Command = 0;      // no command
    }

        . . .

}

#int_TIMER2
TIMER2_isr()
{                                       // update PWMs during rollover for
                                        // glitch-free operation
    if(!Brake_Command)
            // don't allow both Brake and Drive at the same time!!
        set_pwm1_duty(Drive_Command);     // put x% duty on Drive Signal
    else
        set_pwm1_duty(0);                 // put 0% duty on Drive Signal

    if(!Drive_Command)
            // don't allow both Brake and Drive at the same time!!
        set_pwm1E_duty(Brake_Command);    // put x% duty on Brake Signal
    else
        set_pwm1E_duty(0);                // put 0% duty on Brake Signal
}
```

Figure 5.21 *The PWM and PID Control Loop (continued)*

Battery Health and Motor Current Monitoring

The A/D converter monitors the braking input, battery voltage, and motor current. The PIC18F458 has 10-bit A/Ds, which provide more than adequate resolution for these parameters. The battery voltage and motor current measurements are passed to the CAN bus using the *out_data* structure. The drive unit itself uses the battery voltage to determine if the foot switch is made. The *BATT_9V* definition is computed using the resistor divider values and the full-scale voltage (5 volts) and maximum counts (1023) of the A/D. The *BATT_9V* reference is also used to determine the lowest operating voltage for the drive. The display unit, utilizing its own reference calculations, handles the low voltage indication to the user.

In the case of the motor current, the important issue for the drive to deal with is a stalled motor or an overcurrent situation. A separate path is used to generate an interrupt in the case of a stalled motor. This allows the drive to respond to the fault condition quickly, greatly reducing the possibility of a catastrophic failure. If it were up to the software and the A/D converter to figure out that the motor stalled or was shorted, many milliseconds could go by with more than 100 amperes flowing! Using a more hardware-based interrupt path, this time is reduced to microseconds, allowing for a nearly immediate response (Figure 5.22).

Cruise and Brake Control

The handbrake controller input is responsible for much of the scooter's functionality and personality from the rider's point of view. There are several "points of interest" along the entire travel of the handbrake control as the operator is squeezing it. If there is no input *(BRAKE_MIN)*, the scooter is either idle or cruising. If the input is between *CRUISE_SET* and *BRAKE_APPLIED*, and if the scooter is in motion to at least 3 mph, then a cruise control set point is established. If the control is squeezed beyond *BRAKE_APPLIED* and up to *BRAKE_MAX*, then the applied amount of input is provided in full scale to the electronic braking PWM, such that *BRAKE_APPLIED* provides 0% of the braking PWM and *BRAKE_MAX* provides 97% of the braking PWM (Figure 5.23).

As was mentioned in Section 5.6.4, "Hardware Design – Drive Unit," the braking PWM is coupled using a gate-driving transformer, which means that a PWM of more than 97% will not provide a usable gate signal to the MOSFET. Because the handbrake controller only provides a percentage of the total travel available to the potentiometer, we get the added benefit of a brake fault signal. If the brake input should fall below *BRAKE_MIN* or rise above *BRAKE_MAX*, we know that something has happened such as a broken brake control cable, a loose return spring, a broken wire, or a short in the potentiometer or its wiring. As shown in Figure 5.23, a fault is generated if any of these conditions should be detected at the braking input.

CAN Communications

The CAN communication link used in the scooter is a popular form of connecting controllers in noisy environments, such as automobiles, heavy equipment, motorcycles, factory automation, and commercial lighting systems, just to name a few. Some of the benefits of CAN that were discussed previously had to do with its electronic resilience. Another, perhaps even more important feature is its diligence in delivering a message. The CAN controller has a fairly complex set of registers that control features of the communications such as bit rate, where and how the data is sampled, the level of filtering and identification required of a message, and the number of times the message can fail before the delivery attempts should cease.

CCS PICC provides a basic library set that makes setting up this complex controller much easier. The basic setup and handlers are in a file, "can-18xxx8.c," and the header file, "can-

```
enum { BATTERY,BRAKE,MOTOR_I,AUX_INPUT }; // analog data channels
#define LAST_CHANNEL AUX_INPUT
int a2d_chan;
int16 analog_data[LAST_CHANNEL+1];          // analog data queue

/* A/D counts when battery is at 9Volts */
#define BATT_9V       (int16)((((4.7/14.7)*9)/5)*1023)


#int_EXT
EXT_isr()
{
      set_pwm1_duty (0);  // overcurrent on motor driver.. 0 all PWMs..
      set_pwm1E_duty(0);

      flags.system_fault = 1;       // set fault flag
      flags.cruising = 0;
      flags.error = Motor_Over_I;   // a stalled motor was detected
}

void get_analog_data(void)
{
   analog_data[a2d_chan] = read_adc();  // read last value
   a2d_chan++;
   if(a2d_chan > LAST_CHANNEL)
      a2d_chan=0;                       // select the next channel
   set_adc_channel(a2d_chan);           // circularly read the analog
                                        // channels
}

main()
{
      . . .

      get_analog_data();          // circularly service the A/D channels

      . . .

      if(analog_data[BATTERY] > BATT_9V)
        flags.footswitch = 1;    // if valid battery voltage is present
                                 // then
      else                        // the footswitch is engaged
        flags.footswitch = 0;
```

Figure 5.22 *A/D Service and Stall Detection (continues)*

```
    . . .

}
```

Figure 5.22 *A/D Service and Stall Detection (continued)*

```
/* brake voltage min 1.75V */
#define BRAKE_MIN      (int16)((1.75/5.0)*1023)
/* 3.25V is the most the brake should be */
#define BRAKE_MAX      (int16)((3.25/5.0)*1023)

/* brake is applied at 5% above min */
#define CRUISE_SET     (int16)(((BRAKE_MAX-BRAKE_MIN)*0.05)+BRAKE_MIN)
/* brake is applied at 15% above min */
#define BRAKE_APPLIED (int16)(((BRAKE_MAX-BRAKE_MIN)*0.15)+BRAKE_MIN)

main()
{

      . . .

      // Make desicion about the mode.. coast, cruise, brake...

      if((analog_data[BRAKE] > CRUISE_SET) &&
                            (analog_data[BRAKE] < BRAKE_APPLIED))
      {
         flags.cruising = 0;            // unlock the cruise..
         target_counts = delta_counts; // establish setpoint
      }
      else if((analog_data[BRAKE] > BRAKE_APPLIED) &&
                                  (analog_data[BRAKE] < BRAKE_MAX))
      {
         flags.cruising = 0;
         target_counts = 0;             // reset the cruise setpoint
         Drive_Command = 0;

         // apply the remaining brake range to then entire PWM range
         T_val = ((analog_data[BRAKE] - BRAKE_APPLIED)*
                                  (MAXB_PWM/(BRAKE_MAX-BRAKE_APPLIED)));

         if(T_val < MIN_PWM)       // limit the output to prevent PWM
                                   // rollover
```

Figure 5.23 *Handbrake Control Input Interpretation (continues)*

```
            T_val = MIN_PWM;
        if(T_val > MAXB_PWM)      // limit the output to prevent PWM
                                  //rollover
            T_val = MAXB_PWM;
        disable_interrupts(global);    // ensure that value is not half
                                       // updated
        Brake_Command = (int16)T_val;
        enable_interrupts(global);
    }
    else if((target_counts > MIN_PULSES) &&
                (flags.footswitch) && (!flags.system_fault))
    {
        Brake_Command = 0;
        flags.cruising = 1;   // if not breaking and just memorized a
                              // speed..
    }                         // then cruising!
    else
        Brake_Command = 0;


    if((analog_data[BRAKE] < BRAKE_MIN) || (analog_data[BRAKE] >
    BRAKE_MAX))
    {
        flags.system_fault = 1;        // set fault flag
        flags.cruising = 0;
        flags.error = Brake_Fault;     // braking input error detected..
    }

    . . .

}
```

Figure 5.23 *Handbrake Control Input Interpretation (continued)*

18xxx8.h," contains some of the device specific information like special function register and bit locations as well as CAN bus sampling, rate, filtering, and identification parameters.

Using the library functions shown in Figure 5.24 and Figure 5.25, a nearly transparent data link is established between the drive and display units. The CAN controller can send and receive up to 8 bytes of data in a packet. Our system is fairly simple, and everything we need to send out from the drive can easily fit within the 8-byte limit (Figure 5.26).

```
/////////////////////////////////////////////////////////////////////////
////                          can-18xxx8.c                           ////
//// CAN Library routines for Microchip's PIC18Cxx8 and 18Fxx8 line  ////
////                                                                 ////
//// This library provides the following functions:                  ////
////  (for more information on these functions see the comment       ////
////   header above each function)                                   ////
////                                                                 ////
////    can_init - Configures the PIC18xxx8 CAN peripheral           ////
////                                                                 ////
////    can_set_baud - Sets the baud rate control registers          ////
////                                                                 ////
////    can_set_mode - Sets the CAN module into a specific mode      ////
////                                                                 ////
////    can_set_id - Sets the standard and extended ID               ////
////                                                                 ////
////    can_get_id - Gets the standard and extended ID               ////
////                                                                 ////
////    can_putd - Sends a message/request with specified ID         ////
////                                                                 ////
////    can_getd - Returns specified message/request and ID          ////
////                                                                 ////
////    can_kbhit - Returns true if there is data in one of the      ////
////                receive buffers                                  ////
////                                                                 ////
////    can_tbe - Returns true if the transmit buffer is ready to    ////
////              send more data                                     ////
////                                                                 ////
////    can_abort - Aborts all pending transmissions                 ////
////                                                                 ////
//// PIN_B3 is CANRX, and PIN_B2 is CANTX.  You will need a CAN      ////
//// transeiver to connect these pins to CANH and CANL bus lines.    ////
////                                                                 ////
//// CCS provides an example, ex_can.c, which shows how to use this  ////
//// library.                                                        ////
////                                                                 ////
/////////////////////////////////////////////////////////////////////////
////        (C) Copyright 1996,2003 Custom Computer Services         ////
//// This source code may only be used by licensed users of the CCS  ////
//// C compiler.  This source code may only be distributed to other  ////
//// licensed users of the CCS C compiler.  No other use,            ////
//// reproduction or distribution is permitted without written       ////
//// permission.  Derivative programs created using this software    ////
//// in object code form are not restricted in any way.              ////
/////////////////////////////////////////////////////////////////////////
```

Figure 5.24 *CCS "can-18xxx8.c" Library File (continues)*

```
#include "can-18xxx8.h"

//////////////////////////////////////////////////////////////////////////
//
// can_init()
//
// Initializes PIC18xxx8 CAN peripheral.  Sets the RX filter and masks so the
// CAN peripheral will receive all incoming IDs.  Configures both RX buffers
// to only accept valid messages (as opposed to all messages, or all
// extended message, or all standard messages).  Also sets the tri-state
// setting of B2 to output, and B3 to input (apparently the CAN peripheral
// doesn't keep track of this)
//
// The constants (CAN_USE_RX_DOUBLE_BUFFER, CAN_ENABLE_DRIVE_HIGH,
// CAN_ENABLE_CAN_CAPTURE) are given a default define in the can-18xxx8.h
// file.
// These default values can be overwritten in the main code, but most
// applications will be fine with these defaults.
//
////////////////////////////////////////////////////////////////////////////////
void can_init(void) {
   can_set_mode(CAN_OP_CONFIG);//must be in config mode before params can be set
   can_set_baud();

   RXB0CON=0;
   RXB0CON.rxm=CAN_RX_VALID;
   RXB0CON.rxb0dben=CAN_USE_RX_DOUBLE_BUFFER;
   RXB1CON=RXB0CON;

   CIOCON.endrhi=CAN_ENABLE_DRIVE_HIGH;
   CIOCON.cancap=CAN_ENABLE_CAN_CAPTURE;

   can_set_id(&RXM0EIDL, CAN_MASK_ACCEPT_ALL, 0);  //set mask 0
   can_set_id(&RXF0EIDL, 0, 0);  //set filter 0 of mask 0
   can_set_id(&RXF1EIDL, 0, 0);  //set filter 1 of mask 0

   can_set_id(&RXM1EIDL, CAN_MASK_ACCEPT_ALL, 1);  //set mask 1
   can_set_id(&RXF2EIDL, 0, 1);  //set filter 0 of mask 1
   can_set_id(&RXF3EIDL, 0, 1);  //set filter 1 of mask 1
   can_set_id(&RXF4EIDL, 0, 1);  //set filter 2 of mask 1
   can_set_id(&RXF5EIDL, 0, 1);  //set filter 3 of mask 1

   set_tris_b((*0xF93 & 0xFB ) | 0x08);   //b3 is out, b2 is in
```

Figure 5.24 *CCS "can-18xxx8.c" Library File (continues)*

```
//    can_set_mode(CAN_OP_LOOPBACK);
   can_set_mode(CAN_OP_NORMAL);
}

////////////////////////////////////////////////////////////////////////
//
// can_set_baud()
//
// Configures the baud rate control registers.  All the defines here
// are defaulted in the can-18xxx8.h file.  These defaults can, and
// probably should, be overwritten in the main code.
//
////////////////////////////////////////////////////////////////////////
void can_set_baud(void) {
   BRGCON1.brp=CAN_BRG_PRESCALAR;
   BRGCON1.sjw=CAN_BRG_SYNCH_JUMP_WIDTH;

   BRGCON2.prseg=CAN_BRG_PROPAGATION_TIME;
   BRGCON2.seg1ph=CAN_BRG_PHASE_SEGMENT_1;
   BRGCON2.sam=CAN_BRG_SAM;
   BRGCON2.seg2phts=CAN_BRG_SEG_2_PHASE_TS;

   BRGCON3.seg2ph=CAN_BRG_PHASE_SEGMENT_2;
   BRGCON3.wakfil=CAN_BRG_WAKE_FILTER;
}

void can_set_mode(CAN_OP_MODE mode) {
   CANCON.reqop=mode;
   while( (CANSTAT.opmode) != mode );
}


////////////////////////////////////////////////////////////////////////
//
// can_set_id()
//
// Configures the xxxxEIDL, xxxxEIDH, xxxxSIDL and xxxxSIDH registers to
// configure the defined buffer to use the specified ID
//
//    Paramaters:
//       addr - pointer to first byte of ID register, starting with
//       xxxxEIDL.
//             For example, a pointer to RXM1EIDL
//       id - ID to set buffer to
```

Figure 5.24 *CCS "can-18xxx8.c" Library File (continues)*

```
//     ext - Set to TRUE if this buffer uses an extended ID, FALSE if not
//
////////////////////////////////////////////////////////////////////////////
void can_set_id(int* addr, int32 id, int1 ext) {
   int *ptr;

   ptr=addr;

   if (ext) {  //standard
      //eidl
      *ptr=make8(id,0); //0:7

      //eidh
      ptr--;
      *ptr=make8(id,1); //8:15

      //sidl
      ptr--;
      *ptr=make8(id,2) & 0x03;    //16:17
      *ptr|=(make8(id,2) << 3) & 0xE0; //18:20
      *ptr|=0x08;


      //sidh
      ptr--;
      *ptr=((make8(id,2) >> 5) & 0x07 ); //21:23
      *ptr|=((make8(id,3) << 3) & 0xF8);//24:28
   }
   else {   //standard
      //eidl
      *ptr=0;

      //eidh
      ptr--;
      *ptr=0;

      //sidl
      ptr--;
      *ptr=(make8(id,0) << 5) & 0xE0;

      //sidh
      ptr--;
      *ptr=(make8(id,0) >> 3) & 0x1F;
      *ptr|=(make8(id,1) << 5) & 0xE0;
```

Figure 5.24 *CCS "can-18xxx8.c" Library File (continues)*

```
   }
}

////////////////////////////////////////////////////////////////////////
//
// can_get_id()
//
// Returns the ID of the specified buffer.  (The opposite of can_set_id())
// This is used after receiving a message, to see which ID sent the
// message.
//
//   Paramaters:
//     addr - pointer to first byte of ID register, starting with xxxxEIDL.
//            For example, a pointer to RXM1EIDL
//     ext - Set to TRUE if this buffer uses an extended ID, FALSE if not
//
//   Returns:
//     The ID of the buffer
//
////////////////////////////////////////////////////////////////////////
int32 can_get_id(int * addr, int1 ext) {
   int32 ret;
   int * ptr;

   ret=0;
   ptr=addr;

   if (ext) {
      ret=*ptr;  //eidl

      ptr--;     //eidh
      ret|=((int32)*ptr << 8);

      ptr--;     //sidl
      ret|=((int32)*ptr & 0x03) << 16;
      ret|=((int32)*ptr & 0xE0) << 18;

      ptr--;     //sidh
      ret|=((int32)*ptr << 21);

   }
   else {
      ptr-=2;    //sidl
      ret=((int32)*ptr & 0xE0) >> 5;
```

Figure 5.24 *CCS "can-18xxx8.c" Library File (continues)*

```
      ptr--;      //sidh
      ret|=((int32)*ptr << 3);
   }

   return(ret);
}

////////////////////////////////////////////////////////////////////////
//
// can_get_id()
//
// Puts data on a transmit buffer, at which time the CAN peripheral will
// send when the CAN bus becomes available.
//
//    Paramaters:
//       id - ID to transmit data as
//       data - pointer to data to send
//       len - length of data to send
//       priority - priority of message.  The higher the number, the
//                  sooner the CAN peripheral will send the message.
//                  Numbers 0 through 3 are valid.
//       ext - TRUE to use an extended ID, FALSE if not
//       rtr - TRUE to set the RTR (request) bit in the ID, false if NOT
//
//    Returns:
//       If successful, it will return which transmit buffer that data was
//       placed in.  Numbers 0 through 2 are valid.
//       If un-successful, will return 0xFF
//
////////////////////////////////////////////////////////////////////////
int can_putd(int32 id, int * data, int len, int priority, int1 ext, int1
rtr) {
   int i;
   int * txd0;
   int ret;

   txd0=&TXRXBaD0;

    // find emtpy transmitter
    //map access bank addresses to empty transmitter
   if (!TXB0CON.txreq) {
      CANCON.win=CAN_WIN_TX0;
      ret=0;
   }
   else if (!TXB1CON.txreq) {
```

Figure 5.24 *CCS "can-18xxx8.c" Library File (continues)*

```
      CANCON.win=CAN_WIN_TX1;
      ret=1;
   }
   else if (!TXB2CON.txreq) {
      CANCON.win=CAN_WIN_TX2;
      ret=2;
   }
   else
   {
      return(0xFF);
   }

   //set priority.
   TXBaCON.txpri=priority;

   //set tx mask
   can_set_id(&TXRXBaEIDL, id, ext);

   //set tx data count
   TXBaDLC=len;
   TXBaDLC.rtr=rtr;

   for (i=0; i<len; i++)
   {
      *txd0=*data;
      txd0++;
      data++;
   }

   //enable transmission
   TXBaCON.txreq=1;


   CANCON.win=CAN_WIN_RX0;

   return(ret);
}

////////////////////////////////////////////////////////////////////////////
//
// can_getd()
//
// Gets data from a receive buffer, if the data exists
//
//    Returns:
```

Figure 5.24 *CCS "can-18xxx8.c" Library File (continues)*

```
//      id - ID who sent message
//      data - pointer to array of data
//      len - length of received data
//      stat - structure holding some information (such as which buffer
//             received it, ext or standard, etc)
//
//    Returns:
//      Function call returns a TRUE if there was data in a RX buffer,
//      FALSE
//      if there was none.
//
//////////////////////////////////////////////////////////////////////////////
int1 can_getd(int32 & id, int * data, int & len, struct rx_stat & stat)
{
    int i;
    int * ptr;

    if (RXB0CON.rxful) {
        CANCON.win=CAN_WIN_RX0;
        stat.buffer=0;

        CAN_INT_RXB0IF=0;

        stat.err_ovfl=COMSTAT.rx0ovfl;
        COMSTAT.rx0ovfl=0;

        if (RXB0CON.rxb0dben) {
         stat.filthit=RXB0CON.filthit0;
        }
    }
    else if ( RXB1CON.rxful )
    {
        CANCON.win=CAN_WIN_RX1;
        stat.buffer=1;

        CAN_INT_RXB1IF=0;

        stat.err_ovfl=COMSTAT.rx1ovfl;
        COMSTAT.rx1ovfl=0;

        stat.filthit=RXB1CON.filthit;
    }
    else
    {
      return (0);
```

Figure 5.24 *CCS "can-18xxx8.c" Library File (continues)*

```
    }

    len = RXBaDLC.dlc;
    stat.rtr=RXBaDLC.rtr;

    stat.ext=TXRXBaSIDL.ext;
    id=can_get_id(&TXRXBaEIDL,stat.ext);

    ptr = &TXRXBaD0;
    for ( i = 0; i < len; i++ )
    {
        *data = *ptr;
        data++;
        ptr++;
    }

    // return to default addressing
    CANCON.win=CAN_WIN_RX0;

    stat.inv=CAN_INT_IRXIF;
    CAN_INT_IRXIF = 0;

    if (stat.buffer)
    {
      RXB1CON.rxful=0;
    }
    else
    {
      RXB0CON.rxful=0;
    }

    return(1);
}
```

Figure 5.24 *CCS "can-18xxx8.c" Library File (continued)*

```
//////////////////////////////////////////////////////////////////////////
////                          can-18xxx8.h                           ////
////                                                                 ////
//// Prototypes, definitions, defines and macros used for and with   ////
//// the CCS CAN library for PIC18Fxx8 and PIC18Cxx8.                ////
////                                                                 ////
//// (see can-18xxx8.c)                                              ////
////                                                                 ////
//////////////////////////////////////////////////////////////////////////
////        (C) Copyright 1996,2003 Custom Computer Services         ////
//// This source code may only be used by licensed users of the CCS  ////
//// C compiler.  This source code may only be distributed to other  ////
//// licensed users of the CCS C compiler.  No other use,            ////
//// reproduction or distribution is permitted without written       ////
//// permission.  Derivative programs created using this software    ////
//// in object code form are not restricted in any way.              ////
//////////////////////////////////////////////////////////////////////////

#ifndef __CCS_CAN_LIB_DEFINES__
#define __CCS_CAN_LIB_DEFINES__

#IFNDEF CAN_USE_EXTENDED_ID
  #define CAN_USE_EXTENDED_ID         TRUE
#ENDIF

#IFNDEF CAN_BRG_SYNCH_JUMP_WIDTH
  #define CAN_BRG_SYNCH_JUMP_WIDTH  0  //synchronized jump width (def: 1 x Tq)
#ENDIF

#IFNDEF CAN_BRG_PRESCALAR
  #define CAN_BRG_PRESCALAR  8
//baud rate generator prescalar (def: 4) ( Tq = (2 x (PRE + 1))/Fosc )
#ENDIF

#ifndef CAN_BRG_SEG_2_PHASE_TS
 #define CAN_BRG_SEG_2_PHASE_TS   TRUE
//phase segment 2 time select bit (def: freely programmable)
#endif

#ifndef CAN_BRG_SAM
 #define CAN_BRG_SAM 0
//sample of the can bus line
(def: bus line is sampled 1 times prior to sample point)
#endif
```

Figure 5.25 *CCS "can-18xxx8.h" Header File (continues)*

```
#ifndef CAN_BRG_PHASE_SEGMENT_1
 #define CAN_BRG_PHASE_SEGMENT_1  5 //phase segment 1 (def: 6 x Tq)
#endif

#ifndef CAN_BRG_PROPAGATION_TIME
 #define CAN_BRG_PROPAGATION_TIME 2
//propagation time select (def: 3 x Tq)
#endif

#ifndef CAN_BRG_WAKE_FILTER
 #define CAN_BRG_WAKE_FILTER FALSE
//selects can bus line filter for wake up bit
#endif

#ifndef CAN_BRG_PHASE_SEGMENT_2
 #define CAN_BRG_PHASE_SEGMENT_2 5
//phase segment 2 time select (def: 6 x Tq)
#endif

#ifndef CAN_USE_RX_DOUBLE_BUFFER
 #define CAN_USE_RX_DOUBLE_BUFFER FALSE
//if buffer 0 overflows, do NOT use buffer 1 to put buffer 0 data
#endif

#ifndef CAN_ENABLE_DRIVE_HIGH
 #define CAN_ENABLE_DRIVE_HIGH 0
#endif

#ifndef CAN_ENABLE_CAN_CAPTURE
 #define CAN_ENABLE_CAN_CAPTURE 0
#endif

enum CAN_OP_MODE {CAN_OP_CONFIG=4, CAN_OP_LISTEN=3, CAN_OP_LOOPBACK=2,
CAN_OP_DISABLE=1, CAN_OP_NORMAL=0};
enum CAN_WIN_ADDRESS {CAN_WIN_RX0=0, CAN_WIN_RX1=5, CAN_WIN_TX0=4,
CAN_WIN_TX1=3, CAN_WIN_TX2=2};

//can control
struct {
       int1 void0; //0
       CAN_WIN_ADDRESS win:3;    //1:3 //window address bits
       int1 abat;   //4 //abort all pending transmissions
       CAN_OP_MODE reqop:3;//5:7 //request can operation mode bits
} CANCON;
#byte CANCON = 0xF6F
```

Figure 5.25 *CCS "can-18xxx8.h" Header File (continues)*

```
enum CAN_INT_CODE {CAN_INT_WAKEUP=7, CAN_INT_RX0=6, CAN_INT_RX1=5,
CAN_INT_TX0=4, CAN_INT_TX1=3, CAN_INT_TX2=2, CAN_INT_ERROR=1,
CAN_INT_NO=0};

//can status register READ-ONLY
struct {
       int1 void0;  //0
       CAN_INT_CODE icode:3;      //1:3  //interrupt code
       int1 void4;  //4
       CAN_OP_MODE opmode:3;      //5:7  //operation mode status
} CANSTAT;
#byte CANSTAT = 0xF6E

//communication status register READ-ONLY
struct {
       int1 ewarn;           //0 //error warning
       int1 rxwarn;          //1 //receiver warning
       int1 txwarn;          //2 //transmitter warning
       int1 rxbp;    //3 //receiver bus passive
       int1 txbp;    //4 //transmitter bus passive bit
       int1 txbo;    //5    //transmitter bus off
       int1 rx1ovfl; //6    //receive buffer 1 overflow
       int1 rx0ovfl; //7    //receive buffer 0 overflow
} COMSTAT;
#byte COMSTAT=0xF74
#byte COMSTATUS=0xF74

//baud rate control register 1
struct {
       int brp:6;   //0:5  //baud rate prescalar
       int sjw:2;   //6:7  //synchronized jump width
} BRGCON1;
#byte BRGCON1=0xF70

//baud rate control register 2
struct {
       int prseg:3; //0:2 //propagation time select
       int seg1ph:3; //3:5 //phase segment 1
       int1 sam; //6 //sample of the can bus line
       int1 seg2phts; //7 //phase segment 2 time select
} BRGCON2;
#byte BRGCON2=0xF71
```

Figure 5.25 *CCS "can-18xxx8.h" Header File (continues)*

```
//baud rate control register 3
struct {
        int seg2ph:3;//0:2  //phase segment 2 time select
        int void543:3;      //3:5
        int1 wakfil; //6 //selects can bus line filter for wake-up
        int1 void7;  //7
} BRGCON3;
#byte BRGCON3=0xF72

//can i/o control register
struct {
        int void3210:4;  //0:3
        int1 cancap; //4 //can message receive caputre
        int1 endrhi; //5 //enable drive high
        int void76:2;//6:7
} CIOCON;
#byte CIOCON=0xF73

//transmit buffer n control register
struct txbNcon_struct {
        int  txpri:2;//0:1  //transmit priority bits
        int1 void2; //2
        int1 txreq;  //3 transmit request status (clear to request
        message abort)
        int1 txerr;  //4    //transmission error detected
        int1 txlarb; //5    //transmission lost arbitration status
        int1 txabt;  //6    //transmission aborted status
        int1 void7;
};
struct txbNcon_struct TXB0CON;
struct txbNcon_struct TXB1CON;
struct txbNcon_struct TXB2CON;
struct txbNcon_struct TXBaCON;
#byte  TXB0CON=0xF40
#byte  TXB1CON=0xF30
#byte  TXB2CON=0xF20
#byte TXBaCON=0xF60 //txbXcon when in the access bank

#byte  TXB0CONR=0xF40
#byte  TXB1CONR=0xF30
#byte  TXB2CONR=0xF20

//transmit buffer n standard identifier
#byte TXB0SIDH=0xF41
#byte TXB0SIDL=0xF42
```

Figure 5.25 *CCS "can-18xxx8.h" Header File (continues)*

```
#byte TXB1SIDH=0xF31
#byte TXB1SIDL=0xF32
#byte TXB2SIDH=0xF21
#byte TXB2SIDL=0xF22

//transmit buffer n extended identifier
#byte TXB0EIDH=0xF43
#byte TXB0EIDL=0xF44
#byte TXB1EIDH=0xF33
#byte TXB1EIDL=0xF34
#byte TXB2EIDH=0xF23
#byte TXB2EIDL=0xF24

//transmit buffer n data byte m
#byte TXB0D0=0xF46
#byte TXB0D7=0xF4D
#byte TXB1D0=0xF36
#byte TXB1D7=0xF3D
#byte TXB2D0=0xF26
#byte TXB2D7=0xF2D

//transmit buffer n data length
struct txbNdlc_struct {
       int dlc:4;    //0:3
       int void54:2; //4:5
       int1 rtr; //6 //transmission frame remote tranmission
       int1 void7; //7
};
struct txbNdlc_struct TXB0DLC;
struct txbNdlc_struct TXB1DLC;
struct txbNdlc_struct TXB2DLC;
struct txbNdlc_struct TXBaDLC;
#byte TXB0DLC=0xF45
#byte TXB1DLC=0xF35
#byte TXB2DLC=0xF25
#byte TXBaDLC=0xF65  //txbXdlc when in the access bank


//transmit error count register
#byte TXERRCNT=0xF76


enum CAN_RX_MODE {CAN_RX_ALL=3, CAN_RX_EXT=2, CAN_RX_STD=1,
CAN_RX_VALID=0};
```

Figure 5.25 *CCS "can-18xxx8.h" Header File (continues)*

```
//receive buffer 0 control register
struct {
        int1 filthit0;       //0 //filter hit
        int1 jtoff;  //1 //jump table offset
        int1 rxb0dben;       //2 //receive buffer 0 double buffer enable
        int1 rxrtrro;//3 //receive remote transfer request
        int1 void4;  //4
        CAN_RX_MODE rxm:2;  //5:6 //receiver buffer mode
        int1 rxful;  //7 //receive full status
} RXB0CON;
#byte RXB0CON=0xF60

//receive buffer 1 control register
struct {
        int filthit:3;       //0:2
        int1 rxrtrro;//3 //receive remote transfer request
        int1 void4;  //4
        CAN_RX_MODE rxm:2;  //5:6 //receive buffer mode
        int1 rxful;  //7    //receive full
} RXB1CON;
#byte  RXB1CON=0xF50


//receive buffer n standard identifier
#byte  RXB0SIDH=0xF61
#byte  RXB0SIDL=0xF62
#byte  RXB1SIDH=0xF51
#byte  RXB1SIDL=0xF52

//receive buffer n extended identifier
#byte  RXB0EIDH=0xF63
#byte  RXB0EIDL=0xF64
#byte  RXB1EIDH=0xF53
#byte  RXB1EIDL=0xF54

#byte TXRXBaEIDL=0xF64

struct {
    int void012:3; //0:3
    int1 ext;   //extended id
    int1 srr;   //substitute remove request bit
    int void567:3; //5:7
} TXRXBaSIDL;
#byte TXRXBaSIDL=0xF62
```

Figure 5.25 *CCS "can-18xxx8.h" Header File (continues)*

```
//receive buffer n data length code register
struct rxbNdlc_struct {
       int dlc:4;    //0:3 //data length code
       int1 rb0; //4 //reserved
       int1 rb1;     //5 //reserved
       int1 rtr;     //6 //receiver remote transmission request bit
       int1 void7;   //7
};
struct rxbNdlc_struct RXB0DLC;
struct rxbNdlc_struct RXB1DLC;
struct rxbNdlc_struct RXBaDLC;
#byte  RXB0DLC=0xF65
#byte  RXB1DLC=0xF55
#byte  RXBaDLC=0xF65

//receive buffer n data field byte m register
#byte RXB0D0=0xF66
#byte RXB0D7=0xF6D
#byte RXB1D0=0xF56
#byte RXB1D7=0xF5D
#byte TXRXBaD0=0xF66
#byte TXRXBaD7=0xF6D

//receive error count
#byte RXERRCNT=0xF75

//receive acceptance filter n standard indifier
#byte RXF0SIDH=0xF00
#byte RXF0SIDL=0xF01
#byte RXF1SIDH=0xF04
#byte RXF1SIDL=0xF05
#byte RXF2SIDH=0xF08
#byte RXF2SIDL=0xF09
#byte RXF3SIDH=0xF0C
#byte RXF3SIDL=0xF0D
#byte RXF4SIDH=0xF10
#byte RXF4SIDL=0xF11
#byte RXF5SIDH=0xF14
#byte RXF5SIDL=0xF15

//receive acceptance filter n extended indifier
#byte RXF0EIDH=0xF02
#byte RXF0EIDL=0xF03
#byte RXF1EIDH=0xF06
#byte RXF1EIDL=0xF07
```

Figure 5.25 *CCS "can-18xxx8.h" Header File (continues)*

```
#byte RXF2EIDH=0xF0A
#byte RXF2EIDL=0xF0B
#byte RXF3EIDH=0xF0E
#byte RXF3EIDL=0xF0F
#byte RXF4EIDH=0xF12
#byte RXF4EIDL=0xF13
#byte RXF5EIDH=0xF16
#byte RXF5EIDL=0xF17

//receive acceptance mask n standard identifer mask
#byte RXM0SIDH=0xF18
#byte RXM0SIDL=0xF19
#byte RXM1SIDH=0xF1C
#byte RXM1SIDL=0xF1D

//value to put in mask field to accept all incoming id's
#define CAN_MASK_ACCEPT_ALL   0

//receive acceptance mask n extended identifer mask
#byte RXM0EIDH=0xF1A
#byte RXM0EIDL=0xF1B
#byte RXM1EIDH=0xF1E
#byte RXM1EIDL=0xF1F

//can interrupt flags
#bit CAN_INT_IRXIF = 0xFA4.7
#bit CAN_INT_WAKIF = 0xFA4.6
#bit CAN_INT_ERRIF = 0xFA4.5
#bit CAN_INT_TXB2IF = 0xFA4.4
#bit CAN_INT_TXB1IF = 0xFA4.3
#bit CAN_INT_TXB0IF = 0xFA4.2
#bit CAN_INT_RXB1IF = 0xFA4.1
#bit CAN_INT_RXB0IF = 0xFA4.0

//PROTOTYPES and MACROS

struct rx_stat {
   int1 err_ovfl;
   int filthit:3;
   int1 buffer;
   int1 rtr;
   int1 ext;
   int1 inv;
};
```

Figure 5.25 *CCS "can-18xxx8.h" Header File (continues)*

```
void   can_init(void);
void   can_set_baud(void);
void   can_set_mode(CAN_OP_MODE mode);
void   can_set_id(int* addr, int32 id, int1 ext);
int32 can_get_id(int * addr, int1 ext);
int    can_putd(int32 id, int * data, int len, int priority, int1 ext,
int1 rtr);
int1   can_getd(int32 & id, int * data, int & len, struct rx_stat &
stat);

#define can_kbhit()                    (RXB0CON.rxful || RXB1CON.rxful)
#define can_tbe()                      (!TXB0CON.txreq || !TXB1CON.txreq ||
!TXB2CON.txreq)
#define can_abort()                    (CANCON.abat=1)
#endif
```

Figure 5.25 *CCS "can-18xxx8.h" Header File (continued)*

```
#include "can-18xxx8.c"    /* include the CCS CAN library */

// this structure contains the operating system state
struct STAT_FLAGS
{
   int footswitch : 1;      // footswitch made
   int cruising : 1;        // cruise control enabled
   int system_fault :1;     // fault, system off
   int error : 5;           // detail as to fault reason
} flags;

// possible fault conditions
enum  {NO_ERROR, Motor_Over_I, Max_Speed, Brake_Fault, Brake_Time,
Motor_Fault };

// this message structure will be used as-is in the display unit!
// it is monitor package that is sent out once a second
// it contains speed, analog readings, and the system state.
struct CAN_MESSAGE {
   int16     total_pulses;      // wheel pulses in the last 1 second
   int       battery_volts;     // 8-bit A/D reading of battery
   int       brake_volts;       // 8-bit A/D reading of brake controller
   int       motor_current;     // 8-bit A/D reading of motor current
```

Figure 5.26 *CAN Interface Software, Drive Unit (continues)*

```
   int       aux_volts;          // 8-bit A/D reading of aux input
   struct STAT_FLAGS status;     // a copy of the "flags" structure
   int       spare;              // extra byte for future expansion
} out_data;

//send a stream of 8 bytes of data from id 24
int32 tx_id=24;                  // CAN standard ID
int tx_len=8;                    // 8 bytes per package (this is the max)
int tx_pri=3;                    // highest priority setting
int1 tx_rtr=0;                   // no requested data
int1 tx_ext=0;                   // no extended identity


#int_TIMER3
TIMER3_isr()                     // 26ms tick at 10MHz
{
      . . .

   can_time++;
   if(can_time > 37)             // ..once a second..
   {
                                 // convert to 8 bit data for the display
      out_data.battery_volts = analog_data[battery] >> 2;
      out_data.brake_volts = analog_data[brake] >> 2;
      out_data.motor_current = analog_data[motor_i] >> 2;
      out_data.aux_volts = analog_data[aux_input] >> 2;
      out_data.status = flags;

                    // send data structure out onto the CAN bus
      can_putd(tx_id, &out_data, tx_len,tx_pri,tx_ext,tx_rtr);

      out_data.total_pulses = 0; // reset total pulses in this interval
                                 // once it has been sent..
      can_time = 0;              //put data on transmit buffer


      . . .
   }
}
```

Figure 5.26 *CAN Interface Software, Drive Unit (continued)*

The data structure CAN_MESSAGE was carefully constructed to have a total size of 8 bytes. Close attention was paid to the data types used, and the analog values were scaled down so that each fit into 1 byte. No appreciable loss of resolution for the analog values will be noticed by the user at the display. In the CAN Interface Software example code in Figure 5.26, the CAN messages are generated every 1 second within the Timer3 interrupt routine. At each interval, the A/D data is scaled and placed into the *out_data* structure. The function *can_putd()* moves the *out_data* structure contents to the controller's physical output registers. Once the call is made, the CAN controller hardware takes over and does whatever it takes to deliver the message out to the network.

5.6.11 SYSTEM INTEGRATION AND SOFTWARE DEVELOPMENT PHASE, DISPLAY UNIT

The scooter drive is self-contained in that it gets its functional information directly from the wheel speed sensor and the handbrake controller. The drive packages information and provides status on the CAN bus as to what it is doing and why.

The display unit is also a somewhat stand-alone device. It basically listens to the CAN bus for drive-unit generated messages, deciphers them, and presents the data of interest to the rider.

Interfacing with human beings is never a simple task. In all human, or man-machine interfaces, a certain amount of intuitive operation is expected. Whenever possible, things should be marked as clearly as possible. The user should be able to get information or perform a task with as little effort as possible. Displayed data should be put in familiar formats, using familiar units—inches, acres, pounds, seconds, and so on. The display unit of the scooter is no exception. Simplicity is the key. As shown in Section 5.6.7, "Software Design – Drive Unit," the display unit is minimalist by design. There is an LCD display, a couple of buttons, a beeper, and an LED. In this prototype system, we had to compromise in the area of the display because an off-the-shelf LCD is being used, not allowing the luxury of a clear set of nomenclature, but that just adds to the challenge of trying to make something operate as smoothly and intuitively as possible while working within design constraints.

The Collection and Conversion of the CAN Data

The CAN communications processing is made simple by its structure. The packets can be up to 8 bytes in length as defined by the hardware of the CAN controller. In the display unit, the software looks for messages, 8 bytes in length, which originate from the drive unit, which has an ID number of 24. By mimicking the data structure of the drive unit's packets, the data is automatically separated into its elements as the packets arrive (Figure 5.27).

The collection of the data is paced by the arrival of the data packets. The drive unit sends a packet every second, and the CAN controller hardware does whatever it takes to get the message through. Unless there is a tremendous amount of interference, in a peer-to-peer setup like this scooter, it is nearly a certainty that the data will arrive on time, every time.

```
// this structure contains the operating system state
struct STAT_FLAGS
{
   int footswitch : 1;      // footswitch made
   int cruising : 1;        // cruise control enabled
   int system_fault :1;     // fault, system off
   int error : 5;           // detail as to fault reason
};
// possible fault conditions
enum  {NO_ERROR, Motor_Over_I, Max_Speed, Brake_Fault, Brake_Time,
Motor_Fault };

// this message structure is used as-is in the drive unit!
// it is a monitor package that is sent out once a second
// it contains speed, analog readings, and the system state.
struct CAN_MESSAGE {
   int16     total_pulses;     // wheel pulses in the last 1 second
   int       battery_volts;    // 8-bit A/D reading of battery
   int       brake_volts;      // 8-bit A/D reading of brake controller
   int       motor_current;    // 8-bit A/D reading of motor current
   int       aux_volts;        // 8-bit A/D reading of aux input
   struct STAT_FLAGS status;   // a copy of the "flags" structure
   int       spare;            // extra byte for future expansion
} in_data;

//get a stream of rx_len bytes of data from ID rx_td
struct rx_stat rxstat;
int32 rx_id;
int rx_len;


// pulses per 0.1 mile = (((1ft/pi(4"))*5280ft/mi)*16p/ft)/10)
#define PP1TM  (((12/(3.14159*4)*5280)*16)/10)

// pulses per mile per hour * 10 =
((((1ft/pi(4"))*5280ft/mi)*16p/ft)/3600sec/hr)*10)
#define PPMPHx10  ((((12/(3.14159*4)*5280)*16)/3600)*10)

// millivolts per A/D count
#define MVperADC ((((14.7/4.7)*5)/255)*1000)

// amperes per A/D count:  255 counts = 100A
#define AMPSperADC   (10000/255)

//  A/D counts when battery is at 10 Volts
```

Figure 5.27 *CAN Interface Software, Display Unit*

```
#define BATT_10V (int)((((4.7/14.7)*10)/5)*255)

      . . .

main()
{
      . . .

      if ( can_kbhit() )    //if data is waiting in buffer...
      {
         if(can_getd(rx_id, &in_data, rx_len, rxstat))
         { //...then get data from buffer
            if((rx_id == 24) && (rx_len == 8))
            {
               // if the packet is valid..
               // convert all the data to their display units (*10)
               mph      += (in_data.total_pulses * 100)/PPMPHx10;
               mph      /= 2; // just a basic average to help smooth the data

               kph      = (mph * 88)/550;     // kph as integer - 2 digits

               miles   += (int32)in_data.total_pulses;
               t_val    = miles / (int32)PP1TM;
               m_trip   = (int16) t_val;

               if(m_trip > 199)
               {
                  miles = 0;       // roll over trip meter if > 18.8 miles
                  t_val = 0;
                  m_trip = 0;
               }
               t_val    = (t_val * (int32)88)/(int32)550;
               k_trip   = (int16) t_val;

               t_val    = ((int32)MVperADC*(int32)in_data.battery_volts)/100;
               volts    = (int16) t_val;

               t_val    = ((int32)AMPSperADC*(int32)in_data.motor_current)/100;
               current  = (int16) t_val;
            }
         }
       . . .
}
```

Figure 5–27 *CAN Interface Software, Display Unit*

As the valid packets arrive and are verified by length and ID, a series of conversions takes place. A choice was made to convert all the units every time instead of converting only the parameter being viewed by the rider. This decision was made based on the available amount of computing time, which in this case is a huge amount of time at a 1-second-per-update interval, and the number of total parameters that need conversion.

As shown in Figure 5.27, the function *can_kbhit()* detects when a complete CAN message has been received. The call to the function *can_getd()* pulls the data from the CAN controller's internal receiver buffer and moves the data into the structure *in_data*. The *in_data* structure is of type *CAN_MESSAGE* and is identical to the drive unit's *out_data* structure. Once the data is moved from the hardware buffers into the structure, the various parameters are scaled for display. The #defines *PP1TM*, *PPMPHx10*, *MVperADC*, and *AMPSperADC* are used to form scaling constants, making it easier to perform potential future adjustments as well as making the software easier to read.

Because the LCD display is configured in this design as a 2-1/2 digit display with a single decimal point, the range of numbers that can be displayed are 0.0 to 19.9 or 0 to 199. This had to be factored into the unit calculations. For instance, 18.8 miles would be approximately 30.1 kilometers, which would not display correctly. Therefore, some simple adjustments are made. Miles and mph have a decimal point, whereas kilometers and kph do not. For the same reasons, voltage has a decimal point, having a range of 0.0 to15.0 volts, whereas motor current does not with its range of 0 to 100 amperes. To simplify the display formatting process, a structure *DISPITEM*, shown in Figure 5.28, was created that contains a pointer to the value to be displayed, a flag to indicate if it contains a decimal point, and a units marker configuration that indicates to the operator what parameter is in view.

The *DISPITEM* structure is processed by the *build_LCD()* function. This function uses the standard library *sprintf()* function to convert the numeric information into an ASCII string that can be parsed and formatted into a pattern suitable for the LCD display (Figure 5.29).

The variable *current_item* is used as an index into the *DISPITEM* structure array, indicating what is to be displayed. This makes for quick and easy updates to the display because all that is required is a call to *build_LCD()* with the *current_item* set to the desired parameter. The *build_LCD()* function completely formats the LCD segment information based on the look-up table LCD_LUT. The array LCD_PORT is used to hold the patterns of seven-segment configurations that will be placed on the LCD. In addition to the numerals, there are a few other details that are taken care of, such as displaying the decimal point, the letter 'E', the low battery indication, and blanking the empty digits.

The Buttons, Beeper, and Indicators

The buttons are simple switches that are brought in on input pins and monitored directly by the software. The buttons are "edge detected" in the *main()* loop as shown in Figure 5.30. The flags x and y are used to filter the button action such that only the rising edge causes action. This gives the button a positive feel and the user to see a change as the button is depressed as well as hear the audible confirmation of the beeper.

```
unsigned int16 mph,kph,m_trip,k_trip,volts,current;

#define  MAX_ITEM 5
struct DISPITEM {
   int16 *value;      // pointer to scaled value for display
   int1 dp;           // decimal point?
   int unit_mark;     // segment unit marker pattern
   } disp_table[MAX_ITEM+1] = {
         &mph,     1, 1,
         &kph,     0, 4,
         &m_trip,  1, 3,
         &k_trip,  0, 6,
         &volts,   1, 0,
         &current,0, 7
   };
int current_item;

#define TRIP_M_ITEM 2
#define TRIP_K_ITEM 3
```

Figure 5.28 *DISPITEM Structure Definition*

```
unsigned char dispstr[6];      // LCD Display contents in ASCII i.e. "158"
const unsigned char LCD_LUT[12] = {
         0b00111111,        // '0'
         0b00000110,        // '1'
         0b01011011,        // '2'
         0b01001111,        // '3'
         0b01100110,        // '4'
         0b01101101         // '5'
         0b01111101,        // '6'
         0b00000111,        // '7'
         0b01111111,        // '8'
         0b01101111,        // '9'
         0b00000000,        // ' '
         0b01111001         // 'E'
      };
#define BLANK   10       /* index to blank */
#define E_CHAR 11        /* index to E character */
unsigned char LCD_PORT[4]; // LCD port image.. MSD,D,LSD,UNITS
```

Figure 5.29 *The build_LCD() Data Conversion Function (continues)*

```
void build_LCD(void)  // convert formatted ASCII string into and LCD bitmap
{
   if(in_data.status.system_fault)
   {       // if there is a system fault, then display it..
      sprintf(dispstr," E%d",in_data.status.error);
   }       // otherwise, format the information into an ASCII numeric string
   else
   {
      sprintf(dispstr,"%3lu",*disp_table[current_item].value);
          // and then translate it to a segment pattern for the LCD
   }

   if(dispstr[0] == '1')
      LCD_PORT[0] = 1;  // if 1 in MSD then turn on the half-digit
   else
      LCD_PORT[0] = 0;  // if not a 1 then nothing at all

   if(dispstr[1] == ' ')
      LCD_PORT[1] = LCD_LUT[BLANK];  // map segments into the LCD port image
   else if(dispstr[1] == 'E')
      LCD_PORT[1] = LCD_LUT[E_CHAR];
   else
      LCD_PORT[1] = LCD_LUT[dispstr[1] - '0'];

   if((disp_table[current_item].dp) && (dispstr[1] != 'E'))
      LCD_PORT[1] |= 0x80;        // turn on decimal point if required


   if(dispstr[2] == ' ')
      LCD_PORT[2] = LCD_LUT[BLANK];
   else if(dispstr[2] == 'E')
      LCD_PORT[2] = LCD_LUT[E_CHAR];
   else
      LCD_PORT[2] = LCD_LUT[dispstr[2] - '0'];

   if(low_battery)        // turn on low battery indicator if required
      LCD_PORT[2] |= 0x80;

   if(dispstr[1] != 'E')
      LCD_PORT[3] = disp_table[current_item].unit_mark;
   else
      LCD_PORT[3] = 0;  // if there is an error -- turn off units..
}
```

Figure 5.29 *The build_LCD() Data Conversion Function (continued)*

```
#define STATUS_LED       PIN_B4
#define SELECT_BUTTON    PIN_B0
#define RESET_BUTTON     PIN_B1

#define TRIP_M_ITEM 2
#define TRIP_K_ITEM 3

void main()
{
      . . .

   while(1)
   {

      . . .

      if(input(SELECT_BUTTON) && (!sel_is_hi))
      {
         sel_is_hi = 1;                  // on low-to-high transition of
                                         // button
         current_item++;                 // move to next display item
         if(current_item > MAX_ITEM)     // wrap around table
            current_item = 0;
         build_LCD();                    // update display with button
         beep_on();                      // brief beep for button feedback
      }
      delay_ms(10);                      // a little switch de-bounce time
      if(!input(SELECT_BUTTON))          // released finger?
         sel_is_hi = 0;

      if(input(RESET_BUTTON) && (!res_is_hi))
      {
         res_is_hi = 1;                  // on low-to-high transition of
                                         // button
         if((current_item == TRIP_M_ITEM) || (current_item == TRIP_K_ITEM))
         {
            miles = 0;                   // reset trip odometers
            m_trip = 0;
            k_trip = 0;
            build_LCD();                 // update display with button
            beep_on();                   // brief beep for button feedback
         }
      }
      delay_ms(10);                      // a little switch de-bounce time
```

Figure 5.30 *Button Detection and Display Item Selection (continues)*

```
      if(!input(RESET_BUTTON))              // released finger?
         res_is_hi = 0;

   }
}
```

Figure 5.30 *Button Detection and Display Item Selection (continued)*

The SELECT button is used to increment the variable *current_item*, which is used by the *build_LCD()* function and handled by the display routines. The RESET button is used to reset or clear the trip odometer. If the *current_item* variable is indexed to the *TRIP_M_ITEM* or the *TRIP_K_ITEM* of the *disp_table[]* array, then pressing the RESET button will reset the associate variables, clearing the odometer.

The beeper is driven from a PWM. The CCP1 and Timer2 peripherals of the PIC are used to create the tone. The beeper is turned on and off using the following macro definitions:

```
#define beep_off() set_pwm1_duty(0)
#define beep_on() {set_pwm1_duty(512); beep_time = 10;}
```

The beeper is controlled by the Timer3 interrupt service routine, the same routine that is used to drive the LCD display itself. The macro *beep_on()* sets the PWM output to 50%, forming a square wave at the piezo element at the PWM output frequency. The same macro also sets the variable *beep_time* to a value, which is decremented in the Timer3 service routine. When *beep_time* reaches 0, the *beep_off()* macro is used to set the PWM duty cycle to 0%, turning the beeper off.

```
#int_TIMER3
TIMER3_isr()                    // 26ms tick (38Hz) at 10MHz
{
   if((temp & 0x04) &&
            ((in_data.status.system_fault) || (low_battery)))
      output_bit(STATUS_LED,1);      // status light on
   else
      output_bit(STATUS_LED,0);      // status light off

   temp++;
   . . .

   beep_time-;               // time down beeper from button press
   if(beep_time == 0)
      beep_off();             // set beeper to off
}
```

The LED indicator is also driven by the Timer3 service routine. If a fault is present or a low battery value is detected, then the STATUS_LED flashes. Low battery conditions are tested in the main loop by the following code:

```
//  A/D counts when battery is at 10 Volts
#define BATT_10V (int)((((4.7/14.7)*10)/5)*255)

//  A/D counts when battery is at 6 Volts (footswitch not made)
#define BATT_OFF (int)((((4.7/14.7)*6)/5)*255)


if((in_data.battery_volts < BATT_10V) &&
          (!in_data.status.system_fault) &&
               (in_data.battery_volts > BATT_OFF))
  low_battery = 1;
else
  low_battery = 0;   // detect and display low battery!
```

The BATT_10V definition computes the number of A/D counts available on a 5-volt full-scale, 8-bit A/D, based on the resistor divider used in the drive unit (10 K and 4.7 K ohms) at 10 volts. Again, this type of definition may look like a lot of typing, but it can be a tremendous time-saver if a resistor value changes, the low voltage point changes, or even if the width of the data were to change from 8 to 10 bits. The flag *low_battery* is also used by the *build_LCD()* function to turn on the "LOW BATT" segment of the LCD display.

Driving the LCD

The LCD, as discussed in Section 5.6.6, "Hardware Design - Display Unit," depends on two things in order to operate: an AC waveform and a difference in potential across any segment that is to be "lit." Here again, the Timer3 interrupt service routine is being invoked every 26 milliseconds, or approximately 38 hertz. As mentioned previously, the *build_LCD()* function preformats the LCD segment information into an easy-to-manage array. During each Timer3 interrupt, the data is moved from the *LCD_PORT[]* array to the actual output ports of the PIC. For each pass through the routine, as shown in Figure 5.31 the data is complemented with respect to the COM signal (PIN_C0 of the PIC). Any segment that is high when COM is low, or vice versa, will be visible. Any segment that is at the same potential as the COM signal will be off and invisible.

5.6.12 SYSTEM TEST PHASE

At this point in the project, the tests specified for system test during the Test Definition Phase of the project are carried out to give the user confidence that the unit performs as defined by the specifications. As stated earlier, the tests may be as extensive as needed to convince a customer, or they may be as simple as comparing the results to a competitive device.

For the purposes of this textbook, extensive testing is 'above and beyond' the scope of the requirements and would contribute little to expanding your knowledge of project development.

However, one important topic to be addressed here is what to do if the project does not meet specifications (fails a system test). Problems such as incorrect calibration of the speedome-

```
#int_TIMER3
TIMER3_isr()                    // 26ms tick (38Hz) at 10MHz
{

     . . .

   if(temp++ & 1)
   {                                    // DATA OUT RIGHT SIDE UP..
      output_d(LCD_PORT[1]);    // 3rd SEGMENT (digit and decimal point)

      output_a(LCD_PORT[2]);    // 2nd SEGMENT (digit and low_batt)
      output_bit(PIN_A5,LCD_PORT[2] & 0x10);
      output_e((LCD_PORT[2]>>5));

      output_c((LCD_PORT[3]<<3));       // 1st SEGMENT (units markers)
      output_bit(PIN_C1,LCD_PORT[0]);   // 4TH SEGMENT (half digit)

      output_bit(PIN_C0,0);             // COM pin opposite..
   }
   else
   {                                    // DATA OUT UPSIDE DOWN..
      output_d(~LCD_PORT[1]);    // 3rd SEGMENT (digit and decimal point)

      output_a(~LCD_PORT[2]);    // 2nd SEGMENT (digit and low_batt)
      output_bit(PIN_A5,!(LCD_PORT[2] & 0x10));
      output_e(~(LCD_PORT[2]>>5));

      output_c(~(LCD_PORT[3]<<3));      // 1st SEGMENT (units markers)
      output_bit(PIN_C1,!LCD_PORT[0]);  // 4TH SEGMENT (half digit)

      output_bit(PIN_C0,1);             // COM pin opposite..
   }

      . . .

}
```

Figure 5.31 *LCD Drive Process Example (Timer3 ISR)*

ter results, lack of linearity in the braking input, incorrect measurement of the motor current, and so on, will all be spotted during system test. The important issue is what to do about it, and there are really only two choices: fix it or change the specification. Changing the specification should only be done as a very last resort and should ***not*** be required if the Definition Phase was carried out properly. So 'fixing it' is the only *real* choice.

As an example, consider the handbrake controller input. Its intended calibration was based on an empirically derived number from the manufacturer (90° of travel, which is approximately 30% of the range of the potentiometer), and the design allows for some variation in the actual results, allowing for losses due to friction, method of mounting, and so on, so that the handbrake could be properly calibrated.

For example, suppose that when we measure the output of the handbrake we notice an error of 4%. This means that we just did not get quite as much voltage out of the system as we expected over the entire range of the handbrake. So when we plot the output voltage versus the percentage of movement, our graph will look something like "Measured A" shown in Figure 5.32.

In this case, even though there is some error, the result is linear, so a simple adjustment of the scaling would be in order. This could be accomplished by changing the scaling constant in the software from

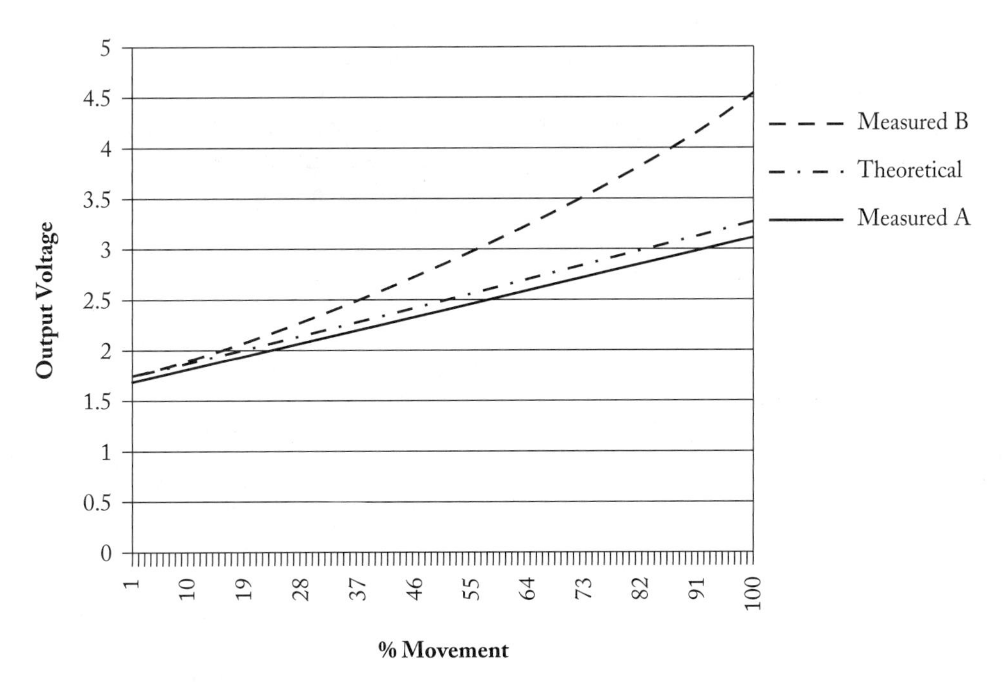

Figure 5.32 *Example, Measured Braking Input*

```
/* brake voltage min 1.75V */
#define BRAKE_MIN       (int16)((1.75/5.0)*1023)

/* 3.25V is the most the brake should be */
#define BRAKE_MAX       (int16)((3.25/5.0)*1023)
```

to

```
/* brake voltage min 1.69V */
#define BRAKE_MIN       (int16)((1.69/5.0)*1023)

/* 3.12V is the most the brake should be */
#define BRAKE_MAX       (int16)((3.12/5.0)*1023)
```

This easy adjustment shows the value of putting conversion constants and calibration factors in as constant variables stored in FLASH memory or as macro definitions. You can adjust these easily if necessary without looking through the program to find where the constants or calibration factors are used, and perhaps missing one or two, if they are used in multiple places. Examination of the program will show that these are inserted in the program and really do not increase or decrease code size by their use, so they are a no-cost convenience.

If the result were nonlinear, like the "Measured B" data in Figure 5.32, a more serious type of correction might be in order, requiring a complex algebraic expression or perhaps the simpler approach of using a look-up table and some interpolation. The choice of method, in this case, greatly depends on the expected accuracy, the precision of the result, and the available computing time. Overall, a look-up table (LUT) is a tough approach to beat no matter how great someone is with "that fancy math." In real systems like the one we are describing, it is not uncommon at all to come across some really strange-looking curves and bends in the data that are collected from a device. As long as there is curve to the earth, there is gravity pulling us down, and the sun rises every day to cause a constant change in temperature, the data that is collected is going have a shape, and it is probably not going to be a straight line.

So we can convert the "Measured B" data to corrected values, more in-line with the "ideal data" using a LUT and linear-interpolation approach. One of the best parts of the LUT approach is that you can use the data you actually collected. Table 5.5 shows measured voltages.

If the data is more straight than it is curved or S-shaped, fewer points can be used in the actual table, and the linear interpolation can fill in the values in between, as they are needed. With ten voltage measurements from our handbrake control testing, an array is formed to create the LUT.

```
const int16 Bvolts[11] = /* A/D counts */
        {360,391,430,474,524,577,638,704,774,850,932};
```

The following function, *calc_braking()*, shows how a voltage measured by the A/D and passed to the function as the parameter *braking*, in A/D counts, is checked against the values of the LUT. If the value is lower than the LUT, it is less than 1.76 volts and nothing should be computed. If the value is higher than the LUT, it is greater than 4.55 volts and

% Braking Input	Output Voltage	A/D Counts
0	1.76	360
10	1.91	391
20	2.10	430
30	2.32	474
40	2.56	524
50	2.82	577
60	3.12	638
70	3.44	704
80	3.78	774
90	4.15	850
100	4.55	932

Table 5.5 *Measured % of Braking Input vs. Voltage and A/D Counts*

nothing can be computed either, so an error should display like: "E3," which means "Brake Input Fault." This can be accomplished by returning a bogus value like 9999 to the caller.

Two values are selected from the LUT table: one above braking, *t*, and one below braking, *b*, and the index of the lower value is kept as *x*. The delta ($t - b$) is calculated for the two table values to determine the slope of what is effectively a short linear segment within the long, curved data. The delta from the measured value *braking* to the lower selected value *b* is used to compute at what point (in percentage) the measured data falls on the short line segment by dividing (*braking-b*) by (*t-b*). The base *braking* percentage movement value is then computed by multiplying our *braking* percentage by the difference in percentage movement between the points associated with *t* and *b* and adding back the percentage of movement associated with the point *b*. By formatting the data table in even 10% increments, the software is simplified. The index *x* is used to form the mantissa of the resulting computed percentage, and the percentage calculated for the short line segment, between points *t* and *b*, becomes the fractional portion of the resulting computed percentage.

```
/* brake voltage min 1.75V */
#define BRAKE_MIN      (int16)((1.75/5.0)*1023)

/* 3.25V is the most the brake should be */
#define BRAKE_MAX      (int16)((3.25/5.0)*1023)

int16 calc_braking(int16 braking) // return the adjusted braking
{
  int16 t,b,x; // temp values, top, bottom, and index
  int32 v1,v2,v3;     // temp values for computation

  if(braking < Bvolts [0])
       return (9999); // too low to measure,
```

```
    if(braking > Bvolts [10])
         return (9999); // too high to measure,
                        // "9999" is an error message!!

    for(x=0; x < 9; x++)
    {
         if(braking < Bvolts[x+1])    // find where the counts fall
         {                                        // in the table
            t = Bvolts[x+1];
            b = Bvolts[x];        // top and bottom values
            break;                // x will be our base braking % / 10
         }
    }   // now calculate the percentage between top and bottom
    v1 = ((int32)braking - (int32)b) * (int32)100;
    v2 = (int32)t - (int32)b;          // the two values
    v3 = (v1/v2);                // percentage of braking diff * 10

    v1 = (int32)x * (int32)100;        // make x into braking input
                                       // * 100
    v1 += v3;             // now add in percentage of difference
    v1 /= (int32)10;      // and scale to a whole percentage
                          // v1 is a % of full scale
                          // now apply this % to the ideal..


    v2 = (((int32)(BRAKE_MAX-BRAKE_MIN) * v1) /(int32)100);
    v2 +=(int32)BRAKE_MIN;

    return (int16)v2; // v2 is new interpolated value
}
```

Therefore, in this example, if *braking* was measured at 600 A/D counts, the values as a result of the **for** loop for the index *x* and the top and bottom values of the linear segment, *t* and *b*, would be

$$x = 5$$

$$t = \text{Bvolts}[5+1] = 638$$

$$b = \text{Bvolts}\ [5] = 577$$

Stepping through the program, the computations would go like this:

```
    v1 = (int32)braking -(int32)b * (int32)100;
```

result: v1 = (600 – 577) * 100 = 2300

```
    v2 = (int32)t - (int32)b;    // the two values
```

result: v2 = 638 – 577 = 61

```
v3 = (v1/v2);   // percentage of braking diff * 10
```

result: v3 = (2300/61) = 37, which is 37% * 10, or 3.7%

```
v1 = (int32)x * (int32)100;  // make x into braking * 100
                             // (fraction!)
```

result: v1 = 5 * 100 = 500, which is fixed point for 50% applied brake (2.5 volts)

```
v1 += v3;// now add in percentage of difference
```

result: v1 = 500 + 37 = 537, again, which is fixed point for 53.7% applied

```
v1 = v1 / (int32)10; // now scale to whole %
```

v1 = 537 / 10 = **53% braking**

finally, the ideal count range is from 358 to 665 counts (1.75 to 3.25 volts) as defined by *BRAKE_MIN and BRAKE_MAX*:

v2 = (((BRAKE_MAX-BRAKE_MIN) * v1) / 100);

v2 + = BRAKE_MIN;

result:

v2 = (((665-358) * 53) / 100) + 358 = *520 Counts*

In this example, we simply throw the fractional portion away. In this instance, a simple LUT and linear interpolation can be applied effectively to linearize a nonlinear result so that the project will meet its specifications.

As shown in the preceding example, the purpose of the system test is to ensure, to whatever degree necessary, that the system performs to specifications. After the tests are completed and any necessary adjustments are made, the project is ready for use with a high degree of confidence in the results.

5.7 CHALLENGES

Following are various features that could be added to or changed in the software to enhance the electronic scooter operation and/or performance:

1. Data log the scooter battery, motor on time, and distance traveled, and predict the remaining time and distance. Add this "remaining time" or "distance to go" to the display.
2. Modify the system for a faster, 24 V DC unit. Be sure to make all necessary hardware and software changes.
3. Double the resolution of the wheel sensor, and make the appropriate software changes.
4. Make the display wireless. This will make a drive more suitable to a conventional skateboard.

5. Add brake lights and turn signals to the scooter.
6. Make the turn signals automatic, using some sort of tilt sensing.
7. Add a dusk-to-dawn headlamp for night riding.

5.8 CHAPTER SUMMARY

In this chapter, project development has been approached as a *process*, an orderly set of steps, which, when followed, will virtually always lead to a successful project. The process has been demonstrated by the development of an electronic scooter drive based on the Microchip PIC18F458 microcontroller.

5.9 EXERCISES

EXERCISE

1. List each of the steps of the process of project development and give an example of the activities that would take place during that step (Section 5.4).
2. In which step of the project development process would each of the following occur (Section 5.4)?
   a. Bread boarding a sensor and its conditioning circuitry
   b. Simulating a circuit's operation
   c. Creating detailed specifications for a project
   d. Drawing a project schematic
   e. Testing individual software functions
   f. Writing software flowcharts
   g. Testing the final project to its specifications
   h. Doing proof of concept testing on a questionable circuit
3. Using the data shown in Table 5.6, write a program to perform a look-up table and linear-interpolation operation to return a compensated temperature from the A/D reading (Section 5.6.12).

A/D Values, 10 Bits (0–1023)	*Temperature °C*
123	-20
344	0
578	20
765	40
892	60
1002	80

Table 5.6 *Linear-Interpolation Exercise*

5.10 LABORATORY ACTIVITY

The only laboratory activity that is really appropriate to this chapter is to demonstrate the process of project development in developing a project. Some suggested projects are shown below. Any of these ideas could be modified or expanded through the use of displays and/or additional sensors or input devices.

1. A robotic 'mouse' that can follow black electrical tape on a whitish vinyl floor. This project will require controlling the speed and steering as well as sensing the black line.
2. A device composed of the microcontroller and a video camera that can detect simple shapes such as a square, a triangle, or a circle of black paper on a white surface.
3. A device using motors, sensors, and a cardboard arrow to point to and follow a heat source as it moves about the room. This device would make excellent use of stepper motors.
4. A replacement for a furnace thermostat.
5. A security system for an automobile with sensors to detect doors being opened (including the gas tank cover, the trunk, and the hood) and which would detect when the vehicle is being moved.

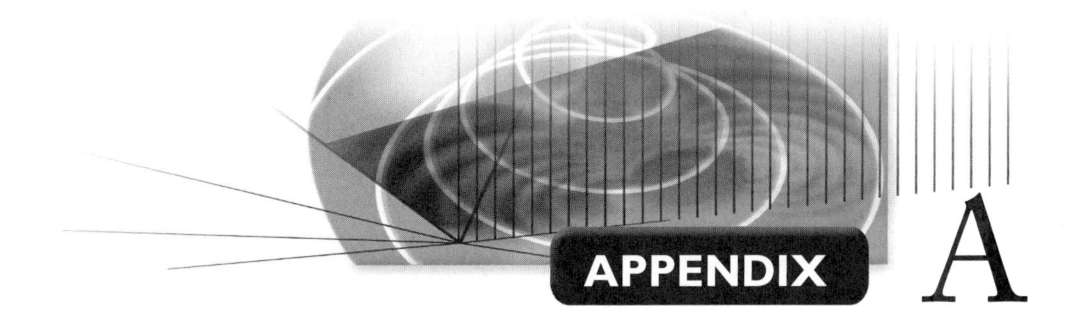

Library Functions Reference

INTRODUCTION

Much of the programming power and convenience of the C language lies in its built-in or library functions. These are routines that accomplish many of the more common tasks that face a C programmer.

As with all C functions, the library functions are called from your code by using the function name and passing values to the function or receiving values returned by the function or both. In order to make use of the functions, the programmer needs to know what parameters the function is expecting to be passed to it and the nature of any values returned from it. This information is made clear by the prototype for the function. The function prototypes are contained in a set of header files that the programmer "includes" into his or her code.

There are many, many library functions available in most C compilers. In order to avoid including a whole host of unneeded function prototypes into every program, the library functions are gathered into groups by function. That is, the function prototypes for the group of routines concerned with the standard input and output routines are contained in a header file called *stdio.h*. If programmers want to use some of these functions, they would put the following statement into the beginning of their code:

```
#include <stdio.h>
```

Having included the header file, the functions concerned with standard I/O such as the *printf* function (referred to in earlier chapters) are made available to the programmer.

The following is a comprehensive list of library functions available in CCS-PICC grouped by their header file names. This reference section describes each of the library functions and provides an example of its use.

FUNCTIONS LISTED BY CATEGORY

Function Prototype	*Required Input*
RS232 I/O Functions	
void assert(*condition*);	#include <assert.h> #use rs232
char fgetc(*stream identifier*);	#use rs232
void fgets(char *str, *stream identifier*);	#use rs232
void fprintf(*stream identifier*, char *fmtstr [, arg1, arg2, ...])	#use rs232
void fputc(char c, *stream identifier*);	#use rs232
void fputs(char *str, *stream identifier*);	#use rs232
char getc();	#use rs232
void gets(char *str);	#use rs232
char kbhit();	#use rs232
void printf([*funtion name*],*stream identifier*, char *fmtstr [, arg1, arg2, ...]);	#use rs232
void putc(char c);	#use rs232
void puts(char *str);	#use rs232
void set_uart_speed(const int32 baud);	#use rs232
SPI I/O	
void setup_spi(const *modes*);	
char spi_data_is_in();	
char spi_read([char c]);	
void spi_write(char c);	
Discrete I/O	
void output_low(const *pin*);	
void output_high(const *pin*);	
void output_float(const *pin*);	
void output_bit(const *pin*, char value);	
char input(const *pin*);	
void output_*port*(char value);	
char input_*port()*	
void port_b_pullups(char value);	
void set_tris_*port*(char value);	

Function Prototype	***Required Input***
Capture/Compare/PWM	
void setup_ccpx(const *mode*);	
void set_pwmx_duty(int16 *value*);	
Delays	
void delay_us(unsigned char c); void delay_us(const int16 c);	
void delay_ms(unsigned char c); void delay_ms(const int16 c);	
void delay_cycles(cont unsigned char c);	
Parallel Slave I/O	
void setup_psp(*mode*);	
int1 psp_input_full();	
int1 psp_output_full();	
int1 psp_overflow();	
I^2C I/O	
void i2c_start();	#use i2c
void i2c_stop();	#use i2c
int8 i2c_read([int1 ack]);	#use i2c
int1 i2c_write(int8 data);	#use i2c
int1 i2c_poll();	#use i2c
Processor Controls	
void sleep();	
void reset_cpu();	
char restart_cause();	
void disable_interrupts(*int_level*);	#int_xxx
void enable_interrupts(*int_level*);	#int_xxx
void ext_int_edge([*source*],*edge*);	
int8 read_bank(int8 bank, int16 offset);	
void write_bank(int8 bank, int16 offset, int8 value);	
int16 label_address(*program_label*);	
void goto_address(int16 addr); void goto_address(int32 addr);	

Function Prototype	***Required Input***
Bit/Byte Manipulation	
int1 shift_right(char *addr, char bytes, int1 value);	
int1 shift_left(char *addr, char bytes, int1 value);	
void rotate_right(char *addr, char bytes);	
void rotate_left(char *addr, char bytes);	

void bit_clear(int8 var, char c);
void bit_clear(int16 var, char c);
void bit_clear(int32 var, char c);	

void bit_set(int8 var, char c);
void bit_set(int16 var, char c);
void bit_set(int32 var, char c);	

void bit_test(int8 var, char c);
void bit_test(int16 var, char c);
void bit_test(int32 var, char c);	

void swap(int8 x);	
int8 = make8(int16 var, int8 offset); int8 = make8(int32 var, int8 offset);	
int16 = make16(int8 var_high, int8 var_low);	
int32 = make32(int8 var1[, int8 var2, int8 var3, int8 var4]);	
Analog Comparator	
void setup_comparator(*mode*);	
Standard C Math	

int8 abs(int8 x);
int16 abs(int16 x);
int32 abs(int32 x);
	#include <stdlib.h>

float acos(float *x*);	#include <math.h>
float asin(float x);	#include <math.h>
float atan(float x);	#include <math.h>
float ceil(float x);	#include <math.h>
float cos(float *x*);	#include <math.h>
float exp(float x);	#include <math.h>
float floor(float x);	#include <math.h>
int16 labs(int16 x);	#include <stdlib.h>
float sinh(float x);	#include <math.h>
float log(float x);	#include <math.h>
float log10(float x);	#include <math.h>
float pow(float x, float y);	#include <math.h>
float sin(float x);	#include <math.h>
float cosh(float x);	#include <math.h>
float tanh(float x);	#include <math.h>
float fabs(float x);	#include <math.h>
float fmod(float x, float y);	#include <math.h>
float atan2(float y, float x);	#include <math.h>

Function Prototype	***Required Input***
Standard C Math	
float frexp(float x, int *expon);	#include <math.h>
float ldexp(float x, int expon);	#include <math.h>
float modf(float x, float *ipart);	#include <math.h>
float sqrt(float x);	#include <math.h>
float tan(float x);	#include <math.h>
A/D Conversions	
void setup_adc_ports(*mode*);	
void setup_adc(*mode*);	
void set_adc_channel(int8 channel);	
int8 read_adc(); int16 read_adc();	
Standard C Char	
int8 atoi(char *str);	#include <stdlib.h>
int32 atoi32(char *str);	#include <stdlib.h>
int16 atol(char *str);	#include <stdlib.h>
float atof(char *str);	#include <stdlib.h>
char tolower(char c);	
char toupper(char c);	
unsigned char isalnum(char c);	#include <ctype.h>
unsigned char isalpha(char c);	#include <ctype.h>
unsigned char isamoung(char c, const char *str);	
unsigned char isdigit(char c);	#include <ctype.h>
unsigned char islower(char c);	#include <ctype.h>
unsigned char isspace(char c);	#include <ctype.h>
unsigned char isupper(char c);	#include <ctype.h>
unsigned char isxdigit(char c);	#include <ctype.h>
unsigned char iscntrl (char c);	#include <ctype.h>
unsigned char isgraph(char c);	#include <ctype.h>
unsigned char isprint(char c);	#include <ctype.h>
unsigned char ispunct(char c);	#include <ctype.h>
int8 strlen(char *str);	#include <string.h>
char *strcpy(char *dest, char *src);	#include <string.h>
char *strncpy(char *dest, char *src, unsigned char n);	#include <string.h>
signed int8 strcmp(char *str1, char *str2);	#include <string.h>
signed char stricmp(char *str1, char *str2);	#include <string.h>
signed char strncmp(char *str1, char *str2, unsigned char n);	#include <string.h>
char *strcat(char *str1, char *str2);	#include <string.h>
char *strstr(char *str1, char *str2);	#include <string.h>
char *strchr(char *str, char c);	#include <string.h>

Function Prototype	***Required Input***
Standard C Char	
char *strrchr(char *str, char c);	#include <string.h>
char *strtok(char *str1, char flash *str2);	#include <string.h>
unsigned char strspn(char *str, char *set);	#include <string.h>
unsigned char strcspn(char *str, char *set);	#include <string.h>
char *strpbrk(char *str, char *set);	#include <string.h>
char *strlwr(char *str);	#include <string.h>
void sprintf(char *str, char flash *fmtstr [, arg1, arg2, ...]) ;	
Standard C Memory	
void memset(void *buf, unsigned char c, unsigned char n);	
void memcpy(void *dest, void *src, unsigned char n);	
int8 offsetof(*structure type*, *field*);	#include <stddef.h>
int8 offsetofbit(*structure type*, *field*);	#include <stddef.h>
Timers	

void setup_timer_0(*mode*);
void setup_timer_1(*mode*);
void setup_timer_2(*mode*, unsigned int8 period, int8 postscale);
void setup_timer_3(*mode*);	

void set_rtcc(int8 value);
void set_rtcc(int16 value);
void set_timer0(int8 value);
void set_timer0(int16 value);
void set_timer1(int16 value);
void set_timer2(int8 value);
void set_timer3(int16 value);	

int8 get_rtcc(); int8 get_timer0();
int16 get_rtcc(); int16 get_timer0();
int16 get_timer1(); int8 get_timer2();
int16 get_timer3();	

void setup_wdt(*mode*);	#fuses WDT
void restart_wdt();	#fuses WDT
Internal EEPROM	
int8 read_eeprom(int8 addr);	
void write_eeprom(int8 addr, int8 value);	
int16 read_program_eeprom(int16 addr); int8 read_program_eeprom(int32 addr);	
void write_program_eeprom(int16 addr, int16 data); void write_program_eeprom(int32 addr, int8 data);	

abs

```
#include <stdlib.h>
int8 abs(int8 x);
int16 abs(int16 x);
int16 labs(int16 x);
int32 abs(int32 x);
```

The *abs* function returns the absolute value of the integer *x*.

Returns: Absolute value of *x* returned as the same variable type as *x*

```
#include <16F877.h>
#include <stdlib.h>
#fuses HS,NOWDT          // a Microchip PIC16F877
#use delay(clock=10000000)

void main()
{
    int16 int_pos_val;
    int32 long_pos_val;

    int_pos_val = abs(-19574);    // get absolute value of an integer
    long_pos_val = abs(-125000); // get absolute value of a long

    while(1)
        ;
}
```

Results: int_pos_val = 19564

long_pos_val = 125000

acos

```
#include <math.h>
float acos(float x);
```

The *acos* function calculates the arc cosine of the floating point number *x*. The result is in the range of 0 to π. The variable *x* must be in the range of −1 to 1.

Returns: acos(x) in the range of 0 to π, where *x* is in the range of −1 to 1

```
#include <16F877.h>
#include <math.h>
#fuses HS,NOWDT          // a Microchip PIC16F877
#use delay(clock=10000000)
#use rs232(baud=9600, xmit=PIN_C6, rcv=PIN_C7)
```

```
void main()
{
    float new_val;

    new_val = acos(0.875);
    printf("%f",new_val);
    while(1)
        ;
}
```

Results: new_val = 0.505360

---

asin

#include <math.h>

float asin(float x);

The *asin* function calculates the arc sine of the floating point number *x*. The result is in the range of $-\pi/2$ to $\pi/2$. The variable *x* must be in the range of -1 to 1.

Returns: asin(x) in the range of $-\pi/2$ to $\pi/2$, where *x* is in the range of -1 to 1

```
#include <16F877.h>
#include <math.h>
#fuses HS,NOWDT           // a Microchip PIC16F877
#use delay(clock=10000000)
#use rs232(baud=9600, xmit=PIN_C6, rcv=PIN_C7)

void main()
{
    float new_val;
    float new_val2;
    new_val = asin(-1);
    new_val2 = asin(1);
    printf("%f",new_val);
    printf("%f",new_val2);
    while(1)
        ;
}
```

Results: new_val = -1.570796

new_val2 = 1.570796

---

assert

#include <assert.h>

#use rs232()

void assert(condition);

The *assert* function tests *condition* (any relational expression) during runtime. If *condition* evaluates to false, then an error message is transmitted to STDERR. (STDERR is the USART defined by the first **#use rs232** statement in the program.) The error message includes the file name and line number of the call to *assert*. No code is generated for the call to *assert* if you include the line **#define NODEBUG** in your program. This allows you to include asserts in your code for testing and quickly eliminate them from the final program.

Returns: Debug statement printed to STDERR

```
#include <16F877.h>
#include <assert.h>

#fuses HS,NOWDT // a Microchip PIC16F877
#use delay(clock=10000000)
#use rs232(baud=9600,parity=N,xmit=PIN_C6,rcv=PIN_C7,bits=8)

void main()
{
    char j;
    j = 10;
    assert(j<9);
    while(1)
          ;
}
```

Results: The following is transmitted on pin C6 at 9600 baud:

```
Assertion failed:j<9 ,file c:\windows \desktop\pic\appen_a.c ,line 12
```

---

atan

#include <math.h>

float atan(float x);

The *atan* function calculates the arc tangent of the floating point number *x*. The result is in the range of $-\pi/2$ to $\pi/2$.

Returns: atan(x) in the range of $-\pi/2$ to $\pi/2$

```
#include <16F877.h>
#include <math.h>
#fuses HS,NOWDT          // a Microchip PIC16F877
#use delay(clock=10000000)
#use rs232(baud=9600, xmit=PIN_C6, rcv=PIN_C7)

void main()
{
    float new_val;
    float new_val2;
```

```
        new_val = atan(-1);
        new_val2 = atan(1);
        printf("%f",new_val);
        printf("%f",new_val2);
        while(1)
            ;
}
```

Results: new_val = -0.785398

new_val2 = 0.785398

atan2

#include <math.h>

float atan2(float y, float x);

The *atan2* function calculates the arc tangent of the floating point numbers *y/x*. The result is in the range of $-\pi$ to π.

Returns: atan(y/x) in the range of $-\pi$ to π

```
#include <16F877.h>
#include <math.h>
#fuses HS,NOWDT          // a Microchip PIC16F877
#use delay(clock=10000000)
#use rs232(baud=9600, xmit=PIN_C6, rcv=PIN_C7)

void main()
{
    float new_val;
    new_val = atan2(2.34, 5.12);
    printf("%f\n\r",new_val);
    while(1)
        ;
}
```

Results: new_val = 0.428685

atoi, atof, atol, atoi32

#include <stdlib.h>

int8 atoi(char *str);

float atof(char *str);

int16 atol(char *str);

int32 atoi32(char *str);

The *atoi*, *atol*, and *atoi32* functions convert the string to an 8-bit, 16-bit or 32-bit integer, respectively. The *atof* function converts the string pointed to by *str* to a floating point number. Numbers may be preceded by + and – signs. Valid numeric characters are the digits *0* through *9* and *A* through *F*. If the number to be converted is hexadecimal, the number should be preceded by '0x'. The decimal point is also accepted by the *atof* function. If the result cannot be represented, the behavior is undefined.

The conversion stops when the first non-numeric character is encountered. All three functions return signed values.

Returns:

- atoi – signed 8-bit integer equivalent of the ASCII string pointed to by *str*
- atof – floating point equivalent of the ASCII string pointed to by *str*
- atol – signed 16-bit integer equivalent of the ASCII string pointed to by *str*
- atoi32 – signed 32-bit integer equivalent of the ASCII string pointed to by *str*

```
#include <16F877.h>
#include <stdlib.h>
#fuses HS,NOWDT          // a Microchip PIC16F877
#use delay(clock=10000000)
#use rs232(baud=9600, xmit=PIN_C6, rcv=PIN_C7)

#define MAX_ENTRY_LENGTH     5
void main(void)
{
    char mystr[MAX_ENTRY_LENGTH+1];
    int myint16;
    char c;

    while (1)
    {
       c = 0;
       printf("Enter a signed integer number followed by !\n\r");
       while (c < MAX_ENTRY_LENGTH)
       {
            mystr[c++] = getchar(); // wait for a character
                    // if it is our terminating character, then quit!
            if (mystr[c-1] == '!')
                    break;
       }
       mystr[c] = '\0';        // null terminate the string!

       myint16 = atol(mystr);        // convert

       printf("Your integer value is: %d\n\r",myint16);
```

```
    }
}
```

Results:

The following is transmitted by the UART :

```
Enter a signed integer number followed by !
```

Once '!' is received, the string is converted and the result is transmitted to the UART.

---

bit_clear

void bit_clear(int8 var, char c);

void bit_clear(int16 var, char c);

void bit_clear(int32 var, char c);

The *bit_clear* function clears bit *c* of the variable *var*. *var* may be an 8-, 16-, or 32-bit variable. *c* may be any bit of *var*. The least significant bit is 0, and the most significant bits are 7, 15, and 31, respectively. This function is the same as

var &= ~(1 << bit);

Returns: None

```
#include <16F877.h>
#fuses HS,NOWDT             // a Microchip PIC16F877
#use delay(clock=10000000)
#use rs232(baud=9600, xmit=PIN_C6, rcv=PIN_C7)

void main()
{
    int8 a;

    a = 0xAA;
    printf("Starting a: %02X\n\r",a);
    bit_clear(a,3);
    printf("Cleared a:  %02X\n\r",a);
    bit_set(a,2);
    printf("Set a:      %02X\n\r",a);
    while(1)
    {
    }
}
```

Results: Transmits at 9600 baud the following three lines:

```
Starting a: AA
Cleared a:  A2
Set a:      A6
```

bit_set

void bit_set(int8 var, char c);

void bit_set(int16 var, char c);

void bit_set(int32 var, char c);

The *bit_set* function sets bit *c* of the variable *var*. *var* may be an 8-, 16-, or 32-bit variable. *c* may be any bit of *var*. The least significant bit is 0, and the most significant bits are 7, 15, and 31, respectively. This function is the same as

```
var |= (1 << bit);
```

Returns: None

See *bit_clear* for an example.

bit_test

void bit_test(int8 var, char c);

void bit_test(int16 var, char c);

void bit_test(int32 var, char c);

The *bit_test* function returns bit *c* of the variable *var*. *var* may be an 8-, 16-, or 32-bit variable. *c* may be any bit of *var*. The least significant bit is 0, and the most significant bits are 7, 15, and 31, respectively.

Returns: 1 if bit *c* of *var* is a 1 or 0 if it is a 0

```
#include <16F877.h>
#fuses HS,NOWDT          // a Microchip PIC16F877
#use delay(clock=10000000)
#use rs232(baud=9600, xmit=PIN_C6, rcv=PIN_C7)

void main()
{
    int8 a;

    a = 0xAA;
    if (bit_test(a,3) != 0)
        printf("bit 3 of a is set!\n\r");
    else
        printf("bit 3 of a is clear!\n\r");
    while(1)
    {
    }
}
```

Results: Transmits the following at 9600 baud:

```
bit 3 of a is set!
```

ceil

#include <math.h>

float ceil(float x);

The *ceil* function returns the smallest integer value that is not less than the floating point number *x*. In other words, the *ceil* function rounds *x* up to the next integer value and returns that value.

Returns: Smallest integer value that is not less than the floating point number x

```
#include <16F877.h>
#include <math.h>
#fuses HS,NOWDT              // a Microchip PIC16F877
#use delay(clock=10000000)
#use rs232(baud=9600, xmit=PIN_C6, rcv=PIN_C7)

void main()
{
    float new_val;
    new_val = ceil(2.531);
    printf("%f\n\r",new_val);
    while(1)
          ;
}
```

Results: new_val = 3.000000

cos

#include <math.h>

float cos(float x);

The *cos* function calculates the cosine of the floating point number *x*. The angle *x* is expressed in radians.

Returns: cos(x)

```
#include <16F877.h>
#include <math.h>
#fuses HS,NOWDT              // a Microchip PIC16F877
#use delay(clock=10000000)
#use rs232(baud=9600, xmit=PIN_C6, rcv=PIN_C7)

void main()
{
```

```
        float new_val;
        new_val = cos(5.121);
        printf("%f\n\r",new_val);
        while(1)
            ;
    }
```

Results: new_val = 0 .397336

cosh

#include <math.h>

float cosh(float x);

The *cosh* function calculates the hyperbolic cosine of the floating point number *x*. The angle *x* is expressed in radians.

Returns: cosh(x)

```
#include <16F877.h>
#include <math.h>
#fuses HS,NOWDT          // a Microchip PIC16F877
#use delay(clock=10000000)
#use rs232(baud=9600, xmit=PIN_C6, rcv=PIN_C7)

void main()
{
    float new_val;
    new_val = cosh(5.121);
    printf("%f\n\r",new_val);
    while(1)
        ;
}
```

Results: new_val = 83.754363

delay_cycles

void delay_cycles(cont unsigned char c);

The *delay_cycles* function creates a delay of the specified number of instruction clocks, *c*. An instruction clock is equal to four oscillator clocks. *c* must be a constant between 1 and 255. This function works by executing the precise number of instructions to cause the requested delay. It does not use any timers. If interrupts are enabled, the time spent in an interrupt routine is not counted toward the time. As a result, the delay time may be longer than requested if an interrupt is serviced during the delay.

Returns: None

```
#include <16F877.h>
```

```
#fuses HS,NOWDT           // a Microchip PIC16F877
#use delay(clock=10000000)
#use fixed_io(c_outputs = PIN_C2)

void main()
{
    int16 mycnt;
    while(1)
    {
        output_bit(PIN_C2,0);
        for (mycnt=0;mycnt<5000;mycnt++)
            delay_cycles(250);
        output_bit(PIN_C2,1);
        delay_ms(500);
    }
}
```

Results: On a 10 MHz system, toggles Port C, pin 2 at 500-millisecond intervals

delay_ms

void delay_ms(unsigned char c);

void delay_ms(const int16 c);

The *delay_ms* function creates a delay of the specified number of milliseconds, *c*. If *c* is a variable, it must be between 0 and 255. If *c* is a constant, it must be between 0 and 65535. This function requires the **#use delay** preprocessor directive and works by executing a precise number of instructions to cause the requested delay. It does not use any timers. If interrupts are enabled, the time spent in an interrupt routine is not counted toward the time. As a result, the delay time may be longer than requested if an interrupt is serviced during the delay.

Returns: None

See *delay_cycles* for an example.

delay_us

void delay_us(unsigned char c);

void delay_us(const int16 c);

The *delay_us* function creates a delay of the specified number of microseconds, *c*. If *c* is a variable, it must be between 0 and 255. If *c* is a constant, it must be between 0 and 65535. This function requires the **#use delay** preprocessor directive and works by executing a precise number of instructions to cause the requested delay. It does not use any timers. If interrupts are enabled, the time spent in an interrupt routine is not counted toward the time. As a result, the delay time may be longer than requested if an interrupt is serviced during the delay.

Returns: None

See *setup_ccpx* for an example.

---

disable_interrupts

#int_xxx

void disable_interrupts(int_level);

The *disable_interrupts* function disables the interrupt at the given level, *int_level. int_level* is a constant representing an interrupt type. The constants for a particular device can be found in the device *.h* file. Disabling interrupts with the **GLOBAL** constant globally disables all interrupts for the device at the global level without affecting the individual interrupt enable flags. To use a specific *int_level* constant, the associated preprocessor directive **#int_xxx** must be used in the program.

Returns: None

```
#include <16F877.h>
#fuses HS,NOWDT          // a Microchip PIC16F877
#use delay(clock=10000000)
#use rs232(baud=9600, xmit=PIN_C6, rcv=PIN_C7)

#define INTS_PER_SECOND 38      // (10000000/(4*256*256))

byte seconds;   // A running seconds counter
byte int_count; // Number of interrupts left before a second
                // has elapsed

#int_rtcc                  // This function is called every time
void clock_isr()           // the RTCC (timer0) overflows (255->0).
{                          // For this program this is apx 38 times
   if(--int_count==0)      // per second.
   {
      ++seconds;
      int_count=INTS_PER_SECOND;
   }

}


void main()
{
   int_count=INTS_PER_SECOND;
   set_timer0(0);
   setup_counters( RTCC_INTERNAL, RTCC_DIV_256);
   enable_interrupts(GLOBAL);
```

```
    while(1)
    {

        printf("Press any key to begin.\n\r");
        getc();
        seconds = 0;
        enable_interrupts(INT_RTCC);
        printf("Press any key to stop.\n\r");
        getc();
        disable_interrupts(INT_RTCC);
        printf("%u seconds.\n\r",seconds);
    }
}
```

Results: The microcontroller prompts the user to "Press any key to begin." and waits for the user to send a character. Once a character is received, the Timer0 (also called **INT_RTCC**) interrupt is enabled and starts counting seconds. The user is prompted to "Press any key to stop." Once a second character is received, the Timer0 interrupt is disabled and the number of seconds between the two characters is transmitted to the user.

---

enable_interrupts

#int_xxx

void enable_interrupts(int_level);

The *enable_interrupts* function enables the interrupt at the given level, *int_level. int_level* is a constant representing an interrupt type. The constants for a particular device can be found in the device *.h* file. Enabling interrupts with the **GLOBAL** constant globally enables interrupts for the device at the global level without affecting the individual interrupt enable flags. To use a specific *int_level* constant, the associated preprocessor directive **#int_xxx** must be used in the program.

Returns: None

See *disable_interrupts* for an example.

---

exp

#include <math.h>

float exp(float x);

The *exp* function calculates the natural (base *e*) exponential of the variable *x*.

Returns: e^x

```
#include <16F877.h>
```

```
#include <math.h>
#fuses HS,NOWDT          // a Microchip PIC16F877
#use delay(clock=10000000)
#use rs232(baud=9600, xmit=PIN_C6, rcv=PIN_C7)

void main()
{
    float new_val;
    new_val = exp(5);
    printf("%f\n\r",new_val);
    while(1)
        ;
}
```

Results: new_val = e^5 = 148.413158

---

ext_int_edge

void ext_int_edge([*source*],*edge*);

The *ext_int_edge* function establishes whether the rising edge or falling edge of an external interrupt input causes the interrupt. *source* is only available for the PIC18 devices where the external interrupt source can be specified. *edge* is either rising, L_TO_H, or falling, H_TO_L. This is only available on devices with external interrupts, and the external interrupts must be enabled appropriately.

Returns: None

See *sleep* for an example.

---

fabs

#include <math.h>

float fabs(float x);

The *fabs* function returns the absolute value of the floating point number *x*.

Returns: Absolute value of *x* as a floating point number

```
#include <16F877.h>
#include <math.h>
#fuses HS,NOWDT          // a Microchip PIC16F877
#use delay(clock=10000000)
#use rs232(baud=9600, xmit=PIN_C6, rcv=PIN_C7)

void main()
{
    float new_val;
    new_val = fabs(-105.45);
```

```
        printf("%f\n\r",new_val);
        while(1)
            ;
    }
```

Results: new_val = 105.449995

fgetc

#use rs232()

char fgetc(*stream identifier*);

The *fgetc* function waits for a character to be received by the USART identified by *stream identifier*. *stream identifier* is a USART defined by a **#use rs232** preprocessor directive. If a hardware USART is being utilized, the hardware can buffer three characters. If a software USART is being used, then the *fgetc* routine (or *getchar*) must be active while the character is being received.

Returns: Character received by the USART identified by *stream identifier*

```
#include <16F877.h>
#fuses HS,NOWDT // a Microchip PIC16F877
#use delay(clock=10000000)
#use rs232(baud=9600,parity=N,xmit=PIN_C6,rcv=PIN_C7,stream=OUT_STREAM)

void main()
{
    char j;
    fputs("Echo Characters:",OUT_STREAM);
    while(1)
    {
        j = fgetc(OUT_STREAM);
        fputc(j,OUT_STREAM);
    }
}
```

Results: The following is transmitted on pin C6 at 9600 baud:

```
Echo Characters:
```

Then any character received by the PIC on pin C7 is echoed back to pin C6 until power is removed from the device.

fgets

#use rs232()

void fgets(char *str, *stream identifier*);

The *fgets* function waits for a string of characters to be received by the USART identified by *stream identifier*. (*stream identifier* is a USART defined by a **#use rs232** preprocessor directive.) The string is read into the memory pointed by *str* until the carriage return (13) character is received. The string is terminated with a '\0' (NULL). It is up to the programmer to ensure that adequate memory is allocated to *str* to contain the string read from *stream identifier* plus the NULL terminator.

Returns: String received from the USART identified by *stream identifier* is placed into the memory pointed to by *str*.

```
#include <16F877.h>
#fuses HS,NOWDT // a Microchip PIC16F877
#use delay(clock=10000000)
#use rs232(baud=9600,parity=N,xmit=PIN_C6,rcv=PIN_C7,stream=OUT_STREAM)

#define MAX_STR_SIZE  20

void main()
{
    char j[MAX_STR_SIZE+1];    // add one for null!
    char cnt;

    fputs("Echo String -  ",OUT_STREAM);
    fputs("Enter < 20 Characters",OUT_STREAM);
    fputs("    followed by ENTER:",OUT_STREAM);

    while(1)
    {
        fgets(j,OUT_STREAM);
        fputs(j,OUT_STREAM);
    }
}
```

Results: The following is transmitted on pin C6 at 9600 baud:

```
Echo String -
Enter < 20 Characters
   followed by ENTER:
```

Then when the keys '1', '2', '3', '4', '5', '6', and 'ENTER' are pressed and transmitted via a serial program to pin C7 of the PIC, the following is transmitted back by the PIC:

```
123456
```

---

floor

```
#include <math.h>
float floor(float x);
```

The *floor* function returns the integer value of the floating point number *x*. This is the largest whole number that is less than or equal to *x*.

Returns: Integer portion of *x*

```
#include <16F877.h>
#include <math.h>
#fuses HS,NOWDT          // a Microchip PIC16F877
#use delay(clock=10000000)
#use rs232(baud=9600, xmit=PIN_C6, rcv=PIN_C7)

void main()
{
    float new_val;
    new_val = floor(2.531);
    printf("%f\n\r",new_val);
    while(1)
        ;
}
```

Results: new_val = 2.000000

---

fmod

#include <math.h>

float fmod(float x, float y);

The *fmod* function returns the remainder of *x* divided by *y*. This is a modulo function specifically designed for float-type variables. The modulo operator, %, is used to perform the modulo operation on the integer variable types.

Returns: Remainder of *x* divided by *y*

```
#include <16F877.h>
#include <math.h>
#fuses HS,NOWDT          // a Microchip PIC16F877
#use delay(clock=10000000)
#use rs232(baud=9600, xmit=PIN_C6, rcv=PIN_C7)

void main()
{
    float new_val;
    new_val = fmod(25.6,8);
    printf("%f\n\r",new_val);
    while(1)
        ;
}
```

Results: remains = 1.600000

fprintf

#use rs232()

void fprintf(*stream identifier*, char *fmtstr [, arg1, arg2, ...]) ;

The *fprintf* function transmits formatted text according to the format specifiers in the *fmtstr* string to the USART identified by *stream identifier*. *fmtstr* can be made up of constant characters to be output directly as well as special format commands or specifications. If no variables, and, therefore, no special format commands, are used, the string, *fmtstr*, may also be an array of characters. If format specifications are used, then *fmtstr* must be a constant string. *fmtstr* is processed by the *fprintf()* function. As *fprintf()* is processing the *fmtstr*, it outputs the characters and expands each argument according to the format specifications that may be embedded within the string. A percent sign (%) is used to indicate the beginning of a format specification. Each format specification is then related to an argument, *arg1*, *arg2*, and so on, in sequence. There should always be an argument for each format specification and vice versa.

The described implementation of the *fprintf()* format specifications is a reduced version of the standard C function. This is done to meet the minimum requirements of an embedded system. A full "ANSI-C" implementation would require a large amount of memory space, which in most cases would make the functions useless in an embedded application.

In the CCS-PICC library, the format specifications take the generic form *%wt*. *w* is optional and may be **1-9** to specify how many characters are to be output or **01-09** to indicate leading zeros. For floating point, *w* may be **1.1-9.9**. *t* is the type and may be one of the types in Table A.1.

Specification	*Format of Argument*
c	Outputs the next argument as an ASCII character
s	Outputs the next argument as a null terminated character string
u	Outputs the next argument as an unsigned decimal integer
x	Outputs the next argument as an unsigned hexadecimal integer using lowercase letters
X	Outputs the next argument as an unsigned hexadecimal integer using uppercase letters
d	Outputs the next argument as a decimal integer
e	Outputs the next argument in exponential format
f	Outputs the next argument as a floating point number
Lx	Outputs the next argument as an unsigned hexadecimal long integer using lowercase letters
LX	Outputs the next argument as an unsigned hexadecimal long integer using lowercase letters
lu	Outputs the next argument as an unsigned decimal long integer
ld	Outputs the next argument as a signed decimal long integer
%	Outputs the % character

Table A.1 *printf Format Specifiers*

```
#include <16F877.h>
#fuses HS,NOWDT // a Microchip PIC16F877
#use delay(clock=10000000)
#use rs232(baud=9600,parity=N,xmit=PIN_C6,rcv=PIN_C7,stream=OUT_STREAM)

void main()
{
    int16 j;
    char c;


    for (j=0;j<=500;j+=250)
    {
        // print the current value of j
        fprintf(OUT_STREAM,"Decimal: %Lu\tHexadecimal: %LX\n\r",j,j);
        fprintf(OUT_STREAM,"Zero Padded Decimal: %0Lu\n\r",j);
        fprintf(OUT_STREAM,"Four Digit Lower Case Hex: %04Lx\r\n\n",j);
    }

    while (1)
    {
        /* receive the character */
        c=getchar();
        /* and echo it back */
        fprintf(OUT_STREAM,"The received character was %c\n\r",c);

    }
}
```

Results:

The following is output by the microcontroller to the UART at start-up.

```
Decimal: 0        Hexadecimal: 0000
Zero Padded Decimal: 00000
Four Digit Lower Case Hex: 0000

Decimal: 250      Hexadecimal: 00FA
Zero Padded Decimal: 00250
Four Digit Lower Case Hex: 00fa

Decimal: 500      Hexadecimal: 01F4
Zero Padded Decimal: 00500
Four Digit Lower Case Hex: 01f4
```

Then any character received by the UART is transmitted back preceded by the specified string. For example, if the character 'c' is received, this text is sent to the transmitter.

```
The received character was c
```

fputc

#use rs232()

void fputc(char c, *stream identifier*);

The *fputc* function transmits the character *c* to the USART identified by *stream identifier*. *stream identifier* is a USART defined by a **#use rs232** preprocessor directive.

Returns: None

```
#include <16F877.h>
#fuses HS,NOWDT // a Microchip PIC16F877
#use delay(clock=10000000)
#use rs232(baud=9600,parity=N,xmit=PIN_C6,rcv=PIN_C7,stream=OUT_STREAM)

void main()
{
    char j;
    fputs("Echo Characters:",OUT_STREAM);
    while(1)
    {
        j = fgetc(OUT_STREAM);
        fputc(j,OUT_STREAM);
    }
}
```

Results: The following is transmitted on pin C6 at 9600 baud:

```
Echo Characters:
```

Then any character received by the PIC on pin C7 is echoed back to pin C6 until power is removed from the device.

fputs

#use rs232()

void fputs(char *str, stream identifier);

The *fputs* function transmits the string *str* to the USART identified by *stream identifier*. (*stream identifier* is a USART defined by a **#use rs232** preprocessor directive.) Once the string is transmitted, a carriage return (13) and a line feed (10) character are transmitted.

Returns: None

```
#include <16F877.h>
#fuses HS,NOWDT // a Microchip PIC16F877
#use delay(clock=10000000)
#use rs232(baud=9600,parity=N,xmit=PIN_C6,rcv=PIN_C7,stream=OUT_STREAM)
#define MAX_STR_SIZE  20
```

```
void main()
{
    char j[MAX_STR_SIZE+1];    // add one for null!
    char cnt;

    fputs("Echo String - ",OUT_STREAM);
    fputs("Enter < 20 Characters",OUT_STREAM);
    fputs("    followed by ENTER:",OUT_STREAM);

    while(1)
    {
        fgets(j,OUT_STREAM);
        fputs(j,OUT_STREAM);
    }
}
```

Results: The following is transmitted on pin C6 at 9600 baud:

```
Echo String -
Enter < 20 Characters
   followed by ENTER:
```

Then when the keys '1', '2', '3', '4', '5', '6', and 'ENTER' are pressed and transmitted via a serial program to pin C7 of the PIC, the following is transmitted back by the PIC:

```
123456
```

---

frexp

```
#include <math.h>
float frexp(float x, int *expon);
```

The *frexp* function returns the mantissa of the floating point number *x* or 0 if *x* is 0. The power of 2 exponent of *x* is stored at the location pointed to by *expon*. If *x* is 0, the value stored at *expon* is also 0.

In other words, if the following call is made:

```
y = frexp(x, expon);
```

The relationshp between *expon*, *x*, and the return value *y* can be expressed as:

$$x = y * 2^{expon}$$

Returns: Mantissa of *x* in the range 0.5 to 1.0 or 0 if *x* is 0

```
#include <16F877.h>
#include <math.h>
#fuses HS,NOWDT          // a Microchip PIC16F877
#use delay(clock=10000000)
#use rs232(baud=9600, xmit=PIN_C6, rcv=PIN_C7)
```

```
void main(void)
{
    float x,z;
    int y;
    float a,c;
    int b;
    float d,f;
    int e;

    x = 3.14159;
    z=frexp(x,&y);

    a = 0.14159;
    c=frexp(a,&b);

    d = .707000;
    f=frexp(d,&e);

    printf("x=%f\n\ry=%d\n\rz=%f\n\r",x,y,z);
    printf("a=%f\n\rb=%d\n\rc=%f\n\r",a,b,c);
    printf("d=%f\n\re=%d\n\rf=%f\n\r",d,e,f);
    while(1)
        ;
}
```

Results:

```
x=3.141590
y=2
z= .785397
a= .141589
b=-1
c= .283179
d= .707000
e=0
f= .707000
```

get_rtcc, get_timerX,

On devices with an 8-bit Timer0:

int8 get_rtcc();

int8 get_timer0();

On devices with a 16-bit Timer0:

int16 get_rtcc();

```
int16 get_timer0();
```

All devices that support the appropriate timer:

```
int16 get_timer1();
int8 get_timer2();
int16 get_timer3();
```

The *get_timerx* functions returns the value of the associated timer/counter. The number of bits returned is the same as the size of the timer. All timers count up and roll over to zero after they reach their maximum value. The maximum value depends on the number of bits available on the timer. Availability of these functions is device dependent. RTCC and Timer0 are the same.

Returns: 8- or 16-bit value of the associate timer/counter

See *setup_timer_X* for an example.

---

getc

```
#use rs232()
char getc();
```

The *getc* function waits for a character to be received by the default USART. (The default USART is the USART defined by the last **#use rs232** preprocessor directive encountered before the call to *getc*.) If a hardware USART is being utilized, the hardware can buffer three characters. If a software USART is being used, then the *getc* routine (or *getchar*) must be active while the character is being received.

If more than one USART is being used, it is recommended that the function *fgetc* is used instead of *getc* such that the exact USART to be used can be specified.

Returns: Character received by the default USART

See *fgetc*.

---

gets

```
#use rs232()
void gets(char *str);
```

The *gets* function waits for a string of characters to be received by the default USART. (The default USART is the USART defined by the last **#use rs232** preprocessor directive encountered before the call to *gets*.) The string is read into the memory pointed by *str* until the carriage return (13) character is received. The string is terminated with a '\0' (NULL). It is up to the programmer to ensure that adequate memory is allocated to *str* to contain the string read plus the NULL terminator.

If more than one USART is being used, it is recommended that the function *fgets* be used instead of *gets* such that the exact USART to be used can be specified.

Returns: String received from the default USART is placed into the memory pointed to by *str*.

See *fgets*.

---

goto_address

void goto_address(int16 addr);

void goto_address(int32 addr);

The *goto_address* function causes the program execution to jump to the specified address, *addr*. Jumps outside of the current function should be done only with great caution.

Returns: None

```
#include <16F877.h>
#fuses HS,NOWDT           // a Microchip PIC16F877
#use delay(clock=10000000)
#use rs232(baud=9600, xmit=PIN_C6, rcv=PIN_C7)

void main()
{
    int8 a;

    a = 10;
    goto_address(label_address(skip));
    a = 20;
skip:

    printf("a = %d.\r\n",a);

    while(1)
    {
    }
}
```

Results: Transmits the following at 9600 baud:

```
a = 10.
```

i2c_poll

#use i2c()

int1 i2c_poll();

The *i2c_poll* should only be used when the hardware SSP is used (hardware I^2C interface in slave mode). This function returns true if the hardware has a received byte in the buffer. When a true is returned, a call to *i2c_read* immediately returns the byte that was received.

Returns: True (1) byte has been received by the I^2C interface or false (0) no byte has been received.

See *i2c_read* for an example.

i2c_read

#use i2c()

int8 i2c_read([int1 ack]);

The *i2c_read* function reads a byte over the I^2C interface. In master mode this function generates a clock and in slave mode it waits for the clock. There is no timeout for the slave. Use *i2c_poll* to prevent a lockup when reading on the slave side. Use **RESTART_WDT** in the **#use i2c** preprocessor directive to strobe the watchdog timer in the slave mode while waiting.

ack is a 1 or a 0 indicating whether or not to acknowledge the byte while reading. *ack* defaults to 1 (acknowledge).

Returns: 8-bit value read from the I^2C interface

```
// This code contains master and slave side communication examples
// for the I2C port. A conditional compile is used to select the side
// that is being compiled. To compile for the master side, uncomment
// #define MASTER_DEVICE below.
#include <16F877.h>
#fuses HS,NOWDT          // a Microchip PIC16F877
#use delay(clock=10000000)
#use rs232(baud=9600,parity=N,xmit=PIN_C6,rcv=PIN_C7,bits=8)
#use standard_IO(D)

//#define MASTER_DEVICE       1

#ifdef MASTER_DEVICE
// (master side)
#use i2c(master,sda=PIN_D1, scl=PIN_D0)
void main()
{
    int8 c,d;
```

```
    output_float(PIN_D0);
    output_float(PIN_D1);
    printf("Press any key to start: ");
    while(1)
    {
          c = getchar();
          putchar(c);          // echo back to user
          i2c_start();         // send start command
          i2c_write(0xa0);     // send slave address
          i2c_write(c);        // send data
          i2c_stop();          // send stop command

    }
}

#else
// (slave side)
#use i2c(slave,sda=PIN_C4,scl=PIN_C3,address=0xa0,FORCE_HW)
void main()
{

    int8 c;
    while(1)
    {
          // wait for data ready before reading!
          if (i2c_poll())
          {
                c = i2c_read();              // read the data
                output_D(0xFF ^ c);          // make nice for LEDs
          }
    }
}
#endif
```

Results: Whatever value is received by the USART on the master side is transmitted to the slave via the I^{2}C interface. The slave then outputs the inverted value to Port D. The master also echoes the value back over the USART for user feedback.

i2c_start

#use i2c()

void i2c_start();

The *i2c_start* function issues a start condition when the device is in I^{2}C master mode. After the start condition, the clock is held low until *i2c_write* is called. If another *i2c_start* is called before an *i2c_stop* is called, then a special restart condition is issued.

Returns: None

See *i2c_read* for an example.

---

i2c_stop

```
#use i2c()
void i2c_stop();
```

The *i2c_stop* function issues a stop condition when the device is in I^2C master mode.

Returns: None

See *i2c_read* for an example.

---

i2c_write

```
#use i2c()
int1 i2c_write(int8 data);
```

The *i2c_write* function sends a single byte, *data*, over the I^2C interface. In master mode, this function generates a clock with the data. In slave mode, it waits for the clock from the master. No automatic timeout is provided in this function. This function returns the ACK bit. The LSB of the first write after a start determines the direction of data transfer (0 is master to slave, 1 is slave to master).

Returns: ACK bit from write to the I^2C interface

See *i2c_read* for an example.

---

input

```
char input(const pin);
```

The *input* function returns the current state of *pin*. *pin* is the bit address of an I/O pin, and the definitions can be found in the *.h* file of the particular device in use.

The method of I/O used is dependent on the last **#use** *_IO directive.

Returns: 0 if the pin is low or 1 if the pin is high

See *output_low* for an example.

---

input_*port*

```
char input_A();
char input_B();
char input_C();
char input_D();
char input_E();
```

The *input_port* function reads the entire port, *port*, and returns the 8-bit value. The direction register is changed in accordance with the last specified **#use *_io** directive. By default with standard I/O before the input is done, the data direction is set to input.

Returns: 8-bit value representing *port*

See *output_port* for example.

---

isalnum

#include <ctype.h>

unsigned char isalnum(char c);

The *isalnum* function tests *c* to see if it is an alphanumeric character.

Returns: 1 if *c* is alphanumeric

```
#include <16F877.h>
#include <ctype.h>
#fuses HS,NOWDT             // a Microchip PIC16F877

unsigned char c_alnum_flag, d_alnum_flag;
void main(void)
{
    c_alnum_flag = isalnum('1'); // test the ASCII value of 1 (0x31)
    d_alnum_flag = isalnum(1);   // test the value 1
    while(1)
         ;
}
```

Results: c_alnum_flag = 1

d_alnum_flag = 0

---

isalpha

#include <ctype.h>

unsigned char isalpha(char c);

The *isalpha* function tests *c* to see if it is an alphabetic character (upper- or lowercase 'a' through 'z').

Returns: 1 if *c* is alphabetic

```
#include <16F877.h>
#include <ctype.h>
#fuses HS,NOWDT             // a Microchip PIC16F877

unsigned char c_alpha_flag, d_alpha_flag;
void main(void)
```

```
{
    c_alpha_flag = isalpha('a');    // test the ASCII character 'a'
    d_alpha_flag = isalpha('1');    // test the ASCII character '1'
    while(1)
         ;
}
```

Results: c_alpha_flag = 1

d_alpha_flag = 0

isamoung

unsigned char isamoung(char c, const char *str);

The *isamoung* function tests *c* to see if it is one of the characters in the constant string pointed to by **str*.

Returns: 1 if *c* is in **str*

```
#include <16F877.h>
#include <ctype.h>
#fuses HS,NOWDT             // a Microchip PIC16F877

unsigned char c_is_flag, d_is_flag;
void main(void)
{
    char a;
    a = '%';
    c_is_flag = isamoung(a,"abcdef:!@#");
    d_is_flag = isamoung(a,"%^&*()");
    while(1)
         ;
}
```

Results: c_is_flag = 0

d_is_flag = 1

iscntrl

#include <ctype.h>

unsigned char iscntrl (char c);

The *iscntrl* function tests *c* to see if it is a control character. Control characters range from 0d to 31d.

Returns: 1 if *c* is a control character

```
#include <16F877.h>
#include <ctype.h>
```

```
#fuses HS,NOWDT            // a Microchip PIC16F877

unsigned char c_iscntrl_flag, d_iscntrl_flag;
void main(void)
{
    c_iscntrl_flag = iscntrl('\t'); // test the control
                                    // character, horizontal tab
    d_iscntrl_flag = iscntrl('a');  // test the ASCII character a
    while(1)
          ;
}
```

Results: c_iscntrl_flag = 1

d_iscntrl_flag = 0

---

isdigit

#include <ctype.h>

unsigned char isdigit(char c);

The *isdigit* function tests *c* to see if it is an ASCII representation of a decimal digit ('0' through '9').

Returns: 1 if *c* is a decimal digit

```
#include <16F877.h>
#include <ctype.h>
#fuses HS,NOWDT            // a Microchip PIC16F877

unsigned char c_isdigit_flag, d_isdigit_flag;
void main(void)
{
    c_isdigit_flag = isdigit('1');  // test the ASCII character 1
    d_isdigit_flag = isdigit('a');  // test the ASCII character a
    while(1)
          ;
}
```

Results: c_isdigit_flag = 1

d_isdigit_flag = 0

---

isgraph

#include <ctype.h>

unsigned char isgraph(char c);

The *isgraph* function tests *c* to see if it is a printable character but not the space character. *isgraph* returns a 1 for characters that are between 33d and 127d.

Returns: 1 if *c* is a printable, non-space character

```
#include <16F877.h>
#include <ctype.h>
#fuses HS,NOWDT          // a Microchip PIC16F877

unsigned char c_is_flag, d_is_flag;
void main(void)
{
   c_is_flag = isgraph('A');// test the ASCII character A
   d_is_flag = isgraph(' ');// test a space character
   while(1)
         ;
}
```

Results: c_is_flag = 1

d_is_flag = 0

---

islower

#include <ctype.h>

unsigned char islower(char c);

The *islower* function tests *c* to see if it is a lowercase alphabetic character ('a' through 'z').

Returns: 1 if *c* is a lowercase alphabetic character

```
#include <16F877.h>
#include <ctype.h>
#fuses HS,NOWDT          // a Microchip PIC16F877

unsigned char c_islower_flag, d_islower_flag;
void main(void)
{
   c_islower_flag = islower('A');  // test the ASCII character A
   d_islower_flag = islower('a');  // test the ASCII character a
   while(1)
         ;
}
```

Results: c_islower_flag = 0

d_islower_flag = 1

---

isprint

#include <ctype.h>

unsigned char isprint(char c);

The *isprint* function tests *c* to see if it is a printable character. Printable characters are between 32d and 127d.

Returns: 1 if *c* is a printable character

```
#include <16F877.h>
#include <ctype.h>
#fuses HS,NOWDT          // a Microchip PIC16F877

unsigned char c_isprint_flag, d_isprint_flag;
void main(void)
{
    c_isprint_flag = isprint('A');  // test the ASCII character A
    d_isprint_flag = isprint(0x03); // test a control character
    while(1)
        ;
}
```

Results: c_isprint_flag = 1

d_isprint_flag = 0

---

ispunct

#include <ctype.h>

unsigned char ispunct(char c);

The *ispunct* function tests *c* to see if it is a punctuation character. All characters that are not control characters and not alphanumeric characters are considered to be punctuation characters.

Returns: 1 if *c* is a punctuation character

```
#include <16F877.h>
#include <ctype.h>
#fuses HS,NOWDT          // a Microchip PIC16F877

unsigned char c_ispunct_flag, d_ispunct_flag;
void main(void)
{
    c_ispunct_flag = ispunct('.'); // test the ASCII character, period
    d_ispunct_flag = ispunct('\t');// test the horizontal tab
    while(1)
        ;
}
```

Results: c_ispunct_flag = 1

d_ispunct_flag = 0

isspace

#include <ctype.h>

unsigned char isspace(char c);

The *isspace* function tests *c* to see if it is the space character. *isspace* returns a 1 for the space character and 0 for all other values.

Returns: 1 if *c* is the space character

```
#include <16F877.h>
#include <ctype.h>
#fuses HS,NOWDT          // a Microchip PIC16F877

unsigned char c_is_flag, d_is_flag;
void main(void)
{
    c_is_flag = isspace('A');// test the ASCII character A
    d_is_flag = isspace(' ');// test a space character
    while(1)
          ;
}
```

Results: c_is_flag = 0

d_is_flag = 1

isupper

#include <ctype.h>

unsigned char isupper(char c);

The *isupper* function tests *c* to see if it is an uppercase alphabetic character ('A' through 'Z').

Returns: 1 if *c* is an uppercase alphabetic character

```
#include <16F877.h>
#include <ctype.h>
#fuses HS,NOWDT          // a Microchip PIC16F877

unsigned char c_is_flag, d_is_flag;
void main(void)
{
    c_is_flag = isupper('A');// test the ASCII character A
    d_is_flag = isupper('a');// test the ASCII character a
    while(1)
          ;
}
```

Results: c_is_flag = 1

d_is_flag = 0

---

isxdigit

#include <ctype.h>

unsigned char isxdigit(char c);

The *isxdigit* function tests c to see if it is an ASCII representation of a hexadecimal digit ('0' through '9', 'a' through 'f', or 'A' through 'F').

Returns: 1 if *c* is a hexadecimal digit

```
#include <16F877.h>
#include <ctype.h>
#fuses HS,NOWDT           // a Microchip PIC16F877

unsigned char c_is_flag, d_is_flag;
void main(void)
{
   c_is_flag = isxdigit('1');      // test the ASCII character 1
   d_is_flag = isxdigit('g');      // test the ASCII character g
   while(1)
         ;
}
```

Results: c_is_flag = 1

d_is_flag = 0

---

kbhit

#use rs232()

char kbhit();

The *kbhit* function returns a 0 (false) if *getc* will need to wait for a character to come in, or 1 (true) if a character is ready for *getc*. If the USART is hardware based, then *kbhit* returns true, which means a character has been received and is waiting in the hardware buffer for *getc* to read. If the USART is under software control, this function returns true if the start bit of a character is being sent on the receive pin. When a software USART is in use, *kbhit* should be called at least 10 times the bit rate to ensure incoming data is not lost.

Returns: None

```
#include <16F877.h>
#fuses HS,NOWDT // a Microchip PIC16F877
#use delay(clock=10000000)
#use rs232(baud=9600,parity=N,xmit=PIN_C6,rcv=PIN_C7)
```

```
void main()
{
    char j;
    fputs("Echo Characters:");
    while(1)
    {
        // check before calling getc so that I can
        // do other things while I'm waiting!
        if (kbhit())
        {
            j = getc();
            putc(j);
        }
    }
}
```

Results: The following is transmitted on pin C6 at 9600 baud:

```
Echo Characters:
```

Then any character received by the PIC on pin C7 is echoed back to pin C6 until power is removed from the device.

label_address

int16 label_address(*program_label*);

The *label_address* function obtains the address in ROM of the next instruction after the label, *program_label*.

Returns: 16-bit ROM address of *program_label*

```
#include <16F877.h>
#fuses HS,NOWDT              // a Microchip PIC16F877
#use delay(clock=10000000)
#use rs232(baud=9600, xmit=PIN_C6, rcv=PIN_C7)


void main()
{
    char a, b, c;

start:
    a = (b+c) << 2;
end:
    printf("It takes %lu ROM locations.\r\n",
        label_address(end)-label_address(start));
    while(1)
```

```
        {
        }
    }
```

Results: Transmits at 9600 baud:

```
It takes 8 ROM locations.
```

---

labs

#include <stdlib.h>

int16 labs(int16 x);

Returns: the absolute value of *x*

See *abs*.

---

ldexp

#include <math.h>

float ldexp(float x, int expon);

The *ldexp* function calculates the value of *x* multiplied by the result of 2 raised to the power of *expon*.

Returns: $x * 2^{expon}$.

```
#include <16F877.h>
#include <math.h>
#fuses HS,NOWDT          // a Microchip PIC16F877
#use delay(clock=10000000)
#use rs232(baud=9600, xmit=PIN_C6, rcv=PIN_C7)

void main()
{
    float new_val;
    new_val = ldexp(5,3);
    printf("%f\n\r",new_val);
    while(1)
        ;
}
```

Results: new_val = $5 * 2^3$ = 39.999999

---

log

#include <math.h>

float log(float x);

The *log* function calculates the base *e* or natural logarithm of the floating point value *x*. *x* must be a positive, non-zero value.

Returns: log(x)

```
#include <16F877.h>
#include <math.h>
#fuses HS,NOWDT          // a Microchip PIC16F877
#use delay(clock=10000000)
#use rs232(baud=9600, xmit=PIN_C6, rcv=PIN_C7)
void main()
{
    float new_val;
    new_val = log(5);
    printf("%f\n\r",new_val);
    while(1)
        ;
}
```

Results: new_val = 1.609437

---

log10

#include <math.h>

float log10(float x);

The *log10* function calculates the base 10 logarithm of the floating point value *x*. *x* must be a positive, non-zero value.

Returns: $\log_{10}(x)$

```
#include <16F877.h>
#include <math.h>
#fuses HS,NOWDT          // a Microchip PIC16F877
#use delay(clock=10000000)
#use rs232(baud=9600, xmit=PIN_C6, rcv=PIN_C7)

void main()
{
    float new_val;
    new_val = log10(5);
    printf("%f\n\r",new_val);
    while(1)
        ;
}
```

Results: new_val = 0.698970

make16

int16 = make16(int8 var_high, int8 var_low);

The *make16* function uses 2-byte moves to create a 16-bit value from two 8-bit values, *var_high* and *var_low*. If either *var_high* or *var_low* is larger than 8 bits, only the least significant 8 bits are used.

Returns: 16-bit value as

```
(((var_high&0xFF)<<8) | (var_low & 0xFF))

#include <16F877.h>
#fuses HS,NOWDT          // a Microchip PIC16F877
#use delay(clock=10000000)

int8 a;
int16 b;
int32 c;

void main()
{
    c = 0x01ABCDEF;
    a = 0x53;
    b = make16(a,c);
    while(1)
        ;
}
```

Results: b = 0x53EF

make32

int32 = make32(int8 var1[, int8 var2, int8 var3, int8 var4]);

The *make32* function creates a 32-bit value from up to four 8-bit values, *var1*, *var2*, *var3*, and *var4*. *var2*, *var3*, and *var4* are optional. *var1* is the MSB and the last parameter used is the LSB. If the total number of bits provided in the parameters is less than 32, then zeros are added at the MSB.

Returns: 32-bit value

```
#include <16F877.h>
#fuses HS,NOWDT          // a Microchip PIC16F877
#use delay(clock=10000000)

int8 a;
int16 b;
```

```
int32 c,d;

void main()
{
    a = 0x53;
    b = 0xABCD;
    c = make32(a,b);
    while(1)
         ;
}
```

Results: c = 0x0053ABCD

make8

int8 = make8(int16 var, int8 offset);

int8 = make8(int32 var, int8 offset);

The *make8* function uses a single byte move to create an 8-bit value from a 16- or 32-bit value, *var*. *offset* is a byte offset of 0, 1, 2, or 3 specifying which byte should be returned.

Returns: 8-bit value from *var*

```
#include <16F877.h>
#fuses HS,NOWDT          // a Microchip PIC16F877
#use delay(clock=10000000)

int8 a;
int32 c;

void main()
{
    c = 0x01ABCDEF;
    a = make8(c,2);
    while(1)
         ;
}
```

Results: a = 0xAB

*memcpy

void memcpy(void *dest, void *src, unsigned char n);

The function *memcpy* copies *n* bytes from the memory location pointed to by *src* to the memory location pointed to by *dest*.

Returns: Pointer to *dest*

```
#include <16F877.h>
```

```
#fuses HS,NOWDT          // a Microchip PIC16F877
#use delay(clock=10000000)
#use rs232(baud=9600, xmit=PIN_C6, rcv=PIN_C7)

char inputstr[] = "$11.2#";
char outputstr[6];
void *a;
char outputstrf[6];
void *b;
char hello_str[] = "Hello World!";

void main(void)
{
   memcpy(outputstr,inputstr+1,4);
   outputstr[4] = '\0';     // null terminate our new string
   memcpy(outputstrf,hello_str,5);
   outputstrf[5] = '\0';    // null terminate our new string!
   printf("outputstr: %s\n\r",outputstr);
   printf("outputstrf: %s\n\r",outputstrf);
   while (1)
         ;
}
```

Results: UART transmits at 9600 baud,

```
outputstr: 11.2
outputstrf: Hello
```

---

memset

void memset(void *buf, unsigned char c, unsigned char n);

The *memset* function fills *n* bytes of the memory pointed to by *buf* with the character *c*.

Returns: None

```
#include <16F877.h>
#fuses HS,NOWDT          // a Microchip PIC16F877
#use delay(clock=10000000)
#use rs232(baud=9600, xmit=PIN_C6, rcv=PIN_C7)

char inputstr1[] = "abc1";

void main(void)
{
   // starting after a, fill in with some 2's
   memset(&(inputstr1[1]),'2',3);
```

```
        puts(inputstr1);
        while (1)
              ;
    }
```

Results: UART transmits at 9600 baud,

a222

modf

#include <math.h>

float modf(float x, float *ipart);

The *modf* function splits the floating point number, *x*, into its integer and fractional components. The fractional part of *x* is returned as a signed floating point number. The integer part is stored as a floating point number at *ipart*. Notice that the **address** of the variable to hold the integer portion, not the variable itself, is passed to *modf*. Both the integer and the floating point results have the same sign as *x*.

Returns:

- Fractional portion of the floating point number *x* as a signed floating point number
- Sets the value at the address pointed to by **ipart* to the integer part of *x*

```
#include <16F877.h>
#include <math.h>
#fuses HS,NOWDT           // a Microchip PIC16F877
#use delay(clock=10000000)
#use rs232(baud=9600, xmit=PIN_C6, rcv=PIN_C7)

void main(void)
{
    float integer_portion, fract_portion;
    fract_portion = modf(-45.7, &integer_portion);
    printf("fract_portion = %f\n\r",fract_portion);
    printf("integer_portion = %f\n\r",integer_portion);
    while(1)
          ;
}
```

Results: fract_portion = -.700000

integer_portion = -44.999999

offsetof

#include <stddef.h>

int8 offsetof(*structure type*, *field*);

The *offsetof* function returns the offset, in bytes, of the field *field* in a structure of the type *structure type. structure type* is the type name of a structure, and *field* is the name of one of the fields within the structure.

Returns: Offset, in bytes, of *field* in a structure of the type *structure type*

```
#include <16F877.h>
#include <stddef.h>
#fuses HS,NOWDT            // a Microchip PIC16F877

struct time_structure
{
    int hour, min, sec;
    int1 daylight_savings;    // one bit for flag
    int zone : 7;
} mytime;

void main(void)
{
    int sec_offset,zone_offset_bits,a,b;

    // initialize the structure to all 0s
    memset(&mytime,0x00,sizeof(mytime));

    // find the byte offset of the seconds field
    sec_offset = offsetof(time_structure, sec);
    // assign a value to the seconds field using the offset
    *(&mytime + sec_offset) = 28;
    // copy the value to a
    a = mytime.sec;

    // find the bit offset of the zone field
    zone_offset_bits = offsetofbit(time_structure,zone);
    // assign a value to the zone field
    mytime.zone = 3;
    // read the byte from the structure containing the zone field
    b = *(&mytime + (zone_offset_bits / 8));
    // shift the field down to the lowest bits of the byte
    b >>= (zone_offset_bits % 8);

    while (1)
          ;
}
```

Results: sec_offset = 2

a = 28

zone_offset_bits = 25

b = 3

offsetofbit

```
#include <stddef.h>

int8 offsetofbit(structure type, field);
```

The *offsetofbit* function returns the offset, in bits, of the field *field* in a structure of the type *structure type*. *structure type* is the type name of a structure, and *field* is the name of one of the fields within the structure.

Returns: Offset, in bits, of *field* in a structure of the type *structure type*

See *offsetof* for an example.

output_bit

```
void output_bit(const pin, char value);
```

The *output_bit* function outputs *value* (0 or 1) to the specified pin. *pin* is the bit address of an I/O pin, and the definitions can be found in the *.h* file of the particular device in use.

The method of I/O used is dependent on the last **#use** *_IO directive.

Returns: None

See *output_low* for an example.

output_float

```
void output_float(const pin);
```

The *output_float* function sets the data direction of *pin* to input. *pin* is the bit address of an I/O pin, and the definitions can be found in the *.h* file of the particular device in use.

The method of I/O used is dependent on the last **#use** *_IO directive.

Returns: None

```
#include <16F877.h>
#fuses HS,NOWDT          // a Microchip PIC16F877
#use delay(clock=10000000)
#use standard_IO(D)
#use fixed_IO(c_outputs = PIN_C2)

void main()
{
    char d;
    while(1)
    {
```

```
            output_low(PIN_C2);
            delay_ms(500);
            output_float(PIN_D0);     // port D pin 0 is an input
            d = 0;
            while(d == 0)
                 d = input(PIN_D0);
            output_high(PIN_C2);
            for (d=0;d<10;d++)
            {
                 output_low(PIN_D0);
                 delay_ms(500);
                 output_high(PIN_D0);
                 delay_ms(500);
            }
        }
    }
```

Results: This program clears pin C2 to signal that it is reading pin D0. The program waits until it sees pin D0 go high, then it sets pin C2 and enables pin D0 as an output. Pin D0 is flashed 10 times before starting over and calling output_float to set D0 as an input again.

---

output_high

void output_high(const *pin*);

The *output_high* function sets *pin* to the high state. *pin* is the bit address of an I/O pin, and the definitions can be found in the *.h* file of the particular device in use.

The method of I/O used is dependent on the last **#use *_IO** directive.

Returns: None

See *output_low* for an example.

---

output_low

void output_low(const *pin*);

The *output_low* function sets *pin* to the ground state. *pin* is the bit address of an I/O pin, and the definitions can be found in the *.h* file of the particular device in use.

The method of I/O used is dependent on the last #use *_IO directive.

Returns: None

```
#include <16F877.h>
#fuses HS,NOWDT          // a Microchip PIC16F877

#use fixed_io(d_outputs=PIN_D3,PIN_D2,PIN_D1,PIN_D0)
```

```
void main()
{
    char b;
    port_b_pullups(TRUE);
    while(1)
    {
        b = input(PIN_B0);
        if (b == 0)
            output_low(PIN_D0);
        else
            output_high(PIN_D0);

        b = input(PIN_B1);
        output_bit(PIN_D1,b);

        b = input(PIN_B2);
        output_bit(PIN_D2,b);

        b = input(PIN_B3);
        output_bit(PIN_D3,b);
    }
}
```

Results: The lowest 4 bits of Port B are output to the lowest 4 bits of Port D. (Because the internal pull-ups are enabled for Port B, the pin is high if no signal is applied and low if pulled low externally.)

output_*port*

void output_A(char value);

void output_B(char value);

void output_C(char value);

void output_D(char value);

void output_E(char value);

The *output_port* function outputs *value* to the port specified in the function call. The direction register is changed in accordance with the last specified **#use *_io** directive. The availability of the specific functions is dependent on the device in use.

Returns: None

```
#include <16F877.h>
#fuses HS,NOWDT            // a Microchip PIC16F877

#use fixed_io(d_outputs=PIN_D7,PIN_D6,PIN_D5,PIN_D4,PIN_D3,PIN_D2,PIN_D1,PIN_D0)
```

```
void main()
{
    char b;
    port_b_pullups(TRUE);
    while(1)
    {
        b = input_b();
        output_d(b);
    }
}
```

Results: The value read in from Port B is output to Port D. (Because the internal pull-ups are enabled for Port B, the pin is high if no signal is applied and low if pulled low externally. Also be aware that Port B pins 6 and 7 are used by in-circuit programmers; therefore, their values may be affected by the programmer when they are connected.)

---

port_b_pullups

void port_b_pullups(char value);

The *port_b_pull-ups* function enables or disables the internal pull-up resistors on Port B according to *value*. If *value* is **TRUE** (0), the pull-ups are enabled. If *value* is **FALSE** (1), the pull-ups are disabled.

Returns: None

See *output_port* for an example.

---

pow

#include <math.h>

float pow(float x, float y);

The *pow* function calculates *x* raised to the power of *y*.

Returns: x^y

```
#include <16F877.h>
#include <math.h>
#fuses HS,NOWDT          // a Microchip PIC16F877
#use delay(clock=10000000)
#use rs232(baud=9600, xmit=PIN_C6, rcv=PIN_C7)

void main()
{
    float new_val;
    new_val = pow(2,5);
    printf("%f\n\r",new_val);
    while(1)
```

```
            ;
    }
```

Results: new_val = 31.999999

printf

#use rs232()

void printf([*funtion name*],*stream identifier*, char *fmtstr [, arg1, arg2, ...]);

The *printf* function operates very similarly to the *fprintf* function and was covered extensively in Chapter 3. See Section 3.4.2, "Print Formatted, *printf()*, and File Print Formatted, *fprintf()*" in Chapter 3 and the *fprintf* section in this appendix for more information.

psp_input_full

int1 psp_input_full();

The *psp_input_full* function checks the Parallel Slave Port (PSP) to see if there is data ready to be read. *psp_input_full* returns 0 (false) or 1 (true).

Returns: 0 (false) if there is no data to read from the PSP or 1 (true) if data is available to be read

See *setup_psp()* for an example.

psp_output_full

int1 psp_output_full();

The *psp_output_full* function checks the Parallel Slave Port (PSP) to see if data can be written to it. *psp_output_full* returns 0 (false) or 1 (true).

Returns: 0 (false) if there is no data already written to the PSP or 1 (true) if data has been written to the PSP already

```
#include <16F877.h>
#fuses HS,NOWDT           // a Microchip PIC16F877
#use delay(clock=10000000)
#use rs232(baud=9600,parity=N,xmit=PIN_C6,rcv=PIN_C7,bits=8)

void main()
{
    int8 temp8;
    setup_psp(PSP_ENABLED);
    while(1)
    {
        temp8 = getchar();
        putchar(temp8);        // echo for feedback!
        while (!psp_output_full())
```

```
                psp_data = temp8;
        }
    }
```

Results: Data received by the USART is written to the PSP when the output is not full (previous data already read by external device).

---

psp_overflow

int1 psp_overflow();

The *psp_overflow* function checks the Parallel Slave Port (PSP) to see if an overflow condition exists. An overflow condition is caused by the PSP data being overwritten by an external device before the data is read by the microcontroller itself.

Returns: 0 (false) if an overflow condition does not exist on the PSP or 1 (true) if an overflow condition exists

```
#include <16F877.h>
#fuses HS,NOWDT          // a Microchip PIC16F877
#use delay(clock=10000000)
#use rs232(baud=9600,parity=N,xmit=PIN_C6,rcv=PIN_C7,bits=8)

void main()
{
    int8 temp8;
    setup_psp(PSP_ENABLED);
    while(1)
    {
        // try to cause an overflow condition by being busy!
        output_low(PIN_C2);
        delay_ms(1000);
        output_high(PIN_C2);
        delay_ms(250);
        if (psp_input_full())
        {
            if (psp_overflow() == 0)
            {
                temp8 = psp_data;
                printf("\n\rData: %X", temp8);
            }
            else
                printf("\n\rOVERFLOW!\n\r");
        }
    }
}
```

Results: Transmits to the USART whatever is received by the PSP. Prints "OVERFLOW" when an overflow condition is detected

putc

```
#use rs232()
void putc(char c);
```

The *putc* function transmits the character *c* to the default USART. The default USART is the USART defined by the last **#use rs232** preprocessor directive encountered before the call to *putc*.

If more than one USART is being used, it is recommended that the function *fputc* be used instead of *putc* such that the exact USART to be used can be specified.

Returns: None

See *fputc*.

puts

```
#use rs232()
void puts(char *str);
```

The *puts* function transmits the string, *str*, to the default USART. The default USART is the USART defined by the last **#use rs232** preprocessor directive encountered before the call to *putc*. Once the string is transmitted, carriage return (13) and a line feed (10) characters are transmitted.

If more than one USART is being used, it is recommended that the function *fputs* be used instead of *puts* such that the exact USART to be used can be specified.

Returns: None

See *fputs*.

read_adc

```
int8 read_adc();
int16 read_adc();
```

The *read_adc* function reads the digital value from the analog to digital converter. Calls to *setup_adc*, *setup_adc_ports*, and *set_adc_channel* should be made sometime before this function is called. The range of the return value depends on the number of bits in the chip's A/D converter and the setting in the **#device adc=x** preprocessor directive. The relationship between the chip A/D bits, the **#device adc** directive, and the return value size are shown in Table A.2.

#device	8-bit device	10-bit device	11-bit device	16-bit device
adc=8	0-FF	0-FF	0-FF	0-FF
adc=10	X	0-3FF	X	X
adc=11	X	X	0-7FF	X
adc=16	0-FF00	0-FFC0	0-FFE0	0-FFFF

Table A. 2 *Return Values for read_adc*

Returns: None

See Chapter 2, Figure 2.47 for an example.

---

read_bank

int8 read_bank(int8 bank, int16 offset);

The *read_bank* function reads a data byte from the user RAM area of the specified bank at the specified offset. Valid banks are 1-3 depending on the device, and the offset into the bank starts at zero. This function may be used on some devices where full RAM access by auto variables is not efficient. For example, on the PIC16C57 chip, setting the pointer size to 5 bits will generate the most efficient ROM code; however, auto variables cannot be above 0x1F. Instead of going to 8-bit pointers you can save ROM by using this function to write to the bank addresses beyond the reach of the minimized pointers. In this case with the PIC16C57, the bank may be 1-3 and the offset may be 0-15.

Returns: None

See *write_bank* for an example.

---

read_eeprom

int8 read_eeprom(int8 addr);

The *read_eeprom* function reads 1 byte from the EEPROM at address *addr*. The first byte of the EEPROM is address zero, and the size of the EEPROM depends on the device in use. This function is only available on devices with an internal EEPROM.

Returns: 1 byte of data from the EEPROM at address *addr*

```
#include <16F877.h>
#fuses HS,NOWDT            // a Microchip PIC16F877
#use delay(clock=10000000)
#use rs232(baud=9600, xmit=PIN_C6, rcv=PIN_C7)
#define LAST_CMD_ADDR 0x0A

void main(void)
{
```

```
        char c;
        printf("Last Cmd: %c\n\r",read_eeprom(LAST_CMD_ADDR));
        while (1)
        {
            // get a character from the USART and store it!
            write_eeprom(LAST_CMD_ADDR,getchar());
            c = read_eeprom(LAST_CMD_ADDR);
            printf("Last Cmd: %c\n\r",c);
        }
    }
```

Results: After an initial transmission of the data located at **LAST_CMD_ADDR** in the EEPROM, the device waits for a character to be received by the USART, stores it in the EEPROM, reads it from the EEPROM and transmits it in the string:

```
Last Cmd: h
```

If power is cycled to the device, you will notice that the last command data is still correct because it was stored in the EEPROM.

---

read_program_eeprom

For 14-bit devices:

int16 read_program_eeprom(int16 addr);

For 16-bit devices:

int8 read_program_eeprom(int32 addr);

The *read_program_eeprom* function reads data from the program memory of the device from address *addr*. The size of the return value and of the address is dependent on the device in use as shown above. This function is only available on devices that allow reads from program memory.

Returns: 16-bit integer for 14-bit devices

8-bit integer for 16-bit devices

```
#include <16F877.h>
#fuses HS,NOWDT            // a Microchip PIC16F877
#use delay(clock=10000000)
#use rs232(baud=9600, xmit=PIN_C6, rcv=PIN_C7)

#define ID_LOCATION          0x1C00
#define ID_DATA_LENGTH       10
#define TOTAL_ID_LENGTH      16
#define ID_OVERHEAD          10
#org ID_LOCATION, ID_LOCATION + TOTAL_ID_LENGTH

char const ID[10] = {"123456789"};
```

```
void main(void)
{
    int8 temp_int8;
    int16 temp_int16, addr;
    int8 id[ID_DATA_LENGTH];
    int8 cnt;
    char c;
    printf("'R' to read id.\n\r'W' to write id.\n\r");
    while(1)
    {
        c = getchar();
        // write new id
        if (c == 'W')
        {
            putchar(c);              // echo for user feedback
            for (cnt=0;cnt<(ID_DATA_LENGTH-1);cnt++)
            {
                temp_int8 = getchar();
                putchar(temp_int8); // echo for user feedback
                temp_int16 = temp_int8 + 0x3400;
                addr = ID_LOCATION + cnt + ID_OVERHEAD;
                // add return literal command to data byte and write!
                write_program_eeprom(addr,temp_int16);
            }
            // null terminate the string
            addr = ID_LOCATION + cnt + ID_OVERHEAD;
            write_program_eeprom(addr, 0x3400);
        }
        // read and print id
        else if (c == 'R')
        {
            putchar(c);              // echo for user feedback
            for (cnt=0;cnt<(ID_DATA_LENGTH-1);cnt++)
            {
                addr = ID_LOCATION + cnt + ID_OVERHEAD;
                temp_int16 = read_program_eeprom(addr);
                // we only want to keep the lower byte (upper is
                //  return literal command, lower is data)
                id[cnt] = (int8) (temp_int16 & 0xFF);
            }
            // null terminate the string
            id[cnt] = '\0';
            printf(" ID: %s\n\r",id);
        }
```

```
        else
            putchar('?');     // show we didn't understand command
    }
}
```

Results: The UART transmits a prompt to the user to send 'R' to read the current ID bytes or 'W' to write a new ID. Upon receipt of the command, the device either transmits the existing ID or receives a new ID. Following is a copy of a communication session.

Initial power-up:

```
'R' to read id.
'W' to write id.
```

Send 'R' to the device:

```
R ID: Abcdefhih
```

Send 'W' followed by "abcdefghij" to the device:

```
Wabcdefghi?
```

Send a second 'R' to the device:

```
R ID: abcdefghi
```

---

reset_cpu

void reset_cpu();

The *reset_cpu* function jumps to location 0 to force a device reset.

Returns: None

```
#include <16F877.h>
#fuses HS,NOWDT           // a Microchip PIC16F877
#use delay(clock=10000000)
#use rs232(baud=9600, xmit=PIN_C6, rcv=PIN_C7)

void main()
{
    char c;

    printf("Hello World!");

    while(1)
    {
        c = getchar();
        putchar(c);
        if (c == 'R')
            reset_cpu();
    }
}
```

Results: The microcontroller transmits "Hello World!" at 9600 baud on power-up, then echoes every character it receives. If 'R' is received, then the microcontroller resets and transmits "Hello World!" again.

---

restart_cause

char restart_cause();

The *restart_cause* function returns a value indicating the cause of the last processor reset. The actual values are device dependent. See the *.h* file for the specific values for a specific device. Some example values are: **WDT_FROM_SLEEP, WDT_TIMEOUT, MCLR_FROM_SLEEP,** and **NORMAL_POWER_UP.**

Returns: Cause of the last reset of the microcontroller

```
#include <16F877.h>
#fuses HS,WDT              // a Microchip PIC16F877
#use delay(clock=10000000)
#use rs232(baud=9600, xmit=PIN_C6, rcv=PIN_C7)

void main()
{

   switch ( restart_cause() )
   {
      case WDT_TIMEOUT:
      {
        printf("\r\nRestarted - Watchdog timeout!\r\n");
        break;
      }
      case NORMAL_POWER_UP:
      {
        printf("\r\nNormal power up!\r\n");
        break;
      }
   }

   setup_wdt(WDT_2304MS);

   while(TRUE)
   {
      restart_wdt();
      printf("Hit any key to avoid watchdog timeout.\r\n");
      getc();
   }
}
```

Results: On power-up, "Normal power up!" is transmitted at 9600 baud followed by "Hit any key to avoid watchdog timeout." If no keys are pressed (no characters are received by the microcontroller), the watchdog timer expires and resets the microcontroller, at which time "Restarted – Watchdog timeout!" is transmitted. If keys are pressed, then the key press prompt is transmitted once for each key press.

---

restart_wdt

```
#fuses WDT
void restart_wdt();
```

The *restart_wdt* function restarts the watchdog timer. If the watchdog timer is enabled, this must be called periodically to prevent the processor from resetting. The watchdog timer is used to cause a hardware reset if the software stops. See *setup_wdt* for information on enabling the watchdog timer and establishing its timeout period.

Returns: None

See *restart_cause* for an example.

---

rotate_left

```
void rotate_left(char *addr, char bytes);
```

The *rotate_left* function rotates *bytes* number of bytes pointed to by *addr* to the left by 1 bit. *addr* may be an array identifier or an address to a byte or structure. Bit 0 or the lowest byte in RAM is considered the LSB. The value shifted out at the left is shifted in at the right.

Returns: None

```
#include <16F877.h>
#fuses HS,NOWDT          // a Microchip PIC16F877
#use delay(clock=10000000)
#use rs232(baud=9600, xmit=PIN_C6, rcv=PIN_C7)

void main()
{
    int8 bit_array[3];
    int8 i;
    bit_array[0] = 0x05;
    bit_array[1] = 0x18;
    bit_array[2] = 0x64;
    for (i=0;i<24;i++)
    {
        printf("i=%d: %X %X %X\r\n",i,
            bit_array[2],bit_array[1],bit_array[0]);
        rotate_left(bit_array,3);
    }
```

```
        printf("i=%d: %X %X %X\r\n",i,
            bit_array[2],bit_array[1],bit_array[0]);

        while(1)
        {
        }
}
```

Results: Transmits 25 lines containing the current values of the array as it is rotated to the left 24 times. The following are excerpts from that transmission:

```
i=0: 64 18 05
i=4: 41 80 56
i=8: 18 05 64
i=12: 80 56 41
i=16: 05 64 18
i=20: 56 41 80
i=24: 64 18 05
```

---

rotate_right

void rotate_right(char *addr, char bytes);

The *rotate_right* function rotates *bytes* number of bytes pointed to by *addr* to the right by 1 bit. *addr* may be an array identifier or an address to a byte or structure. Bit 0 or the lowest byte in RAM is considered the LSB. The value shifted out at the right is shifted in at the left.

Returns: None

```
#include <16F877.h>
#fuses HS,NOWDT            // a Microchip PIC16F877
#use delay(clock=10000000)
#use rs232(baud=9600, xmit=PIN_C6, rcv=PIN_C7)

void main()
{
    int8 bit_array[3];
    int8 i;

    bit_array[0] = 0x05;
    bit_array[1] = 0x18;
    bit_array[2] = 0x64;
    for (i=0;i<24;i++)
    {
        printf("i=%d: %X %X %X\r\n",i,
            bit_array[2],bit_array[1],bit_array[0]);
        rotate_right(bit_array,3);
    }
```

```
        printf("i=%d: %X %X %X\r\n",i,
            bit_array[2],bit_array[1],bit_array[0]);

        while(1)
        {
        }
}
```

Results: Transmits 25 lines containing the current values of the array as it is rotated to the right 24 times. The following are excerpts from that transmission:

```
i=0: 64 18 05
i=4: 56 41 80
i=8: 05 64 18
i=12: 80 56 41
i=16: 18 05 64
i=20: 41 80 56
i=24: 64 18 05
```

---

set_adc_channel

void set_adc_channel(int8 channel);

The *set_adc_channel* function selects the *channel* analog channel to use for the next *read_adc* call. The channel numbers start with zero and are labeled in the data sheet as AN0, AN1, and so on. Be aware that you must wait the required acquisition time after changing the channel before you can get a valid read. See the data sheets for the selected device to calculate the required acquisition time.

Returns: None

See Chapter 2, Figure 2.47 for an example.

---

set_pwmx_duty

void set_pwm1_duty(int16 *value*);

void set_pwm2_duty(int16 *value*);

The *set_pwmx_duty* function writes the 10-bit value *value* to the PWM to set the duty cycle. An 8-bit value may be used and will be shifted up to become a 10-bit value. The duty_cycle determines the amount of time the PWM signal is high during each cycle according to:

value(1/clock)*t2div*

where *clock* is oscillator frequency and *t2div* is the Timer2 prescaler (set in the call to *setup_timer2()*).

This function is available only on devices with CCP/PWM hardware.

Returns: None

```
#include <16F877.h>
#fuses HS,NOWDT          // a Microchip PIC16F877
#use delay(clock=10000000)
#use fixed_io(c_outputs = PIN_C2)
void main()
{
    int16 duty_time;
    setup_timer_2(T2_DIV_BY_16,250,1);
    setup_ccp1(CCP_PWM);
    duty_time = 0;
    while(1)
    {
        set_pwm1_duty(duty_time);
        duty_time += 100;
        duty_time &= 0x3FF; // roll over when we reach the top
        delay_ms(500);
    }
}
```

Results: This program creates a PWM output on pin C2 and increases the duty cycle by 100 every 500 ms. Only the lower 10 bits of the *duty_time* variable are kept so that the duty cycle rolls over when it becomes greater than 10 bits.

---

set_rtcc, set_timerX,

On devices with an 8-bit Timer0:

void set_rtcc(int8 value);

void set_timer0(int8 value);

On devices with a 16-bit Timer0:

void set_rtcc(int16 value);

void set_timer0(int16 value);

All devices that support the appropriate timer:

void set_timer1(int16 value);

void set_timer2(int8 value);

void set_timer3(int16 value);

The *set_timerx* functions sets the value of the associated timer/counter to *value*. The size of *value* is the same as the size of the timer. All timers count up and roll over to zero after they reach their maximum value. The maximum value depends on the number of bits available on the timer. Availability of these functions is device dependent. RTCC and Timer0 are the same.

Returns: None

See *setup_timer_X* for an example.

set_tris_port

void set_tris_A(char value);

void set_tris_B(char value);

void set_tris_C(char value);

void set_tris_D(char value);

void set_tris_E(char value);

The *set_tris_port* function sets the I/O port direction register associated with *port* according to *value*. Each bit in *value* corresponds to a pin on the port. If the bit is a 1, then the pin is set up as an input. If the bit is a 0, then the pin is set up as an output.

These functions must be used with **#use fast_io**. When the default standard I/O is used, the built-in functions will set the I/O direction automatically and then override any settings made by the *set_tris_port* functions.

Returns: None

```
#include <16F877.h>
#fuses HS,NOWDT          // a Microchip PIC16F877
#use delay(clock=10000000)
#use fast_IO(D)
#use fixed_IO(c_outputs = PIN_C2)

void main()
{
    char d;
    while(1)
    {
        // sequence of the program is, set C2 low,
        // wait 2 seconds, read port D. Set C2 high,
        // output the inverted portD value,
        //  wait 2 seconds and repeat.
        output_low(PIN_C2);
        delay_ms(5000);
        set_tris_D(0xFF);    // port D is all inputs
        d = input_D();
        output_high(PIN_C2);
        set_tris_d(0x00);
        output_D(d^0xFF);
        delay_ms(5000);
    }
}
```

Results: This program cycles between reading and writing Port D. The program sets pin C2 low to signal a read cycle, sets up Port D as an input, waits 5 seconds, then reads Port D. The program then clears pin C2 to signal a write cycle, sets up Port D as an output, and writes the inverted value back out to Port D. The program waits 5 seconds in the write cycle, then loops back to the read cycle and continues.

set_uart_speed

#use rs232()

void set_uart_speed(const int32 baud);

The *set_uart_speed* function changes the baud rate of the hardware RS232 serial port at run-time. *baud* is a constant value 100 to 115200 representing the bits-per-second or baud rate of the port This function is only available on devices with a hardware USART. Also note that not all baud rates can be achieved at all CPU clock frequencies.

Returns: None

```
#include <16F877.h>
#fuses HS,NOWDT           // a Microchip PIC16F877
#use delay(clock=10000000)
#use rs232(baud=9600,parity=N,xmit=PIN_C6,rcv=PIN_C7)

#use fixed_io(b_outputs=PIN_B7, PIN_B6, PIN_B5)

void main()
{
    char b;
    port_b_pullups(TRUE);
    while(1)
    {
        b = INPUT_B();
        switch (b)
        {
            case 0x00:
                set_uart_speed(2400);
                break;
            case 0x01:
                set_uart_speed(4800);
                break;
           case 0x03:
                set_uart_speed(9600);
                break;
           case 0x07:
                set_uart_speed(19600);
                break;
           case 0x0F:
                set_uart_speed(38400);
```

```
                    break;
                case 0x1F:
                    set_uart_speed(57600);
                    break;
            }
            printf("Baud selection was:%X\n\r",b);
            delay_ms(1000);
        }
    }
```

Results: This example changes the current baud rate of the USART based on the input at the lowest 5 bits of Port B. For example, if on Port B no pins were externally connected (and, therefore, high because internal pull-ups are on), the following is transmitted on pin C7 at 57600 baud:

```
Baud selection was:1F
```

Another example is if Port B pins 1 through 5 are pulled down, then the following is transmitted at 2400 baud:

```
Baud selection was:00
```

setup_adc

void setup_adc(*mode*);

The *setup_adc* function configures the analog to digital converter according to the constant *mode*. Valid *mode* values depend on the device selected and can be found in the *.h* file for the device. For example, the valid options for a PIC16F877 from the *16f877.h* file are:

```
// Constants used for SETUP_ADC() are:
#define ADC_OFF                    0
#define ADC_CLOCK_DIV_2            1
#define ADC_CLOCK_DIV_8         0x41
#define ADC_CLOCK_DIV_32        0x81
#define ADC_CLOCK_INTERNAL      0xc1
```

Returns: None

See Chapter 2, Figure 2.47 for an example.

setup_adc_ports

void setup_adc_ports(*mode*);

The *setup_adc_ports* function configures the ADC pins to a specified combination of analog and digital inputs. *mode* is a constant describing the combination of inputs and also establishes what will be used as the reference voltage. Valid *mode* values depend on the device selected and can be found in the *.h* file for the device. For example, the valid options for a PIC16F877 from the *16f877.h* file are:

```
// Constants used in SETUP_ADC_PORTS() are:
```

```
#define NO_ANALOGS               0x86     // None
#define ALL_ANALOG               0x80     // A0 A1 A2 A3 A5 E0 E1 E2
                                           //    Ref=Vdd
#define ANALOG_RA3_REF           0x81     // A0 A1 A2 A5 E0 E1 E2
                                           //    Ref=A3
#define A_ANALOG                 0x82     // A0 A1 A2 A3 A5 Ref=Vdd
#define A_ANALOG_RA3_REF         0x83     // A0 A1 A2 A5 Ref=A3
#define RA0_RA1_RA3_ANALOG       0x84     // A0 A1 A3 Ref=Vdd
#define RA0_RA1_ANALOG_RA3_REF 0x85       // A0 A1 Ref=A3

#define ANALOG_RA3_RA2_REF           0x88     // A0 A1 A5 E0 E1 E2
                                                //    Ref=A2,A3
#define ANALOG_NOT_RE1_RE2           0x89     // A0 A1 A2 A3 A5 E0
                                                //    Ref=Vdd
#define ANALOG_NOT_RE1_RE2_REF_RA3       0x8A  // A0 A1 A2 A5 E0 Ref=A3
#define ANALOG_NOT_RE1_RE2_REF_RA3_RA2   0x8B  // A0 A1 A5 E0 Ref=A2,A3
#define A_ANALOG_RA3_RA2_REF             0x8C // A0 A1 A5 Ref=A2,A3
#define RA0_RA1_ANALOG_RA3_RA2_REF       0x8D // A0 A1 Ref=A2,A3
#define RA0_ANALOG                       0x8E // A0
#define RA0_ANALOG_RA3_RA2_REF           0x8F // A0 Ref=A2,A3
```

Returns: None

See Chapter 2, Figure 2.47 for an example.

---

setup_ccpx

void setup_ccp1(const *mode*);

void setup_ccp2(const *mode*);

The *setup_ccpx* functions initialize the CCP according to mode. *mode* can be one of the constants in Table A.3.

The CCP counters are accessed using the long variables CCP_1 and CCP_2. The CCP operates in three modes: capture mode, compare mode, and PWM mode. In capture mode, the Timer1 value is copied to CCP_x when the input pin event occurs. In compare mode, an action is triggered when Timer1 matches CCP_*x*. In PWM mode, a square wave is generated on pin CCP*x*.

This function is only available on devices with CCP hardware.

Returns: None

Constant	Description
CCP_OFF	Disable the CCP
CCP_CAPTURE_FE	Capture Timer1 counts on falling edge of CCP*x*
CCP_CAPTURE_RE	Capture Timer1 counts on rising edge of CCP*x*
CCP_CAPTURE_DIV_4	Capture Timer1 counts on every 4th rising edge of CCP*x*
CCP_CAPTURE_DIV_16	Capture Timer1 counts on every 16th rising edge of CCP*x*
CCP_COMPARE_SET_ON_MATCH	Output CCP*x* high on compare match between Timer1 and CCP_*x*
CCP_COMPARE_CLR_ON_MATCH	Output CCP*x* low on compare match between Timer1 and CCP_*x*
CCP_COMPARE_INT	Generate an interrupt on compare match between Timer1 and CCP_*x*
CCP_COMPARE_RESET_TIMER	Reset timer on match between Timer1 and CCP_*x*
CCP_PWM	Enable pulse-width modulation on CCP*x*

Table A. 3 *CCP Mode Options*

```
#include <16F877.h>
  #fuses HS,NOWDT          // a Microchip PIC16F877
  #use delay(clock=10000000)
  #use rs232(baud=9600,parity=N,xmit=PIN_C6,rcv=PIN_C7,bits=8)
  #use fixed_io(c_outputs = PIN_C2)
  #use fixed_io(d_outputs = PIN_D0)
  void main()
  {
      long last_ccp1;
      setup_timer_1(T1_INTERNAL|T1_DIV_BY_8);
      setup_ccp1(CCP_CAPTURE_FE);
      last_ccp1 = ccp_1;
      while(1)
      {
           output_bit(PIN_D0,input(PIN_C2));
           delay_us(1000);
           if (last_ccp1 != ccp_1)
           {
               printf("%lu\n\r",ccp_1);
               last_ccp1 = ccp_1;
           }
      }
}
```

Results: This program transmits at 9600 baud, the Timer1 counts captured on the falling edge of CCP1 – pin C2.

setup_comparator

void setup_comparator(*mode*);

The *setup_comparator* function configures the analog comparator module according to *mode*. *mode* is a constant describing which pairs of analog inputs are to be compared. The available constants can be found in the *.h* file for the selected device. For example, these are the options available for the PIC18F458 copied from the file *18f458.h*:

```
#define A0_A3_A1_A2   4
#define A0_A2_A1_A2   3
#define NC_NC_A1_A2   5
#define NC_NC_NC_NC   7
#define A0_VR_A1_VR   2
#define A3_VR_A2_VR   10
#define A0_A2_A1_A2_OUT_ON_A3_A4 6
#define A3_A2_A1_A2   9
```

In these constants, the four parts represent the inputs C1-, C1+, C2-, C2+.

Returns: None

```
#include <18f458.h>
#fuses HS,NOWDT,NOPROTECT
#use delay(clock=10000000)
#use rs232(baud=9600, xmit=PIN_C6, rcv=PIN_C7)

short safe_conditions=TRUE;

#byte CMCON = 0xFB4
#bit C1_COMP = CMCON.6
#bit C2_COMP = CMCON.7

main()
{
   printf("\nRunning voltage test...\n\n");
   setup_comparator(A0_VR_A1_VR);

   while(TRUE)
   {
      if(!C1_COMP)
       printf("Voltage level is below 3.6 V.                \r");
      else
       printf("WARNING!!  Voltage level is above 3.6 V.    \r");
      delay_ms(500);
   }
}
```

Results: This example demonstrates the use of the hardware voltage comparator. Connect a potentiometer wiper to A0 and connect A1 to GND. The top and bottom of the potentiometer are connected to VCC and GND, respectively. Turn the potentiometer back and forth past its center to change the voltage. The following messages should appear as the 3.6 V point is passed by the wiper. The value of A0 is compared to the internal reference.

```
Voltage level is below 3.6 V.
```

or

```
WARNING!!  Voltage level is above 3.6 V.
```

setup_psp

void setup_psp(*mode*);

The *setup_psp* function initializes the Parallel Slave Port (PSP), where *mode* is either **PSP_ENABLED** or **PSP_DISABLED**. The *set_tris_E* function may be used to set the data direction. The data may be read and written to using the variable *psp_data*. This function is only available for devices with PSP hardware.

Returns: None

```
#include <16F877.h>
#fuses HS,NOWDT            // a Microchip PIC16F877
#use delay(clock=10000000)
#use rs232(baud=9600,parity=N,xmit=PIN_C6,rcv=PIN_C7,bits=8)

void main()
{
    int8 temp8;
    setup_psp(PSP_ENABLED);
    while(1)
    {
        if (psp_input_full())
        {
            temp8 = psp_data;
            printf("\n\rData: %X", temp8);
        }
    }
}
```

Results: Whenever data is received on the PSP, the data is transmitted by the USART in hexadecimal format.

setup_spi

void setup_spi(const *modes*);

The *setup_spi* function initializes the Serial Port Interface (SPI). This is used for 2 or 3 wire serial devices that follow a common clock/data protocol. *modes* are sets of defines that describe how the SPI port should be initialized. The *modes* are grouped into three categories: master/slave mode, clock idle mode, and clock speed. The options from these groups are:

Master/Slave Mode:

SPI_MASTER, SPI_SLAVE, SPI_SS_DISABLED

Clock Idle Mode:

SPI_L_TO_H, SPI_H_TO_L

Clock Speed:

SPI_CLK_DIV_4, SPI_CLK_DIV_16, SPI_CLK_DIV_64, SPI_CLK_T2

In the call to *setup_spi*, the options from the groups are OR'ed together to form a single parameter.

```
setup_spi (SPI_MASTER | SPI_L_TO_H | SPI_CLK_DIV_4);
```

This function is only available on devices with SPI hardware.

Returns: None

See *spi_read* for an example.

---

setup_timer_X

void setup_timer_0(*mode*);

void setup_timer_1*(mode);*

void setup_timer_2(*mode*, unsigned int8 period, int8 postscale);

void setup_timer_3(*mode*);

The *setup_timer_x* function sets up the specified timer according to *mode. mode* is a constant consisting of one or more timer options from several groups OR'ed together. The options available depend on the timer being used and are shown in Table A.4. These include clock source options and timer prescale options. Not all devices support all four timers.

Timer2 also has *period* and *postscale* parameters. *period* is an integer between 0 and 255 that determines when the clock value is reset. *postscale* is a number between 1 and 16 that determines how many timer resets occur before an interrupt. *postscale* equal to 1 means one reset before an interrupt; *postscale* equal to 2 means two timer resets before an interrupt and so on.

Returns: None

Timer	Source Options	Prescale Options	Other Options	Size of Timer
0 (also called RTCC)	RTCC_INTERNAL, RTCC_EXT_L_TO_H, RTCC_EXT_H_TO_L	RTCC_DIV_2, RTCC_DIV_4, RTCC_DIV_8, RTCC_DIV_16, RTCC_DIV_32, RTCC_DIV_64, RTCC_DIV_128, RTCC_DIV_256	(Only available on devices with a 16-bit Timer0) RTCC_OFF, RTCC_8_BIT	8 bits or 16 bits, device dependent
1	T1_DISABLED, T1_INTERNAL, T1_EXTERNAL, T1_EXTERNAL_SYNC	T1_DIV_BY_1, T1_DIV_BY_2, T1_DIV_BY_4, T1_DIV_BY_8	T1_CLK_OUT	16 bits
2		T2_DISABLED, T2_DIV_BY_1, T2_DIV_BY_4, T2_DIV_BY_16		8 bits
3	T3_DISABLED, T3_INTERNAL, T3_EXTERNAL, T3_EXTERNAL_SYNC	T3_DIV_BY_1, T3_DIV_BY_2, T3_DIV_BY_4, T3_DIV_BY_8		16 bits

Table A. 4 *Setup Timer Options*

```
#include <16F877.h>
#fuses HS,NOWDT            // a Microchip PIC16F877
#use delay(clock=10000000)
#use rs232(baud=9600,parity=N,xmit=PIN_C6,rcv=PIN_C7,bits=8)

void main()
{
    int16 timer1_val;

    setup_timer_1(T1_INTERNAL|T1_DIV_BY_8);
    // clear the timer
    set_timer1(0);
    // do something to be timed
    printf("\n\rHello World!\n\r");
    // read the new timer value
    timer1_val = get_timer1();
    // print the results
    printf("It took %ld timer ticks at 8!",timer1_val);

    // change the prescale and get a different count!
    setup_timer_1(T1_INTERNAL|T1_DIV_BY_4);
    set_timer1(0);
```

```
        printf("\n\rHello World!\n\r");
        timer1_val = get_timer1();
        printf("It took %ld timer ticks at 4!",timer1_val);

        while(1)
            ;
    }
```

Results: The following is transmitted by the USART:

```
Hello World!
It took 4534 timer ticks at 8!
Hello World!
It took 10399 timer ticks at 4!
```

setup_wdt

#fuses WDT

void setup_wdt(*mode*);

The *setup_wdt* function sets up the watchdog timer according to the constant, *mode*. The available options for *mode* can be found in the *.h* file for the device. For the watchdog timer to operate, it must be enabled and a timeout time set. These are done differently on the different versions of PICC. Table A.5 describes the methods of operating the watchdog for the different versions of PICC.

Operation Description	PCB/PCM Versions	PCH Version
Enable/Disable	#fuses	setup_wdt()
Timeout Time	setup_wdt()	#fuses
Setup Mode Options	WDT_18MS, WDT_36MS, WDT_72MS, WDT_144MS, WDT_288MS, WDT_576MS, WDT_1152MS, WDT_2304MS	WDT_ON, WDT_OFF
Restart	restart_wdt()	Restart_wdt()

Table A. 5 *Watchdog Timer Methods*

Returns: None

See *restart_cause* for an example.

shift_left

int1 shift_left(char *addr, char bytes, int1 value);

The *shift_left* function shifts a bit into an array or structure. *addr* is a pointer to memory, *bytes* is the number of bytes to work with, and *value* is a 0 or 1 to be shifted in. This function returns the value (0 or 1) shifted out. *addr* may be an array identifier or an address of a structure. Bit 0 of the lowest byte in RAM is treated as the LSB.

Returns: Bit shifted out

```
#include <16F877.h>
#fuses HS,NOWDT            // a Microchip PIC16F877
#use delay(clock=10000000)
#use rs232(baud=9600, xmit=PIN_C6, rcv=PIN_C7)


void main()
{
   int8 bit_array[3];
   int8 i;

   bit_array[0] = 0x05;
   bit_array[1] = 0x18;
   bit_array[2] = 0x64;
   for (i=0;i<24;i++)
   {
      printf("i=%d: %X %X %X\r\n",i,
         bit_array[2],bit_array[1],bit_array[0]);
      shift_left(bit_array,3,0);
   }
   printf("i=%d: %X %X %X\r\n",i,
      bit_array[2],bit_array[1],bit_array[0]);

   while(1)
   {
   }
}
```

Results: Transmits 25 lines containing the current values of the array as it shifts a 0 into the array and shifts the array to the right 24 times. The following are excerpts from that transmission:

```
i=0: 64 18 05
i=4: 41 80 50
i=8: 18 05 00
i=12: 80 50 00
i=16: 05 00 00
i=20: 50 00 00
i=24: 00 00 00
```

shift_right

int1 shift_right(char *addr, char bytes, int1 value);

The *shift_right* function shifts a bit into an array or structure. *addr* is a pointer to memory, *bytes* is the number of bytes to work with, and *value* is a 0 or 1 to be shifted in. This function returns the value (0 or 1) shifted out. *addr* may be an array identifier or an address of a structure. Bit 0 of the lowest byte in RAM is treated as the LSB.

Returns: Bit shifted out

```
#include <16F877.h>
#fuses HS,NOWDT            // a Microchip PIC16F877
#use delay(clock=10000000)
#use rs232(baud=9600, xmit=PIN_C6, rcv=PIN_C7)

void main()
{
    int8 bit_array[3];
    int8 i;
    bit_array[0] = 0x05;
    bit_array[1] = 0x18;
    bit_array[2] = 0x64;
    for (i=0;i<24;i++)
    {
        printf("i=%d: %X %X %X\r\n",i,
            bit_array[2],bit_array[1],bit_array[0]);
        shift_right(bit_array,3,1);
    }
    printf("i=%d: %X %X %X\r\n",i,
        bit_array[2],bit_array[1],bit_array[0]);

    while(1)
    {
    }
}
```

Results: Transmits 25 lines containing the current values of the array as it shifts a 1 into the array and shifts the array to the right 24 times. The following are excerpts from that transmission:

```
i=0: 64 18 05
i=4: F6 41 80
i=8: FF 64 18
i=12: FF F6 41
i=16: FF FF 64
i=20: FF FF F6
i=24: FF FF FF
```

sin

```
#include <math.h>
float sin(float x);
```

The *sin* function calculates the sine of the floating point number *x*. The angle *x* is expressed in radians.

Returns: sin(x)

```
#include <16F877.h>
#include <math.h>
#fuses HS,NOWDT             // a Microchip PIC16F877
#use delay(clock=10000000)
#use rs232(baud=9600, xmit=PIN_C6, rcv=PIN_C7)

void main()
{
    float new_val;
    new_val = sin(5.121);
    printf("%f\n\r",new_val);
    while(1)
          ;
}
```

Results: new_val = -0.917673

sinh

```
#include <math.h>
float sinh(float x);
```

The *sinh* function calculates the hyperbolic sine of the floating point number *x*. The angle *x* is expressed in radians.

Returns: sinh(x)

```
#include <16F877.h>
#include <math.h>
#fuses HS,NOWDT             // a Microchip PIC16F877
#use delay(clock=10000000)
#use rs232(baud=9600, xmit=PIN_C6, rcv=PIN_C7)

void main()
{
    float new_val;
    new_val = sinh(5.12);
    printf("%f\n\r",new_val);
    while(1)
```

```
              ;
    }
```

Results: new_val = 83.664702

---

sleep

void sleep();

The *sleep* function issues a **SLEEP** instruction. Details of this function are device specific; however, in general the part will enter low power mode and halt program execution until woken by specific external events. Upon waking up, execution continues after the sleep instruction.

Returns: None

```
#include <16F877.h>
#fuses HS,NOWDT          // a Microchip PIC16F877
#use delay(clock=10000000)
#use rs232(baud=9600, xmit=PIN_C6, rcv=PIN_C7)

// global flag to send processor into sleep mode
short sleep_mode;

// external interrupt when button pushed and released
#INT_EXT
void ext_isr()
{
   static short button_pressed=FALSE;

   if(!button_pressed)          // if button action and was not pressed
   {
      button_pressed=TRUE;     // the button is now down
      sleep_mode=TRUE;         // activate sleep
      printf("The processor is now sleeping.\r");
      ext_int_edge(L_TO_H);    // change so interrupts on release
   }
   else                         // if button action and was pressed
   {
      button_pressed=FALSE;    // the button is now up
      sleep_mode=FALSE;        // reset sleep flag
      ext_int_edge(H_TO_L);    // change so interrupts on press
   }

   if(!input(PIN_B0))           // keep button action synchronized with
      button_pressed=TRUE;     // button flag
```

```
        delay_ms(100);                // debounce button
    }
    // main program that increments counter every second unless sleeping
    void main()
    {
        long counter;

        sleep_mode=FALSE;             // init sleep flag
        port_b_pullups(TRUE);
        ext_int_edge(H_TO_L);         // init interrupt triggering
                                      //   for button press

        enable_interrupts(INT_EXT);// turn on interrupts
        enable_interrupts(GLOBAL);

        printf("\n\n");

        counter=0;                    // reset the counter
        while(TRUE)
        {
            if(sleep_mode)            // if sleep flag set
                sleep();              // make processor sleep
            printf("The count value is:  %5ld     \r",counter);
            counter++;                // display count value and increment
            delay_ms(1000);           // every second
        }
    }
```

Results: The microcontroller increments the counter and displays its value while Port B pin 0 is high and goes to sleep while Port B pin 0 is low. The external interrupts are used to trigger the wake-up from sleep. When counting, the microcontroller transmits at 9600 baud,

```
The processor is now sleeping.
```

Just before sleeping, the microcontroller transmits at 9600 baud,

```
The count value is:     59
```

---

spi_data_is_in

char spi_data_is_in();

The *spi_data_is_in* function returns TRUE if data has been received over the SPI port. This function is only available on devices with SPI hardware.

Returns: True (1) or false (0)

See *spi_read* for an example.

spi_read

char spi_read([char c]);

The *spi_read* function returns a value read from the SPI port. This function is only available on devices with SPI hardware.

If this device is the master, *spi_read* clocks out the value *c* as a new value is received. The received value is returned. If there is no data to send but a read needs to be performed, pass a zero to *spi_read*. If *spi_write* is used to write data to the SPI port, then the data clocked in during the write can be read by calling *spi_read* without passing any parameters to it.

If this device is the slave, then *spi_read* waits for the other device to send clock and data to the SPI port. As the new data is clocked in, the value *c* is clocked out to the other device. To avoid waiting for a clock from the other device, use *spi_data_is_in()* to determine if data is ready.

Returns: Value read from the SPI port

```
// This code contains master and slave side communication examples
// for the SPI port. A conditional compile is used to select the side
// that is being compiled. To compile for the master side, uncomment
// #define MASTER_DEVICE below.

#include <16F877.h>
#fuses HS,NOWDT            // a Microchip PIC16F877
#use delay(clock=10000000)
#use rs232(baud=9600,parity=N,xmit=PIN_C6,rcv=PIN_C7,bits=8)
#use standard_IO(D)

//#define MASTER_DEVICE      1

#ifdef MASTER_DEVICE
// (master side)
void main()
{
    int8 c,d;
    setup_spi(SPI_MASTER|SPI_L_TO_H|SPI_CLK_DIV_16);
    printf("Press any key to start: ");
    while(1)
    {
        c = getchar();
        output_D(0xFF ^ c);   // make nice for LEDs

        // the next two lines represent one method
        spi_write(c);         // data sent using write
        d = spi_read();       // data read using read
```

```
            // alternate method on the next line
            // d = spi_read(c); // write and read data using read

            putchar(d);
        }
    }

#else
// (slave side)
void main()
{
    int8 c;
    setup_spi(SPI_SLAVE|SPI_L_TO_H|SPI_CLK_DIV_16);
    while(1)
    {
        // flash C2 to show when actively running
        output_low(PIN_C2);
        delay_ms(100);
        output_high(PIN_C2);
        delay_ms(100);

        // one method - check for data ready before reading
        if (spi_data_is_in())
        {
            c = spi_read();          // read the data
            output_D(0xFF ^ c);      // make nice for LEDs
            spi_write(c);            // write the data back (and wait
                                     // for clock to send

        }
        // alternate method:
        //c = spi_read(c);           // write last data and read new
        //output_D(0xFF ^ c);
    }
}
#endif
```

Results: This example is meant to be run on two microcontrollers with their SPI clocks and data lines tied together. One should be programmed as the master and one as the slave. (Remember, data out of one should go to data in of the other and vice versa.) The master side transmits to the SPI port whatever is received by the USART and transmits to the USART whatever is received on the SPI port. The master also sends the data to Port D.

The slave side echoes back to the SPI port whatever it receives from the SPI port and displays the received data on Port D. The slave side also flashes pin C2 to show when it is actively running and not waiting for SPI clocks from the master. In one method on the slave side, the *spi_data_is_in* function is used to determine when data is ready to be read from the SPI port without waiting on SPI clocks. If this method is used and the *spi_write* is commented out, C2 continuously flashes. (*spi_write* causes the microcontroller to wait for clocks to send the data.) The alternate method just calls *spi_read* and spends most of its time waiting for clocks.

One other note: There is a one-cycle delay between what is received by the USART and what is transmitted to the USART. This delay is caused by the SPI bus interface. The slave side receives the byte on one data cycle then transmits it back to the master during the next data cycle.

---

spi_write

void spi_write(char c);

The *spi_write* function sets the next value to be written to the SPI port. This function is only available on devices with SPI hardware. If this device is the master, then a clock signal is generated and the value is sent immediately. If this device is the slave in the system, then this value is not sent until a clock is generated by the master in the system.

Returns: None

See *spi_read* for an example.

---

sprintf

void sprintf(char *str, char flash *fmtstr [, arg1, arg2, ...]) ;

The *sprintf* function copies the formatted text according to the format specifiers in the *fmtstr* string to the string *str*. A null termination is appended to the end of *str* after the formatted text is copied to it. The memory for *str* should be large enough to accommodate the copied text and the null termination.

The format specifiers are the same as those used for *printf*. See *printf* for more information on the format specifiers.

```
#include <16F877.h>
#include <string.h>
#fuses HS,NOWDT           // a Microchip PIC16F877
#use delay(clock=10000000)
#use rs232(baud=9600, xmit=PIN_C6, rcv=PIN_C7)

unsigned int16 j;
char mystr1[40];        // make room for the results!
char mystr2[40];
char mystr3[40];

void main(void)
```

```
{
    for (j=0;j<=500;j+=250)
    {
        //copy the formatted strings
        sprintf(mystr1,"\n\rDecimal: %lu\tHexadecimal: %LX\n\r",j,j);
        sprintf(mystr2, "Zero Padded Decimal: %01d\n\r",j);
        sprintf(mystr3, "Four Digit Hexadecimal: %04lx\r\n",j);
        // now send the first two as one long string to the UART
        printf("%s %s",mystr1,mystr2);
        // and send the last one separately with an extra line feed
        puts(mystr3);
    }

    while (1)
        ;
}
```

Results: The following is output by the microprocessor to the UART at start-up.

```
Decimal: 0        Hexadecimal: 0000
 Zero Padded Decimal: 00000
Four Digit Hexadecimal: 0000


Decimal: 250      Hexadecimal: 00FA
 Zero Padded Decimal: 00250
Four Digit Hexadecimal: 00fa


Decimal: 500      Hexadecimal: 01F4
 Zero Padded Decimal: 00500
Four Digit Hexadecimal: 01f4
```

---

sqrt

#include <math.h>

float sqrt(float x);

The *sqrt* function returns, as a floating point variable, the square root of the positive floating point variable *x*. If *x* is negative, the behavior is undefined.

Returns: The square root of the positive floating point variable *x*

```
#include <16F877.h>
#include <math.h>
#fuses HS,NOWDT          // a Microchip PIC16F877
#use delay(clock=10000000)
#use rs232(baud=9600, xmit=PIN_C6, rcv=PIN_C7)
```

```
void main(void)
{
    float my_sqrt;

    my_sqrt = sqrt(6.4);        // get the square root of a float

    printf("my_sqrt = %f\n\r",my_sqrt);
    while(1)
        ;
}
```

Results: my_sqrt = 2.529822

---

*strcat

#include <string.h>

char *strcat(char *str1, char *str2);

The *strcat* function concatenates string *str2* onto the end of string *str1*. The memory allocated for *str1* must be long enough to accommodate the new, longer string plus the null-terminating character or else unexpected results occur. A pointer to *str1* is returned. Neither *str1* or *str2* may be located in flash (cannot be constants).

Returns: **str1* (a pointer to the null-terminated concatenation of strings *str1* and *str2*)

```
#include <16F877.h>
#include <string.h>
#fuses HS,NOWDT             // a Microchip PIC16F877

void main(void)
{
    char stra[] = "abc";
    char strb[10] = "xyz";

    strcat(strb,stra); // add stra to strb

    while(1)
        ;
}
```

Results: strb = xyzabc

---

*strchr

#include <string.h>

char *strchr(char *str, char c);

The *strchr* function locates the first occurrence of the character *c* in the string *str*. If the character *c* is not found within the string, then a null pointer is returned.

Returns: A pointer to the first occurrence character *c* in string *str*. If *c* is not found in *str*, a null pointer is returned.

```
#include <16F877.h>
#include <string.h>
#fuses HS,NOWDT          // a Microchip PIC16F877
#use delay(clock=10000000)
#use rs232(baud=9600, xmit=PIN_C6, rcv=PIN_C7)

void main(void)
{
    char stra[] = "123.45";
    char *strb; // no need to allocate space, pointing
            // into stra - already allocated!


    strb = strchr(stra,'.');
    printf("Full String: %s\n\rNew String: %s\n\r",stra,strb);
    *strb = '?';// replace the decimal point with a ?
    printf("Modified string: %s\n\r",stra);

    while (1)
         ;
}
```

Results: The UART transmits, at 9600 baud,

```
Full String: 123.45
New String: .45
Modified string: 123?45
```

---

strcmp

#include <string.h>

signed int8 strcmp(char *str1, char *str2);

The *strcmp* functions compare string *str1* with string *str2*. The function starts the comparison with the first character in each string. When the character in *str1* fails to match the character in *str2*, the difference in the character values is used to determine the return value of the function.

Returns:

- -1 if *str1* < *str2*
- 0 if *str1* = *str2*
- +1 value if *str1* > *str2*

```
#include <16F877.h>
#include <string.h>
#fuses HS,NOWDT          // a Microchip PIC16F877

char stra[] = "george";
char strb[] = "georgie";
signed int8 result;

void main(void)
{
    result = strcmp(stra, strb);
    while (1)
        ;
}
```

Results: result = -1

---

*strcpy

char *strcpy(char *dest, char *src);

The *strcpy* function copies the string pointed to by *src* to the location pointed to by *dest*. The null-terminating character of the *src* string is the last character copied to the *dest* string. The memory allocated for the *dest* string must be large enough to hold the entire *src* string plus the null-terminating character.

Returns: Pointer to *dest*

See *strtok* for an example.

---

strcspn

#include <string.h>

unsigned char strcspn(char *str, char *set);

The *strcspn* function returns the index of the first character in the string *str* that matches a character in the string *set*. If none of the characters in the string *str* is in the string *set*, the length of *str* is returned. If the first character in the string *str* is in the string *set*, zero is returned.

Returns: Index of the first character in *str* that is in *set*

```
#include <16F877.h>
#include <string.h>
#fuses HS,NOWDT          // a Microchip PIC16F877

void main(void)
{
    char set[] = "1234567890-()";
    char set2[] = ".-()";
```

```
    char stra[] = "1.800.555.1212";
    char index_1;
    char index_2;

    index_1 = strcspn(stra,set);
    index_2 = strcspn(stra,set2);

    while (1)
    ;
}
```

Results: index_1 = 0

index_2 = 1

stricmp

#include <string.h>

signed char stricmp(char *str1, char *str2);

The *stricmp* function compares characters from string *str1* to string *str2*, ignoring the case. The functions start comparing with the first character in each string. When the character in *str1* fails to match the character in *str2* (other than being the upper or lower equivalent), the difference in the character values is used to determine the return value of the function.

Returns:

- -1 if *str1* < *str2*
- 0 if *str1* = *str2*
- +1 if *str1* > *str2*

```
#include <16F877.h>
#include <string.h>
#fuses HS,NOWDT             // a Microchip PIC16F877

char stra[] = "george";
char strb[] = "GEORGE";
char strc[] = "jenny";
signed int8 result;
signed int8 result2;

void main(void)
{
    result = stricmp(stra, strb);
    result2 = stricmp(stra,strc);
    while (1)
        ;
```

```
}
```

Results: result = 0

result2 = -1

strlen

#include <string.h>

int8 strlen(char *str);

The *strlen* function returns the length of the string *str*, not counting the null terminator. The length can be from 0 to 255. The string must be located in RAM and not a constant located in FLASH.

Returns: Length of the string *str*

```
#include <16F877.h>
#include <string.h>
#fuses HS,NOWDT          // a Microchip PIC16F877

void main(void)
{
    char stra[] = "1234567890";
    int8 len1;

    len1 = strlen(stra);
    while(1)
          ;
}
```

Results: len1 = 10

strlwr

#include <string.h>

char *strlwr(char *str);

The *strwlr* function converts *str* to an all lowercase string and returns a pointer to *str*.

Returns: Pointer to *str* converted to lowercase

```
#include <16F877.h>
#include <string.h>
#fuses HS,NOWDT          // a Microchip PIC16F877
#use delay(clock=10000000)
#use rs232(baud=9600, xmit=PIN_C6, rcv=PIN_C7)

void main(void)
{
```

```
    char password[] = "a1CD43b";
    strlwr(password);  // ignore return pointer since original string
                       // is also changed!
    printf("Pass: %s\n\r",password);
    while (1)
        ;
}
```

Results: UART transmits, at 9600 baud,

```
Pass: a1cd43b
```

---

strncmp

#include <string.h>

signed char strncmp(char *str1, char *str2, unsigned char n);

The *strncmp* function compares at most *n* characters from *string str1* to *string str2*. The function starts the comparison with the first character in each string. When the character in *str1* fails to match the character in *str2*, the difference in the character values is used to determine the return value of the function. Any differences between the strings beyond the *nth* character are ignored.

Returns:

- -1 if *str1* < *str2*
- 0 if *str1* = *str2*
- +1 if *str1* > *str2*

```
#include <16F877.h>
#include <string.h>
#fuses HS,NOWDT            // a Microchip PIC16F877

char stra[] = "george";
char strb[] = "georgie";
signed int8 result;

void main(void)
{
    result = strncmp(stra, strb,5);
    while (1)
        ;
}
```

Results: result = 0

*strncpy

#include <string.h>

char *strncpy(char *dest, char *src, unsigned char n);

The *strncpy* function copies up to *n* characters from the string pointed to by *src* to the location pointed to by *dest.* If there are less than *n* characters in the *src* string, then the *src* string is copied to *dest* and a null-terminating character is appended. If *src* string is longer than or equal in length to *n*, then no terminating character is copied or appended to *dest.*

Returns: Pointer to *dest*

```
#include <16F877.h>
#include <string.h>
#fuses HS,NOWDT          // a Microchip PIC16F877
#use delay(clock=10000000)
#use rs232(baud=9600, xmit=PIN_C6, rcv=PIN_C7)

char stra[] = "Hello";
char strb[] = "HELLO";

void main(void)
{
    strncpy(strb,stra,3);           // copy stra to strb
    printf("%s\n\r",strb);
    while (1)
        ;
}
```

Results:

Because 3 is less than the total length of stra, the first three letters are copied over strb, but no terminating null is copied. As a result, the original null termination is used to terminate strb. So, the UART transmits, at 9600 baud,

```
HelLO
```

*strpbrk

#include <string.h>

char *strpbrk(char *str, char *set);

The *strpbrk* function searches the string *str* for the first occurrence of a character from the string *set.* If there is a match, the function returns a pointer to the character in the string *str.* If there is not a match, a null pointer is returned.

Returns: Pointer to the first character in *str* that matches a character in *set*

```
#include <16F877.h>
#include <string.h>
#fuses HS,NOWDT            // a Microchip PIC16F877
#use delay(clock=10000000)
#use rs232(baud=9600, xmit=PIN_C6, rcv=PIN_C7)

void main(void)
{
    char stra[] = "11/25/00";
    char set[] = "/.,!-";
    char alt_set[] = ",.-";
    char strb[] = "November 25, 2000";
    char *pos;
    char *ypos;

    pos = strpbrk(stra,set); // find first
                             // occurrence of something!
    ypos = strpbrk(strb,alt_set);
    printf("Initial Date: %s\n\r",stra);
    printf("String following match: %s\n\r",pos);
    printf("Just the year: %s\n\r",ypos+1);
    while (1)
          ;
}
```

Results: The UART transmits, at 9600 baud,

```
Initial Date: 11/25/00
String following match: /25/00
Just the year:  2000
```

---

*strrchr

```
#include <string.h>
char *strrchr(char *str, char c);
```

The *strrchr* function locates the last occurrence of the character *c* in the string *str*. If the character *c* is not found within the string, then a null pointer is returned.

Returns: Pointer to the last occurrence character *c* in string *str*. If *c* is not found in *str*, a null pointer is returned.

```
#include <16F877.h>
#include <string.h>
#fuses HS,NOWDT            // a Microchip PIC16F877
#use delay(clock=10000000)
#use rs232(baud=9600, xmit=PIN_C6, rcv=PIN_C7)
```

```
void main(void)
{
    char stra[] = "123.45.789";
    char *strb; // no need to allocate space, pointing
            // into stra - already allocated!

    strb = strrchr(stra,'.');
    printf("Full String: %s\n\rNew String: %s\n\r",stra,strb);
    *strb = '6';// replace the decimal point with a 6
    printf("Modified string: %s\n\r",stra);

    while (1)
        ;
}
```

Results: The UART transmits, at 9600 baud,

```
Full String: 123.45.789
New String: .789
Modified string: 123.456789
```

---

strspn

#include <string.h>

unsigned char strspn(char *str, char *set);

The *strspn* function returns the index of the first character in the string *str* that does not match a character in the string *set*. If all characters in the string *str* are in the string *set*, the length of *str* is returned. If no characters in the string *str* are in the string *set*, zero is returned.

Returns: Index of the first character in *str* that is not in *set*

```
#include <16F877.h>
#include <string.h>
#fuses HS,NOWDT          // a Microchip PIC16F877

void main(void)
{
    char set[] = "1234567890-()";
    char set2[] = "1234567890-().";
    char stra[] = "1.800.555.1212";
    char index_1;
    char index_2;

    index_1 = strspn(stra,set);
    index_2 = strspn(stra,set2);

    while (1)
```

```
        ;
    }
```

Results: index_1 = 1

index_2 = 14

---

*strstr

#include <string.h>

char *strstr(char *str1, char *str2);

The *strstr* function searches string *str1* for the first occurrence of string *str2*. If *str2* is found within *str1*, then a pointer to the first character of *str2* in *str1* is returned. If *str2* is not found in *str1*, then a null is returned.

Returns: Pointer to the first character of *str2* in *str1* or null if *str2* is not in *str1*

```
#include <16F877.h>
#include <string.h>
#fuses HS,NOWDT          // a Microchip PIC16F877
#use delay(clock=10000000)
#use rs232(baud=9600, xmit=PIN_C6, rcv=PIN_C7)

void main(void)
{
    char stra[] = "Red Green Blue";
    char strb[] = "Green";
    char strc[] = "B";
    char *ptr;
    char *ptrf;

    // grab a pointer to where Green is
    ptr = strstr(stra,strb);

    // grab a pointer to where the first B is
    ptrf = strstr(stra, strc);

    printf("Starting String: %s\n\r",stra);
    printf("Search String: %s\n\r",strb);
    printf("Results String 1: %s\n\r",ptr);
    printf("Results String 2: %s\n\r",ptrf);
    while (1)
        ;
}
```

Results: The UART transmits, at 9600 baud,

```
Starting String: Red Green Blue
```

```
Search String: Green
Results String 1: Green Blue
Results String 2: Blue
```

*strtok

#include <string.h>

char *strtok(char *str1, char flash *str2);

The function *strtok* scans string *str1* for the first token not contained in the string *str2*. The function expects *str1* to consist of a sequence of text tokens, separated by one or more characters from the string *str2* (token separators). This function may be called repetitively to parse through a string (*str1*) and retrieve tokens that are separated by known characters (*str2*).

The first call to *strtok* with a non-null pointer for *str1* returns a pointer to the first character of the first token in *str1*. The function searches for the end of the token and places a null termination character at the first token separator (character from *str2*) following the token. Subsequent calls to *strtok* with a null (0) passed for *str1* return the next token from *str1* in sequence until no more tokens exist in *str1*. When no more tokens are found, a null is returned.

Note that *strtok* modifies *str1* by placing the null termination characters after each token. To preserve *str1*, make a copy of it before calling *strtok*.

Returns: Pointer to the next token in *str1* or null if no more tokens exist in *str1*

```
#include <16F877.h>
#include <string.h>
#fuses HS,NOWDT           // a Microchip PIC16F877
#use delay(clock=10000000)
#use rs232(baud=9600, xmit=PIN_C6, rcv=PIN_C7)

char mytext[] = "(888)777-2222";
char separators[] = "()-";

void main(void)
{
    char area_code[4];
    char *prefix;
    char *postfix;
    char backup_copy[14];

    // we want to keep the original around too!
    strcpy(backup_copy,mytext);
```

```
        // grab a pointer to the area code
        strcpy(area_code,strtok(mytext, separators));

        // grab a pointer to the prefix
        prefix = strtok(0, separators);

        // grab a pointer to the postfix
        postfix = strtok(0, separators);


        printf("Starting String: %s\n\r",backup_copy);
        printf("Area Code: %s\n\r",area_code);
        printf("Prefix: %s\n\r",prefix);
        printf("Postfix: %s\n\r",postfix);
        while (1)
            ;
    }
```

Results: The UART transmits, at 9600 baud,

```
Starting String: (888)777-2222
Area Code: 888
Prefix: 777
Postfix: 2222
```

---

swap

void swap(int8 x);

The *swap* function swaps the upper nibble and the lower nibble of *x*. This function does not return the new value but, rather, modifies *x* in memory.

Returns: None

```
#include <16F877.h>
#fuses HS,NOWDT          // a Microchip PIC16F877
#use delay(clock=10000000)
#use rs232(baud=9600, xmit=PIN_C6, rcv=PIN_C7)

void main()
{
    int8 a;

    a = 0x49;
    printf("a: %02X\n\r",a);
    swap(a);
    printf("swapped a: %02X\n\r",a);
    while(1)
    {
```

```
        }
    }
```

Results: Transmits, at 9600 baud,

```
a: 49
swapped a: 94
```

---

tan

#include <math.h>

float tan(float x);

The *tan* function calculates the tangent of the floating point number *x*. The angle *x* is expressed in radians.

Returns: tan(x)

```
#include <16F877.h>
#include <math.h>
#fuses HS,NOWDT          // a Microchip PIC16F877
#use delay(clock=10000000)
#use rs232(baud=9600, xmit=PIN_C6, rcv=PIN_C7)

void main()
{
    float new_val;
    new_val = tan(5.121);
    printf("%f\n\r",new_val);
    while(1)
        ;
}
```

Results: new_val = -2.309565

---

tanh

#include <math.h>

float tanh(float x);

The *tanh* function calculates the hyperbolic tangent of the floating point number *x*. The angle *x* is expressed in radians.

Returns: tanh(x)

```
#include <16F877.h>
#include <math.h>
#fuses HS,NOWDT          // a Microchip PIC16F877
#use delay(clock=10000000)
#use rs232(baud=9600, xmit=PIN_C6, rcv=PIN_C7)
```

```
void main()
{
    float new_val;
    new_val = tanh(5.121);
    printf("%f\n\r",new_val);
    while(1)
        ;
}
```

Results: new_val = 0.999928

tolower

char tolower(char c);

The *tolower* function converts the ASCII character *c* from an uppercase character to a lowercase character. If *c* is a lowercase character, *c* is returned unchanged.

Returns: ASCII character *c* as a lowercase character

```
#include <16F877.h>
#fuses HS,NOWDT            // a Microchip PIC16F877
#use delay(clock=10000000)
#use rs232(baud=9600, xmit=PIN_C6, rcv=PIN_C7)

void main(void)
{
    char lower_case_c;
    lower_case_c = tolower('U');
    printf("%c -> %c\n\r",'U',lower_case_c);
    while(1)
        ;
}
```

Results: lower_case_c = 'u'

toupper

char toupper(char c);

The *toupper* function converts the ASCII character *c* from a lowercase character to an uppercase character. If c is an uppercase character, *c* is returned unchanged.

Returns: ASCII character *c* as an uppercase character

```
#include <16F877.h>
#fuses HS,NOWDT            // a Microchip PIC16F877
#use delay(clock=10000000)
#use rs232(baud=9600, xmit=PIN_C6, rcv=PIN_C7)

void main(void)
```

```
{
    char upper_case_c;
    upper_case_c = toupper('r');
    printf("%c -> %c\n\r",'r',upper_case_c);
    while(1)
        ;
}
```

Results: upper_case_c = 'R'

---

write_bank

void write_bank(int8 bank, int16 offset, int8 value);

The *write_bank* function writes the data byte, *value*, to the user RAM area of the specified bank at the specified offset. Valid banks are 1-3 depending on the device, and the offset into the bank starts at zero. This function may be used on some devices where full RAM access by auto variables is not efficient. For example, on the PIC16C57 chip, setting the pointer size to 5 bits will generate the most efficient ROM code; however, auto variables cannot be above 0x1F. Instead of going to 8-bit pointers, you can save ROM by using this function to write to the bank addresses beyond the reach of the minimized pointers. In this case with the PIC16C57, the bank may be 1-3 and the offset may be 0-15.

Returns: None

```
#include <16F877.h>
#fuses HS,NOWDT          // a Microchip PIC16F877
#use delay(clock=10000000)
#use rs232(baud=9600,parity=N,xmit=PIN_C6,rcv=PIN_C7,bits=8)

void main()
{
    int8 i;
    int8 c;

    while(1)
    {
        // until a carriage return is received,
        // read data into bank 1
        i = 0;
        do
        {
            c = getchar();
            write_bank(1,i++,c);
        }while ((c != 13) && (i < 20));

        // move to a new line!
```

```
            printf("\n\r");

            // until the carriage return is read,
            // read data from bank 1 and transmit

            i = 0;
            do
            {
                c = read_bank(1,i++);
                putchar(c);
            }while ((c != 13) && (i < 20));
        }
    }
```

Results: The device stores what is received by the USART in bank 1 until a carriage return is received. Then the device transmits the data back out of the USART.

---

write_eeprom

void write_eeprom(int8 addr, int8 value);

The *write_eeprom* function writes *value* to the EEPROM address *addr*. *value* and *addr* are each 8-bit integers. The first address in the EEPROM is zero and the last address is dependent on the size of the EEPROM in the device in use. This function is only available on devices with an internal EEPROM.

Returns: None

See *read_eeprom* for an example.

---

write_program_eeprom

For 14-bit devices:

void write_program_eeprom(int16 addr, int16 data);

For 16-bit devices:

void write_program_eeprom(int32 addr, int8 data);

The *write_program_eeprom* function writes *data* to the program memory of the device at address *addr*. The size of *data* and *addr* are dependent on the device in use as shown above. This function is only available on devices that allow reads from program memory.

Returns: None

See *read_program_eeprom* for an example.

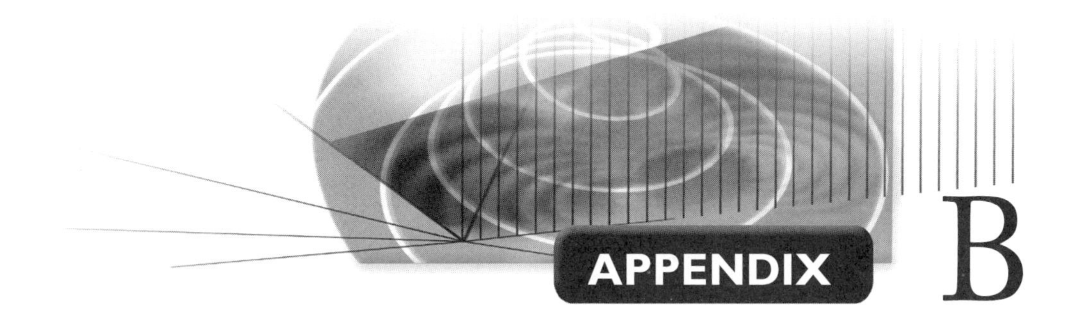

Programming the PIC Microcontrollers

The purpose of this appendix is to introduce the various programming methods available and to provide enough information to troubleshoot programming problems. The information given is not intended to be sufficient to enable the reader to create a programming device; see the microcontroller specification if you wish to create your own programmer or write a *boot loader* program.

All microcontrollers (and microcomputers of all types) require that all or part of the operating program be resident in the computer when it starts up. Microcontrollers and computers cannot do "nothing" (i.e., they must be executing code at all times they are running). If they are in a 'do-nothing' loop, they may be accomplishing nothing, but they are still executing the code that makes up the idle loop.

This means that the operating code must be permanently stored in a nonvolatile section of memory. In the case of the PIC microcontrollers, this memory is composed of FLASH and/or EEPROM memory technology. And, assuming that you are not using an in-circuit emulator or a simulator, the code must be programmed into the FLASH memory in order for the code to be run.

Microchip provides on-board programming via a special synchronous port used exclusively for programming. The program bytes are applied to the synchronous port in a specific sequence and are subsequently stored into the FLASH memory. This code is then executed the next time the microcontroller is reset. The huge advantage to synchronous programming is that it almost always can be done in circuit. The devices do not need to be removed from their application and programmed in a special programming device. This allows for easy field upgrades in the commercial world.

SYNCHRONOUS PORT PROGRAMMING

As an example, consider the method of programming a PIC16F877, which is described below.

The synchronous programming is controlled using specific bits. The bits are shown in Table B.1. In the case of field programming, these bits must be able to be driven or used by the programmer.

(Note: I = input, O = Output, P = Power, V_{DD} and V_{SS} are connected normally during programming.)

Normal Pin Function	Programming Function		
	Function	Input/Output	Pin Description
RB3	PGM	I	Low-Voltage Input
RB6	Clock	I	Clock Input
RB7	Data	I/O	Data Line
$\overline{MCLR}$	$V_{test\ mode}$	P	Program Mode Select

Table B.1 *PIC16F877 Programming Pin Descriptions*

1. After the microcontroller is powered up and running, hold RB6 and RB7 low and raise RB3 to V_{DD} and then apply V_{DD} to $\overline{MCLR}$. This sequence puts the chip into low-voltage programming mode and all other logic into a *RESET* condition.
2. Send a command (see Table B.2) into the synchronous programming port. The command will set how the microcontroller treats the data that follows it.
3. If the command was a *load* or *read* command, transmit or receive the data. Other commands cause different actions to occur.
4. In the case of a multibyte transfer, use the *increment address* command to move the address pointer for the next byte.
5. Repeat steps 3 and 4 until the entire program code is loaded into or read from the FLASH memory.
6. Repeat the entire sequence for the configuration bits and for the EEPROM if it is desired to program these.
7. Use the *Begin Programming Cycle* command as required to actually program the loaded data in the FLASH, the EEPROM, or the configuration register.
8. Pulse RESET low/high while releasing the programming port bits to resume normal operation and to execute the code now loaded into the microcontroller.

Programming operations that may be executed in a PIC16F877 are shown in Table B.2.

The actual programming commands vary somewhat among the various PIC microcontrollers. Check the specification for your particular device for details.

Command	**Function**
Load Configuration	Load data for the configuration register (the 'fuses')
Load Data for FLASH	Load data for the FLASH
Read Data from FLASH	Read data from the FLASH
Increment Address	Increment the address pointer for the next byte
Begin Erase/Programming Cycle	Starts actual programming following *load* commands
Begin Programming only Cycle	Starts actual programming without erase following *load*
Load Data for Memory	Load data for EEPROM
Read Data from Memory	Read data from EEPROM
Bulk Erase 1	Clear memory past certain protection bits
Bulk Erase 2	Clear memory past other protection bits

Table B.2 *PIC16F877 Programming Commands*

COMMERCIAL PROGRAMMERS

There are a number of commercial programmers available to allow programming the FLASH code memory of the microcontrollers. Some, such as ICD-S, made by Custom Computer Services (CCS) and marketed by Progressive Resources, LLC (see http://www.PRLLC.com) and CCS (http://www.CCSINFO.com), use serial communication to the PC. These typically use an Intel Hex file as their source for the code and fuse information. These units have sophisticated PC software to control the programming process and include features such as automatic erase/program/verify sequencing. They also have additional electronics to convert the serial information into appropriate form for the PIC programming port. Overall the commercial programmers make the programming task very easy and convenient.

BOOT LOADER PROGRAMMING

Some of the PIC devices also allow for self-programming via a *boot loader* program. These devices are capable of writing data to the FLASH code memory under the control of an on-board program called a *boot loader*.

The advantage to a boot loader program is that it can be tailored to the application. That is, if you want to be able to update the FLASH code memory from the serial port, then the boot loader can be written to accept data serially (in a standard form like an Intel Hex formatted file, for instance) from a PC or even a PDA, and place the data into FLASH memory. If you want to be able to do the programming via the I^2C port, the boot loader can be tailored to this method as well.

To use a *boot loader*, the loader program must be programmed into the section of FLASH memory called the *boot block*. Control bits are then set so that a RESET causes the *boot loader* to start watching the chosen input device (i.e., the serial port). A certain string of characters, followed by the program code, keys the boot loader to start reprogramming the FLASH section of the device. Receipt of the wrong string of characters or the passage of a set amount of time without receipt of the control string will cause the boot loader to transfer control to the beginning of FLASH code memory to execute the stored program.

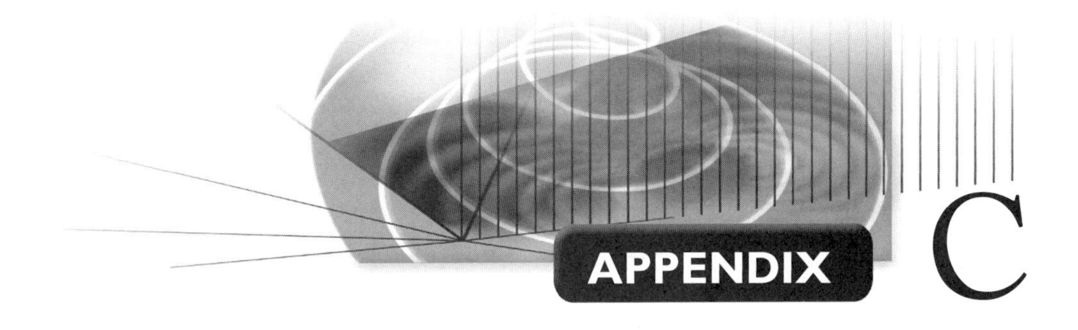

APPENDIX C

CCS ICD-S Serial In-System Programmer/Debugger

The Custom Computer Services (CCS) ICD-S is a complete in-circuit debugging solution for the Microchip PIC16Fxx and PIC18Fxx FlashPIC microcontrollers. ICD-S can debug all PIC16F and PIC18F targets that support debug mode. It also provides in-circuit serial programming (ICSP®) support for all FLASH chips.

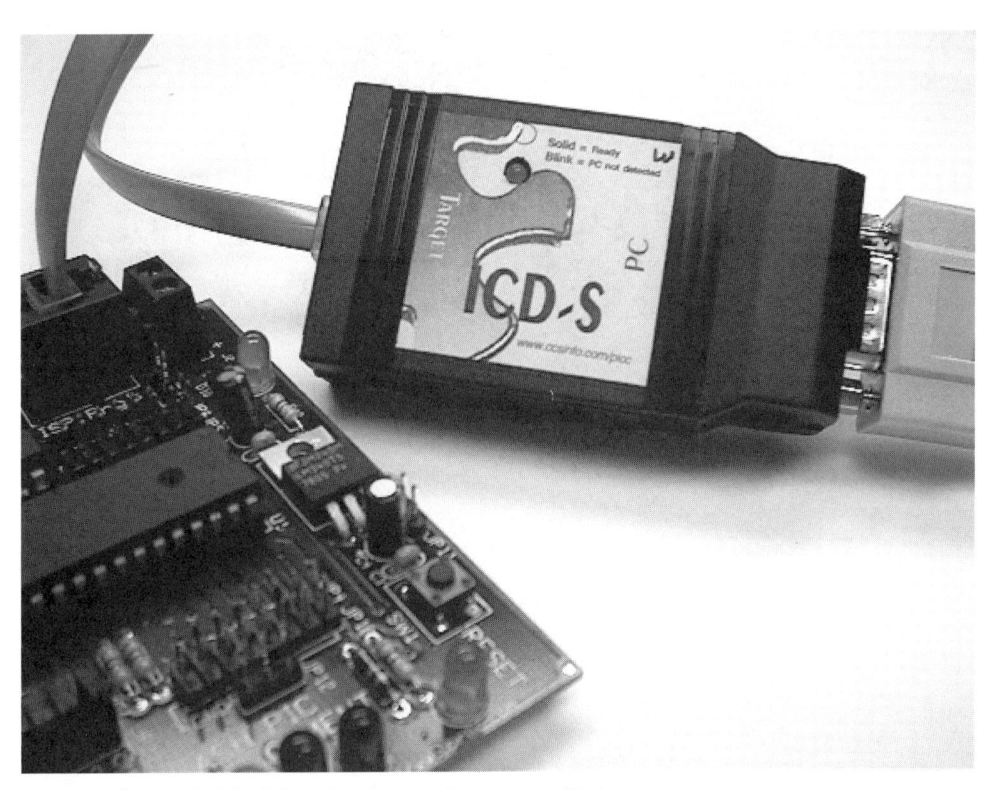

Figure C-1 *CCS ICD-S Serial In-System Programmer/Debugger*

The ICD unit works with the CCS PCW debugger or CCS stand-alone ICD control software. The CCS PCW debugger is a very robust debugger integrated with PCW, and provides very detailed debugging information at the C level. The stand-alone control software allows you to quickly program target chips using the CCS ICD-S in-system serial programmer. The control software also lets you update the ICD-S unit firmware without having to remove the chip from the ICD-S unit.

When MPLAB-ICD firmware is loaded into it, the ICD-S is fully compatible with Microchip's MPLAB IDE and MPLAB ICD1. If you own an MPLAB ICD1, you can load it with CCS-ICD firmware to get PIC18 debugging and ICSP functionality, but this functionality is only provided through CCS software and not Microchip software. This, however, is a great way to upgrade your existing Microchip ICD1 for use with PIC18s.

For more information, refer to the CCS Web site http://www.ccsinfo.com.

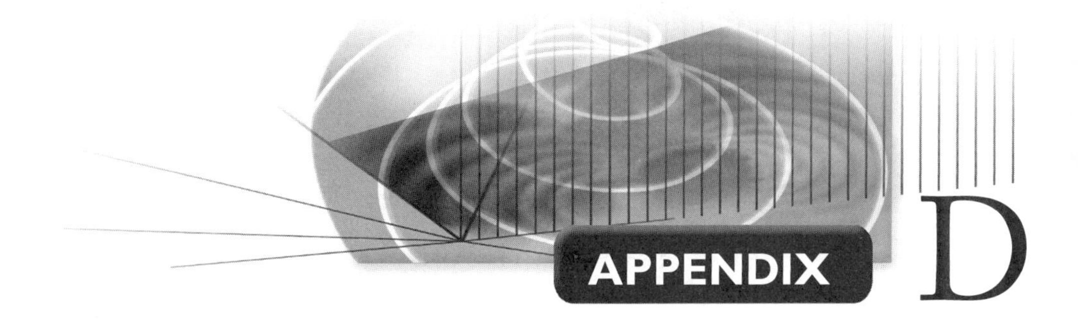

Microchip ICD 2 Serial In-System Programmer/Debugger

Courtesy of Microchip Technology, Inc.

MPLAB® ICD 2 In-Circuit Debugger

DEBUGGER SOLUTION FOR PIC® FLASH PRODUCTS

The MPLAB® ICD 2 (In-Circuit Debugger 2) is the next advanced step for In-Circuit Debugging from Microchip Technology. The MPLAB ICD 2 allows debugging of selected PIC® FLASH microcontrollers using the powerful graphical user interface of the MPLAB Integrated Development Environment (IDE) which is available as a free tool and included with each unit. It is the ideal tool for embedded control designers looking for a low cost alternative to expensive in-circuit emulators.

In-Circuit debugging is achieved using the two dedicated hardware lines (microcontroller pins used only during debugging mode) that allow In-Circuit Serial Programming™ (ICSP™) of the device and debugging capability through proprietary firmware. The MPLAB ICD 2 debug feature is built into the microcontroller and activated by programming the debug code into the target processor. Shared overhead is one stack level, several general purpose file registers and a small bank of program memory when in the debug mode.

The MPLAB ICD 2 firmware is FLASH-based, which allows it to be enhanced to support future microcontroller products and new features, extending the life of the tool – making it a good investment. Firmware downloads are available from the Microchip web site at: www.microchip.com.

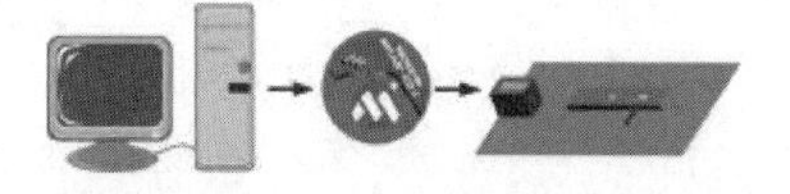

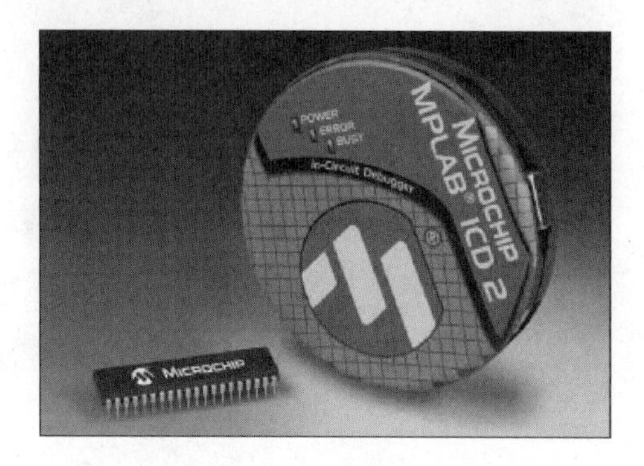

Features

- RS-232 interface to host PC
- Real-time background debugging
- MPLAB IDE compatible (free copy included)
- Built-in over voltage/short circuit monitor
- Firmware upgradeable from PC/web download
- Totally enclosed
- Supports low voltage debug to 2.0 volts
- Diagnostic LED's (Power, Busy, Error)
- Reading/Writing memory space and stack of target microcontroller
- Erase of program memory space with verification
- Freeze-on-Halt

MPLAB ICD 2 In-Circuit Debugger Set-Up

The MPLAB ICD 2 interfaces between the design engineer's PC operating with MPLAB IDE and the product board (target) being developed. The in-circuit debugger acts as an intelligent interface/translator between the two allowing the engineer to look into the active target board's microcontroller permitting real-time viewing of variables and registers using watch windows. A single break point can be set, halting the program at a specific point. Memory read/writes can also be achieved. Additionally, the MPLAB ICD 2 can be used to program or reprogram the PIC FLASH microcontroller while installed on the target board.

PIC® FLASH Products Supported

The PIC FLASH microcontrollers currently supported include: PIC18C601, PIC18C801, PIC18F452, PIC18F248, PIC18F258, PIC18F442, PIC18F448, PIC18F452, PIC18F458, PIC18F6620, PIC18F6720, PIC18F8620 and PIC18F8720.

The MPLAB ICD 2 firmware is continually being updated. A review of the README.ID2 file located in MPLAB IDE is recommended for the most current list of supported parts. As new device firmware becomes available, free downloads are available at www.microchip.com.

Ordering Information:

Model Name: MPLAB® ICD 2

Part Number	Description
DV164005*	ICD 2 Module (Includes ICD 2 Module and USB Cable)
DV164006*	ICD 2 Evaluation Kit (Includes ICD 2 Module, USB Cable, RS-232 Cable, Power Supply and PICDEM™ 2 Plus Demonstration Board - DV163022)
DV164007*	ICD 2 Module ws (Includes ICD 2 Module, USB Cable, RS-232 Cable and Power Supply)
AC162049	Programming Module (Works with DV164005, DV164006 and DV164007 above)
AC162048	RS-232 and Power Supply Kit (Use with DV164005 above for RS-232 communication)
DM163022	PICDEM 2 Plus Demonstration Board (Includes PIC18F452, PIC16F877, LCD 2 x 16 Display, LED's, RS-232 Port, Piezo Sounder, Temperature Sensor, Demonstration Programs, Unassembled Source Code and More)

*USB Ready – To enable the USB feature, please check the Microchip web site for version 6.00 or later of MPLAB IDE. This is a free download at www.microchip.com

Host System Requirements:

- PC-compatible system with a Intel Pentium® class or higher processor, or equivalent
- A minimum of 16 MB RAM
- A minimum of 40 MB available hard drive space
- CD-ROM drive (for use with the accompanying CD)
- Available RS-232 port
- Microsoft® Windows 98, Windows NT® 4.0 or Windows 2000

Customer Support:

Microchip maintains a worldwide network of distributors, representatives, local sales offices, Field Application Engineers and Corporate Application Engineers. Microchip's Internet home page can be reached at: www.microchip.com.

Development Tools from Microchip	
MPLAB® IDE	Integrated Development Environment (IDE)
MPASM™ Assembler	Universal PICmicro® Macro-assembler
MPLINK™ Linker/MPLIB™ Librarian	Linker/Librarian
MPLAB C18	C Compiler for PIC18XXXX MCUs
C Compiler	Sold by Third-party Vendors (HI-TECH, IAR, CCS)
MPLAB SIM Simulator	Software Simulator
MPLAB ICE 2000	Full-featured Modular In-Circuit Emulator
PICSTART® Plus Programmer	Entry-level Development Kit with Programmer
PRO MATE® II Device Programmer	Full-featured, Modular Device Programmer
KEELOQ® Evaluation Kit	Encoder/Decoder Evaluator
KEELOQ Transponder Evaluation Kit	Transmitter/Transponder Evaluator
microID® Developer's Kit	125 kHz and 13.56 MHz RFID Development Tools
MCP2510 CAN Developer's Kit	MCP2510 CAN Evaluation/Development Tool

Americas

Atlanta	(770) 640-0034
Boston	(978) 692-3848
Chicago	(630) 285-0071
Dallas	(972) 818-7423
Detroit	(248) 538-2250
Kokomo	(765) 864-8360
Los Angeles	(949) 263-1888
New York	(631) 273-5305
Phoenix	(480) 792-7966
San Jose	(408) 436-7950
Toronto	(905) 673-0699

Asia/Pacific

Australia	61-2-9868-6733
China – Beijing	86-10-85282100
China – Chengdu	86-28-6766200
China – Fuzhou	86-591-7503506
China – Shanghai	86-21-6275-5700
China – Shenzhen	86-755-2350361
Hong Kong	852-2401-1200
India	91-80-2290061
Japan	81-45-471-6166
Korea	82-2-554-7200
Singapore	65-6334-8870
Taiwan	886-2-2717-7175

Europe

Denmark	45-4420-9895
France	33-1-69-53-63-20
Germany	49-89-627-144-0
Italy	39-039-65791-1
United Kingdom	44-118-921-5869

As of 4/1/02

Microchip Technology Inc. • 2355 W. Chandler Blvd. • Chandler, AZ 85224-6199 USA • (480) 792-7200 • FAX (480) 792-9210

DS51271B

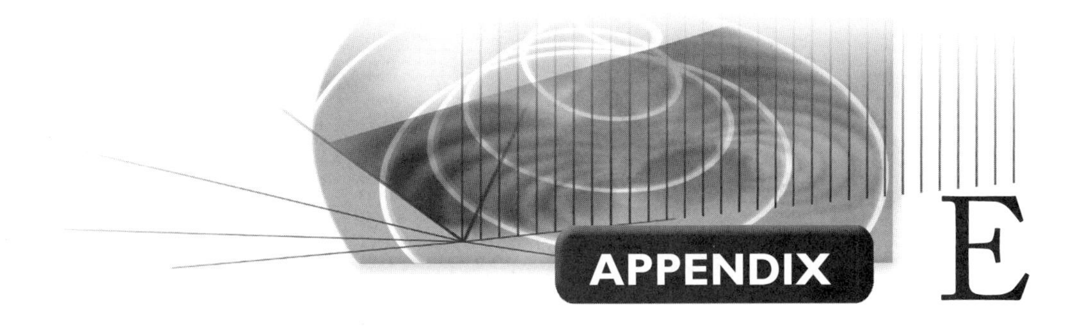

The "*FlashPIC-DEV*" Development Board

The **FlashPIC-DEV** development board is designed for prototyping and laboratory use (Figure E.1). The board supports the PIC16F877, PIC18F458, as well as several other PIC16x and PIC18x microcontrollers in the 40-pin DIP package.

The features of the development board include:

- RS232 through a 9-Pin D-Shell as well as screw terminals and a jumper header
- Up to 32K words of In-System Programmable FLASH memory with up to 256 bytes of EEPROM and up to 1.5K of Internal RAM (depending on processor selection)

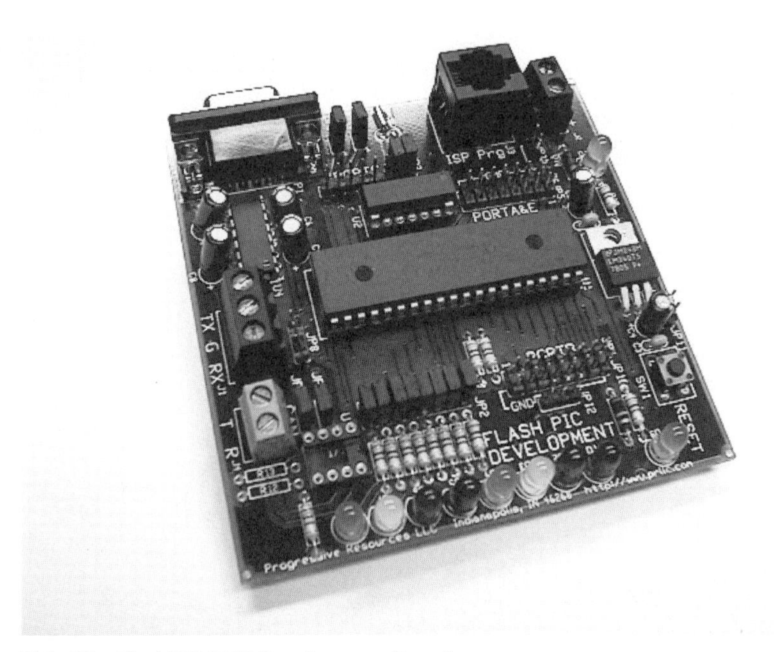

Figure E.1 *The FlashPIC-DEV Development Board*

- Up to eight 10-bit, Analog Inputs, using either internal or user supplied reference
- Nine I/O controlled LEDs, eight of which are jumper selectable
- 32 KHz "watch" crystal for on-board Real-Time operations
- A universal clock socket that allows for "canned oscillators," as well as a variety of crystals, ceramic resonators, and passive terminations
- 0.1-inch centered headers providing simple connection to the processor special function pins and I/O
- A 6-pin, ICD connection provided for in-system programming and debugging. This connection is directly compatible with the Microchip ICD, ICD2, and CCS ICD-S programming hardware. FLASH PICs can also be programmed through RS232 using an appropriate boot loader application.
- On-board regulation that allows for power inputs from 8-38 V DC with an LED power indicator
- Termination provided for 5 V DC output at 250 ma

For more information refer to the Progressive Resources Web site http://www.prllc.com.

SPECIFICATIONS

Voltage range	8 V to 38 V DC
Power consumption	250 mW (nominal)
Dimensions	W 3.7 inches x H 3.7 inches
Mounting	Rubber feet, 4 places
Weight	~3 oz
Operating temperature	0°C to +60°C
Storage temperature	0°C to +85°C
Humidity	0% to 95% at +50°C (noncondensing)

Table E.1 *Specifications*

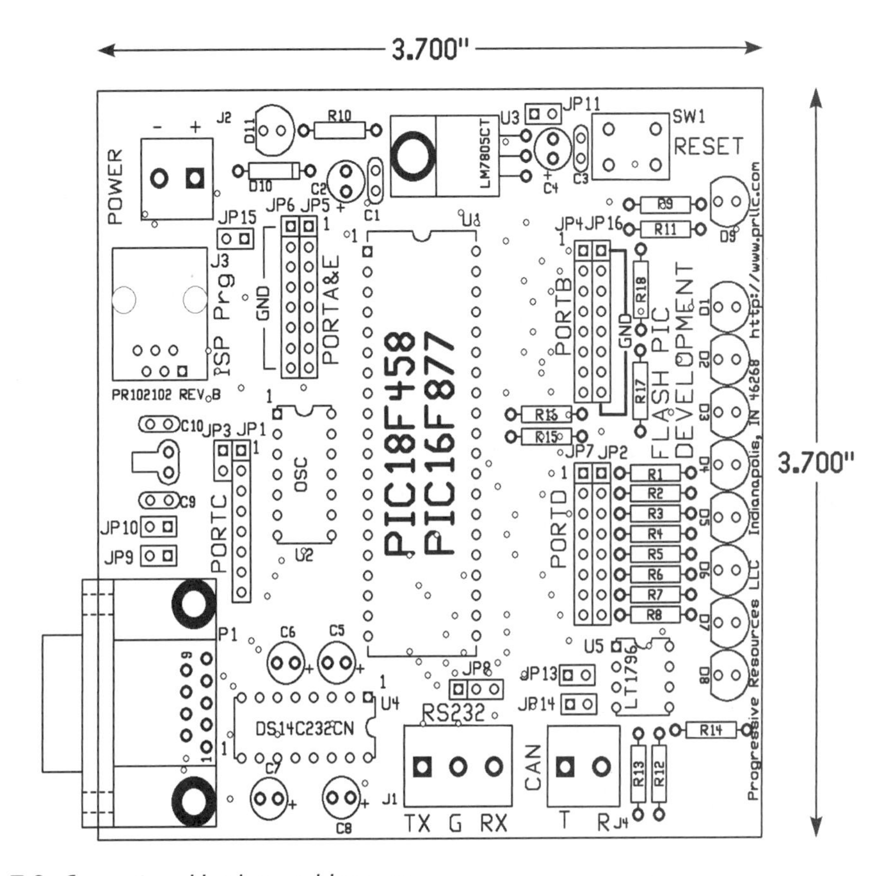

Figure E.2 *Connections, Headers, and Jumpers*

APPLICATION NOTES

POWER

J2 (screw terminal connector) is the power input point (Figure E.2 and Figure E.3). Acceptable voltages are 8–38 V DC (J1-1 is +, J1-2 is ground) (see Table E.1).

JP11 is an output of the regulated 5 V DC that may be used to power other devices. Note, however, that the LM7805 does not have a heat sink and so the actual available power output is somewhat limited, depending on the input voltage and power being consumed. Check the LM7805 regulator specification for details.

SERIAL CONNECTION

P1 is a standard DB-9 connector usually used to connect to a PC. The TX signal is on P1-2, and the RX signal is on P1-3. These are RS-232 level signals.

J1 and JP8 also provide connections for the RS-232 level serial signals. On each connector, pin 1 is the TX signal, pin 3 is the Rx signal, and pin 2 is ground.

JP9 and JP10 are jumpers, which connect the processor serial signals (RXD and TXD) to or from the RS-232 driver chip. These jumpers must be in place for the RS-232 serial connections to work. Removing the jumpers allows use of the Port D bits 0 and 1 for TTL-level I/O.

SPI CONNECTION

Use of the SPI bus is by making connections to the SPI bus signals on Port C (JP1-4, 5, and 6).

PARALLEL PORTS

Parallel ports A/E, B, C, and D are connected to JP5, JP4, JP1, and JP7, respectively (and labeled clearly on the board). Bits are connected sequentially, with bit 0 on pin 1, bit 1 on pin2, etc. These are normal TTL-level signals with or without pull-ups depending on the port initialization setup in the software.

Port A/E (JP5) has a parallel row of ground pins next to it (JP6) providing a convenient ground reference when measuring analog voltages with the internal A/D converter (ADC).

Port B (JP4) has a parallel row of ground pins next to it (JP16) so that enabling the built-in pull-up resistors and then using two-pin jumpers to ground any pins that need a logic 0 applied for input purposes can effect simple input signals.

Port C (JP1) has a parallel set of pins (JP3) at positions 1 and 2. A clock crystal (32.768 KHz) is connected to JP3. JP3 is located adjacent to JP1 (Port C) to provide an easy connection of the clock crystal to Port C bits 0 and 1 for use as a real-time clock.

Port D (JP7) has a parallel row of pins (JP2), each of which is connected to an LED through a 510-ohm series resistor to +5 V DC. Jumping any of the pins of JP7 to the corresponding pin of JP2 allows the use of the on-board LEDs as an output. Because the LEDs are connected to +5 V DC and the port is sinking the LED current, the LED will be on for any pin that outputs logic 0.

SYSTEM CLOCK

As supplied, the system clock is 10 MHz. U2 contains the crystal and caps necessary for the oscillator. Replacing U2 with a TTL, crystal, ceramic resonator, or RC oscillator, or a different integrated oscillator unit allows changing the system clock if necessary.

CAN INTERFACE

A CAN interface driver socket is provided (U5) for a Linear Technologies LT1796 CAN bus interface. Jumpers JP13 and JP14 connect the CAN interface to the appropriate pins on the controller (CANTX and CANRX). The CAN interface is intended for use with the PIC18F45x microcontrollers that have an actual CAN transceiver built in. Refer to the Microchips Web site and data sheets for details on the CAN bus controller and its features.

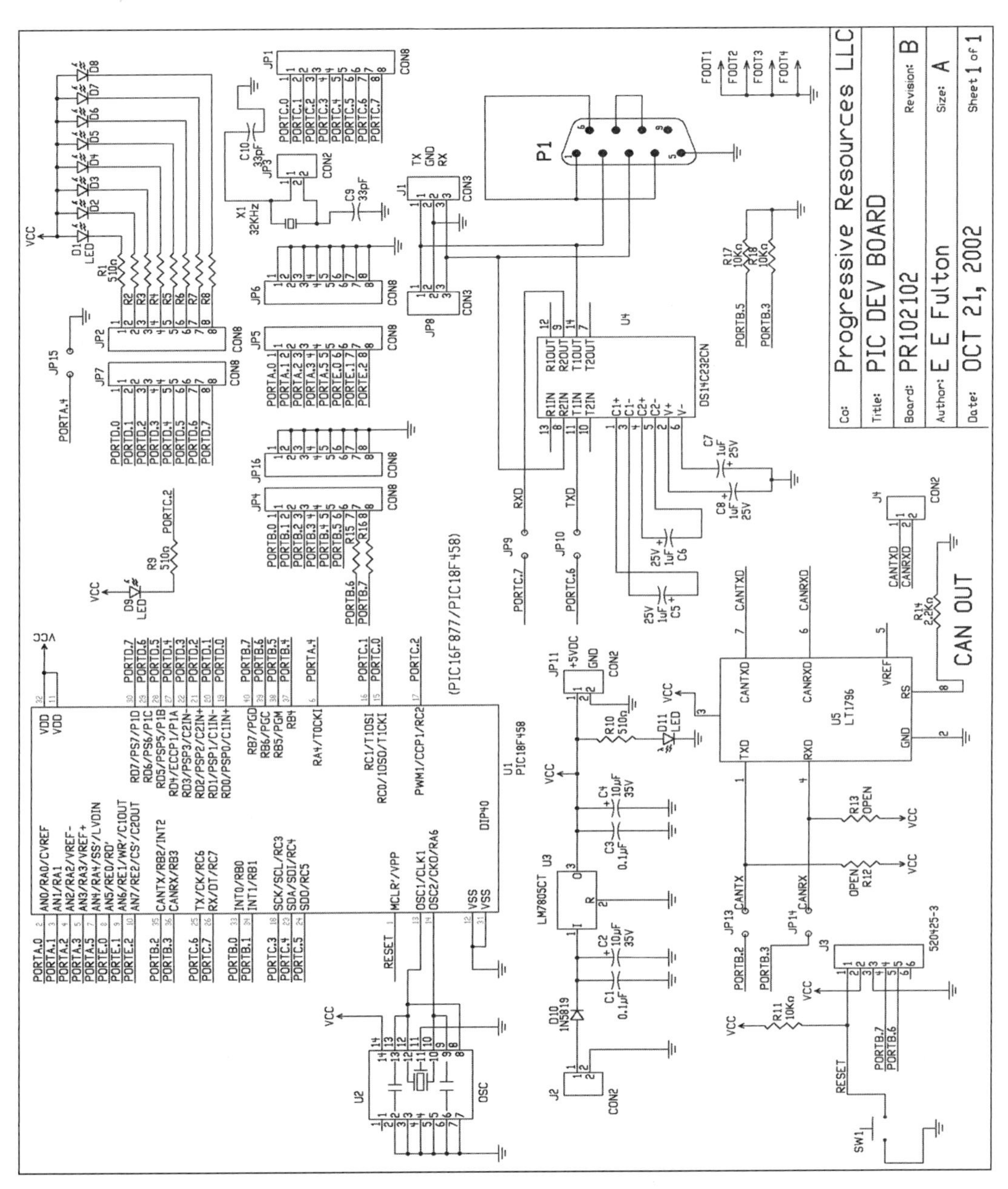

Figure E.3 *Schematic*

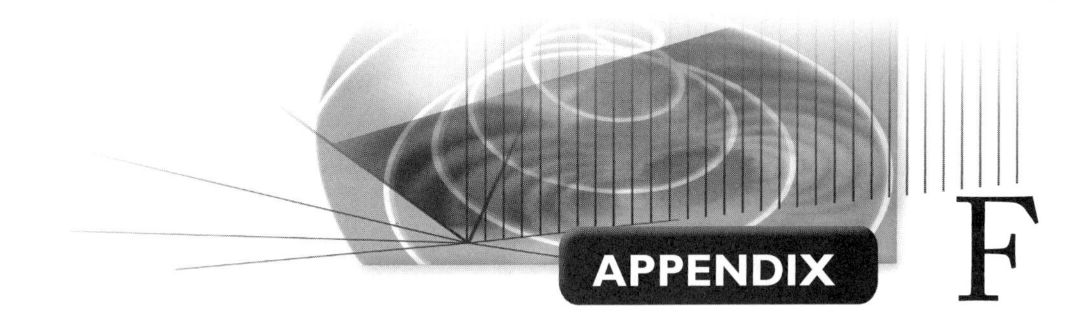

APPENDIX F

ASCII Table

AMERICAN STANDARD CODE FOR INFORMATION INTERCHANGE

Decimal	*Octal*	*Hex*	*Binary*	*Value*
000	000	000	00000000	NUL
001	001	001	00000001	SOH
002	002	002	00000010	STX
003	003	003	00000011	ETX
004	004	004	00000100	EOT
005	005	005	00000101	ENQ
006	006	006	00000110	ACK
007	007	007	00000111	BEL
008	010	008	00001000	BS
009	011	009	00001001	HT
010	012	00A	00001010	LF
011	013	00B	00001011	VT
012	014	00C	00001100	FF
013	015	00D	00001101	CR
014	016	00E	00001110	SO
015	017	00F	00001111	SI
016	020	010	00010000	DLE
017	021	011	00010001	DC1

Decimal	Octal	Hex	Binary	Value
018	022	012	00010010	DC2
019	023	013	00010011	DC3
020	024	014	00010100	DC4
021	025	015	00010101	NAK
022	026	016	00010110	SYN
023	027	017	00010111	ETB
024	030	018	00011000	CAN
025	031	019	00011001	EM
026	032	01A	00011010	SUB
027	033	01B	00011011	ESC
028	034	01C	00011100	FS
029	035	01D	00011101	GS
030	036	01E	00011110	RS
031	037	01F	00011111	US
032	040	020	00100000	SP(Space)
033	041	021	00100001	!
034	042	022	00100010	“
035	043	023	00100011	#
036	044	024	00100100	$
037	045	025	00100101	%
038	046	026	00100110	&
039	047	027	00100111	‘
040	050	028	00101000	(
041	051	029	00101001	)
042	052	02A	00101010	*
043	053	02B	00101011	+
044	054	02C	00101100	,
045	055	02D	00101101	-

Decimal	**Octal**	**Hex**	**Binary**	**Value**
046	056	02E	00101110	.
047	057	02F	00101111	/
048	060	030	00110000	0
049	061	031	00110001	1
050	062	032	00110010	2
051	063	033	00110011	3
052	064	034	00110100	4
053	065	035	00110101	5
054	066	036	00110110	6
055	067	037	00110111	7
056	070	038	00111000	8
057	071	039	00111001	9
058	072	03A	00111010	:
059	073	03B	00111011	;
060	074	03C	00111100	<
061	075	03D	00111101	=
062	076	03E	00111110	>
063	077	03F	00111111	?
064	100	040	01000000	@
065	101	041	01000001	A
066	102	042	01000010	B
067	103	043	01000011	C
068	104	044	01000100	D
069	105	045	01000101	E
070	106	046	01000110	F
071	107	047	01000111	G
072	110	048	01001000	H
073	111	049	01001001	I

Decimal	Octal	Hex	Binary	Value
074	112	04A	01001010	J
075	113	04B	01001011	K
076	114	04C	01001100	L
077	115	04D	01001101	M
078	116	04E	01001110	N
079	117	04F	01001111	O
080	120	050	01010000	P
081	121	051	01010001	Q
082	122	052	01010010	R
083	123	053	01010011	S
084	124	054	01010100	T
085	125	055	01010101	U
086	126	056	01010110	V
087	127	057	01010111	W
088	130	058	01011000	X
089	131	059	01011001	Y
090	132	05A	01011010	Z
091	133	05B	01011011	[
092	134	05C	01011100	\
093	135	05D	01011101	]
094	136	05E	01011110	^
095	137	05F	01011111	_
096	140	060	01100000	`
097	141	061	01100001	a
098	142	062	01100010	b
099	143	063	01100011	c
100	144	064	01100100	d
101	145	065	01100101	e

Decimal	Octal	Hex	Binary	Value
102	146	066	01100110	f
103	147	067	01100111	g
104	150	068	01101000	h
105	151	069	01101001	i
106	152	06A	01101010	j
107	153	06B	01101011	k
108	154	06C	01101100	l
109	155	06D	01101101	m
110	156	06E	01101110	n
111	157	06F	01101111	o
112	160	070	01110000	p
113	161	071	01110001	q
114	162	072	01110010	r
115	163	073	01110011	s
116	164	074	01110100	t
117	165	075	01110101	u
118	166	076	01110110	v
119	167	077	01110111	w
120	170	078	01111000	x
121	171	079	01111001	y
122	172	07A	01111010	z
123	173	07B	01111011	{
124	174	07C	01111100	|
125	175	07D	01111101	}
126	176	07E	01111110	~
127	177	07F	01111111	DEL

PIC16F877 Instruction Set Summary

Courtesy of Microchip Technology, Inc.

13.0 INSTRUCTION SET SUMMARY

Each PIC16F87X instruction is a 14-bit word, divided into an OPCODE which specifies the instruction type and one or more operands which further specify the operation of the instruction. The PIC16F87X instruction set summary in Table 13-2 lists **byte-oriented**, **bit-oriented**, and **literal and control** operations. Table 13-1 shows the opcode field descriptions.

For **byte-oriented** instructions, 'f' represents a file register designator and 'd' represents a destination designator. The file register designator specifies which file register is to be used by the instruction.

The destination designator specifies where the result of the operation is to be placed. If 'd' is zero, the result is placed in the W register. If 'd' is one, the result is placed in the file register specified in the instruction.

For **bit-oriented** instructions, 'b' represents a bit field designator which selects the number of the bit affected by the operation, while 'f' represents the address of the file in which the bit is located.

For **literal and control** operations, 'k' represents an eight or eleven bit constant or literal value.

TABLE 13-1: OPCODE FIELD DESCRIPTIONS

Field	Description
f	Register file address (0x00 to 0x7F)
W	Working register (accumulator)
b	Bit address within an 8-bit file register
k	Literal field, constant data or label
x	Don't care location (= 0 or 1). The assembler will generate code with x = 0. It is the recommended form of use for compatibility with all Microchip software tools.
d	Destination select; d = 0: store result in W, d = 1: store result in file register f. Default is d = 1.
PC	Program Counter
TO	Time-out bit
PD	Power-down bit

The instruction set is highly orthogonal and is grouped into three basic categories:

- **Byte-oriented** operations
- **Bit-oriented** operations
- **Literal and control** operations

All instructions are executed within one single instruction cycle, unless a conditional test is true or the program counter is changed as a result of an instruction. In this case, the execution takes two instruction cycles with the second cycle executed as a NOP. One instruction cycle consists of four oscillator periods. Thus, for an oscillator frequency of 4 MHz, the normal instruction execution time is 1 μs. If a conditional test is true, or the program counter is changed as a result of an instruction, the instruction execution time is 2 μs.

Table 13-2 lists the instructions recognized by the MPASM™ assembler.

Figure 13-1 shows the general formats that the instructions can have.

> **Note:** To maintain upward compatibility with future PIC16F87X products, <u>do not use</u> the OPTION and TRIS instructions.

All examples use the following format to represent a hexadecimal number:

0xhh

where h signifies a hexadecimal digit.

FIGURE 13-1: GENERAL FORMAT FOR INSTRUCTIONS

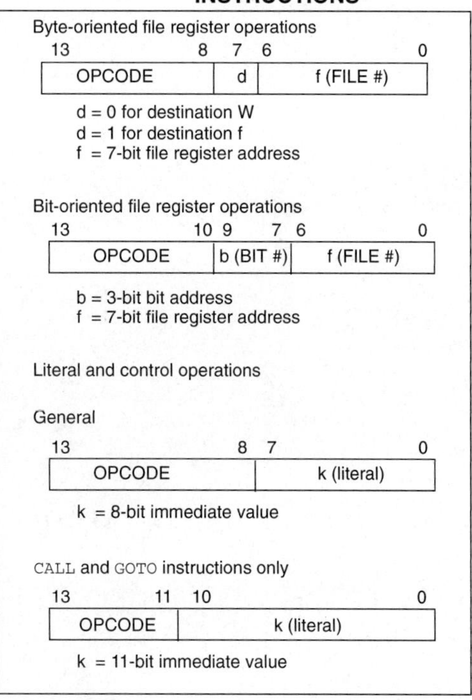

A description of each instruction is available in the PICmicro™ Mid-Range Reference Manual, (DS33023).

TABLE 13-2: PIC16F87X INSTRUCTION SET

Mnemonic,	Operands	Description	Cycles	14-Bit Opcode MSb — LSb				Status Affected	Notes
BYTE-ORIENTED FILE REGISTER OPERATIONS									
ADDWF	f, d	Add W and f	1	00	0111	dfff	ffff	C,DC,Z	1,2
ANDWF	f, d	AND W with f	1	00	0101	dfff	ffff	Z	1,2
CLRF	f	Clear f	1	00	0001	lfff	ffff	Z	2
CLRW	-	Clear W	1	00	0001	0xxx	xxxx	Z	
COMF	f, d	Complement f	1	00	1001	dfff	ffff	Z	1,2
DECF	f, d	Decrement f	1	00	0011	dfff	ffff	Z	1,2
DECFSZ	f, d	Decrement f, Skip if 0	1(2)	00	1011	dfff	ffff		1,2,3
INCF	f, d	Increment f	1	00	1010	dfff	ffff	Z	1,2
INCFSZ	f, d	Increment f, Skip if 0	1(2)	00	1111	dfff	ffff		1,2,3
IORWF	f, d	Inclusive OR W with f	1	00	0100	dfff	ffff	Z	1,2
MOVF	f, d	Move f	1	00	1000	dfff	ffff	Z	1,2
MOVWF	f	Move W to f	1	00	0000	lfff	ffff		
NOP	-	No Operation	1	00	0000	0xx0	0000		
RLF	f, d	Rotate Left f through Carry	1	00	1101	dfff	ffff	C	1,2
RRF	f, d	Rotate Right f through Carry	1	00	1100	dfff	ffff	C	1,2
SUBWF	f, d	Subtract W from f	1	00	0010	dfff	ffff	C,DC,Z	1,2
SWAPF	f, d	Swap nibbles in f	1	00	1110	dfff	ffff		1,2
XORWF	f, d	Exclusive OR W with f	1	00	0110	dfff	ffff	Z	1,2
BIT-ORIENTED FILE REGISTER OPERATIONS									
BCF	f, b	Bit Clear f	1	01	00bb	bfff	ffff		1,2
BSF	f, b	Bit Set f	1	01	01bb	bfff	ffff		1,2
BTFSC	f, b	Bit Test f, Skip if Clear	1 (2)	01	10bb	bfff	ffff		3
BTFSS	f, b	Bit Test f, Skip if Set	1 (2)	01	11bb	bfff	ffff		3
LITERAL AND CONTROL OPERATIONS									
ADDLW	k	Add literal and W	1	11	111x	kkkk	kkkk	C,DC,Z	
ANDLW	k	AND literal with W	1	11	1001	kkkk	kkkk	Z	
CALL	k	Call subroutine	2	10	0kkk	kkkk	kkkk		
CLRWDT	-	Clear Watchdog Timer	1	00	0000	0110	0100	$\overline{TO}$,$\overline{PD}$	
GOTO	k	Go to address	2	10	1kkk	kkkk	kkkk		
IORLW	k	Inclusive OR literal with W	1	11	1000	kkkk	kkkk	Z	
MOVLW	k	Move literal to W	1	11	00xx	kkkk	kkkk		
RETFIE	-	Return from interrupt	2	00	0000	0000	1001		
RETLW	k	Return with literal in W	2	11	01xx	kkkk	kkkk		
RETURN	-	Return from Subroutine	2	00	0000	0000	1000		
SLEEP	-	Go into standby mode	1	00	0000	0110	0011	$\overline{TO}$,$\overline{PD}$	
SUBLW	k	Subtract W from literal	1	11	110x	kkkk	kkkk	C,DC,Z	
XORLW	k	Exclusive OR literal with W	1	11	1010	kkkk	kkkk	Z	

Note 1: When an I/O register is modified as a function of itself (e.g., `MOVF PORTB, 1`), the value used will be that value present on the pins themselves. For example, if the data latch is '1' for a pin configured as input and is driven low by an external device, the data will be written back with a '0'.

2: If this instruction is executed on the TMR0 register (and, where applicable, d = 1), the prescaler will be cleared if assigned to the Timer0 module.

3: If Program Counter (PC) is modified, or a conditional test is true, the instruction requires two cycles. The second cycle is executed as a `NOP`.

Note: Additional information on the mid-range instruction set is available in the PICmicro™ Mid-Range MCU Family Reference Manual (DS33023).

ADDLW	Add Literal and W
Syntax:	[*label*] ADDLW k
Operands:	$0 \le k \le 255$
Operation:	$(W) + k \rightarrow (W)$
Status Affected:	C, DC, Z
Description:	The contents of the W register are added to the eight bit literal 'k' and the result is placed in the W register.

ADDWF	Add W and f
Syntax:	[*label*] ADDWF f,d
Operands:	$0 \le f \le 127$ $d \in [0,1]$
Operation:	$(W) + (f) \rightarrow$ (destination)
Status Affected:	C, DC, Z
Description:	Add the contents of the W register with register 'f'. If 'd' is 0, the result is stored in the W register. If 'd' is 1, the result is stored back in register 'f'.

ANDLW	AND Literal with W
Syntax:	[*label*] ANDLW k
Operands:	$0 \le k \le 255$
Operation:	(W) .AND. (k) $\rightarrow$ (W)
Status Affected:	Z
Description:	The contents of W register are AND'ed with the eight bit literal 'k'. The result is placed in the W register.

ANDWF	AND W with f
Syntax:	[*label*] ANDWF f,d
Operands:	$0 \le f \le 127$ $d \in [0,1]$
Operation:	(W) .AND. (f) $\rightarrow$ (destination)
Status Affected:	Z
Description:	AND the W register with register 'f'. If 'd' is 0, the result is stored in the W register. If 'd' is 1, the result is stored back in register 'f'.

BCF	Bit Clear f
Syntax:	[*label*] BCF f,b
Operands:	$0 \le f \le 127$ $0 \le b \le 7$
Operation:	0 $\rightarrow$ (f<b>)
Status Affected:	None
Description:	Bit 'b' in register 'f' is cleared.

BSF	Bit Set f
Syntax:	[*label*] BSF f,b
Operands:	$0 \le f \le 127$ $0 \le b \le 7$
Operation:	1 $\rightarrow$ (f<b>)
Status Affected:	None
Description:	Bit 'b' in register 'f' is set.

BTFSS	Bit Test f, Skip if Set
Syntax:	[*label*] BTFSS f,b
Operands:	$0 \le f \le 127$ $0 \le b < 7$
Operation:	skip if (f<b>) = 1
Status Affected:	None
Description:	If bit 'b' in register 'f' is '0', the next instruction is executed. If bit 'b' is '1', then the next instruction is discarded and a `NOP` is executed instead, making this a 2T_{CY} instruction.

BTFSC	Bit Test, Skip if Clear
Syntax:	[*label*] BTFSC f,b
Operands:	$0 \le f \le 127$ $0 \le b \le 7$
Operation:	skip if (f<b>) = 0
Status Affected:	None
Description:	If bit 'b' in register 'f' is '1', the next instruction is executed. If bit 'b', in register 'f', is '0', the next instruction is discarded, and a `NOP` is executed instead, making this a 2T_{CY} instruction.

CALL — Call Subroutine

Syntax:	[*label*] CALL k
Operands:	$0 \leq k \leq 2047$
Operation:	(PC)+ 1→ TOS, k → PC<10:0>, (PCLATH<4:3>) → PC<12:11>
Status Affected:	None
Description:	Call Subroutine. First, return address (PC+1) is pushed onto the stack. The eleven-bit immediate address is loaded into PC bits <10:0>. The upper bits of the PC are loaded from PCLATH. `CALL` is a two-cycle instruction.

CLRWDT — Clear Watchdog Timer

Syntax:	[*label*] CLRWDT
Operands:	None
Operation:	00h → WDT 0 → WDT prescaler, 1 → $\overline{TO}$ 1 → $\overline{PD}$
Status Affected:	$\overline{TO}$, $\overline{PD}$
Description:	`CLRWDT` instruction resets the Watchdog Timer. It also resets the prescaler of the WDT. Status bits $\overline{TO}$ and $\overline{PD}$ are set.

CLRF — Clear f

Syntax:	[*label*] CLRF f
Operands:	$0 \leq f \leq 127$
Operation:	00h → (f) 1 → Z
Status Affected:	Z
Description:	The contents of register 'f' are cleared and the Z bit is set.

COMF — Complement f

Syntax:	[*label*] COMF f,d
Operands:	$0 \leq f \leq 127$ $d \in [0,1]$
Operation:	($\bar{f}$) → (destination)
Status Affected:	Z
Description:	The contents of register 'f' are complemented. If 'd' is 0, the result is stored in W. If 'd' is 1, the result is stored back in register 'f'.

CLRW — Clear W

Syntax:	[*label*] CLRW
Operands:	None
Operation:	00h → (W) 1 → Z
Status Affected:	Z
Description:	W register is cleared. Zero bit (Z) is set.

DECF — Decrement f

Syntax:	[*label*] DECF f,d
Operands:	$0 \leq f \leq 127$ $d \in [0,1]$
Operation:	(f) - 1 → (destination)
Status Affected:	Z
Description:	Decrement register 'f'. If 'd' is 0, the result is stored in the W register. If 'd' is 1, the result is stored back in register 'f'.

DECFSZ	**Decrement f, Skip if 0**
Syntax:	[*label*] DECFSZ f,d
Operands:	$0 \le f \le 127$ $d \in [0,1]$
Operation:	(f) - 1 $\rightarrow$ (destination); skip if result = 0
Status Affected:	None
Description:	The contents of register 'f' are decremented. If 'd' is 0, the result is placed in the W register. If 'd' is 1, the result is placed back in register 'f'. If the result is 1, the next instruction is executed. If the result is 0, then a `NOP` is executed instead making it a 2TCY instruction.

GOTO	**Unconditional Branch**
Syntax:	[*label*] GOTO k
Operands:	$0 \le k \le 2047$
Operation:	k $\rightarrow$ PC<10:0> PCLATH<4:3> $\rightarrow$ PC<12:11>
Status Affected:	None
Description:	`GOTO` is an unconditional branch. The eleven-bit immediate value is loaded into PC bits <10:0>. The upper bits of PC are loaded from PCLATH<4:3>. `GOTO` is a two-cycle instruction.

INCF	**Increment f**
Syntax:	[*label*] INCF f,d
Operands:	$0 \le f \le 127$ $d \in [0,1]$
Operation:	(f) + 1 $\rightarrow$ (destination)
Status Affected:	Z
Description:	The contents of register 'f' are incremented. If 'd' is 0, the result is placed in the W register. If 'd' is 1, the result is placed back in register 'f'.

INCFSZ	**Increment f, Skip if 0**
Syntax:	[*label*] INCFSZ f,d
Operands:	$0 \le f \le 127$ $d \in [0,1]$
Operation:	(f) + 1 $\rightarrow$ (destination), skip if result = 0
Status Affected:	None
Description:	The contents of register 'f' are incremented. If 'd' is 0, the result is placed in the W register. If 'd' is 1, the result is placed back in register 'f'. If the result is 1, the next instruction is executed. If the result is 0, a `NOP` is executed instead, making it a 2TCY instruction.

IORLW	**Inclusive OR Literal with W**
Syntax:	[*label*] IORLW k
Operands:	$0 \le k \le 255$
Operation:	(W) .OR. k $\rightarrow$ (W)
Status Affected:	Z
Description:	The contents of the W register are OR'ed with the eight bit literal 'k'. The result is placed in the W register.

IORWF	**Inclusive OR W with f**
Syntax:	[*label*] IORWF f,d
Operands:	$0 \le f \le 127$ $d \in [0,1]$
Operation:	(W) .OR. (f) $\rightarrow$ (destination)
Status Affected:	Z
Description:	Inclusive OR the W register with register 'f'. If 'd' is 0 the result is placed in the W register. If 'd' is 1 the result is placed back in register 'f'.

MOVF	**Move f**
Syntax:	[*label*] MOVF f,d
Operands:	$0 \le f \le 127$ $d \in [0,1]$
Operation:	(f) → (destination)
Status Affected:	Z
Description:	The contents of register f are moved to a destination dependant upon the status of d. If d = 0, destination is W register. If d = 1, the destination is file register f itself. d = 1 is useful to test a file register, since status flag Z is affected.

MOVLW	**Move Literal to W**
Syntax:	[*label*] MOVLW k
Operands:	$0 \le k \le 255$
Operation:	k → (W)
Status Affected:	None
Description:	The eight bit literal 'k' is loaded into W register. The don't cares will assemble as 0's.

MOVWF	**Move W to f**
Syntax:	[*label*] MOVWF f
Operands:	$0 \le f \le 127$
Operation:	(W) → (f)
Status Affected:	None
Description:	Move data from W register to register 'f'.

NOP	**No Operation**
Syntax:	[*label*] NOP
Operands:	None
Operation:	No operation
Status Affected:	None
Description:	No operation.

RETFIE	**Return from Interrupt**
Syntax:	[*label*] RETFIE
Operands:	None
Operation:	TOS → PC, 1 → GIE
Status Affected:	None

RETLW	**Return with Literal in W**
Syntax:	[*label*] RETLW k
Operands:	$0 \le k \le 255$
Operation:	k → (W); TOS → PC
Status Affected:	None
Description:	The W register is loaded with the eight bit literal 'k'. The program counter is loaded from the top of the stack (the return address). This is a two-cycle instruction.

RLF	**Rotate Left f through Carry**
Syntax:	[*label*] RLF f,d
Operands:	0 ≤ f ≤ 127 d ∈ [0,1]
Operation:	See description below
Status Affected:	C
Description:	The contents of register 'f' are rotated one bit to the left through the Carry Flag. If 'd' is 0, the result is placed in the W register. If 'd' is 1, the result is stored back in register 'f'.

RETURN	**Return from Subroutine**
Syntax:	[*label*] RETURN
Operands:	None
Operation:	TOS → PC
Status Affected:	None
Description:	Return from subroutine. The stack is POPed and the top of the stack (TOS) is loaded into the program counter. This is a two-cycle instruction.

RRF	**Rotate Right f through Carry**
Syntax:	[*label*] RRF f,d
Operands:	0 ≤ f ≤ 127 d ∈ [0,1]
Operation:	See description below
Status Affected:	C
Description:	The contents of register 'f' are rotated one bit to the right through the Carry Flag. If 'd' is 0, the result is placed in the W register. If 'd' is 1, the result is placed back in register 'f'.

SLEEP	
Syntax:	[*label*] SLEEP
Operands:	None
Operation:	00h → WDT, 0 → WDT prescaler, 1 → $\overline{TO}$, 0 → $\overline{PD}$
Status Affected:	$\overline{TO}$, $\overline{PD}$
Description:	The power-down status bit, $\overline{PD}$ is cleared. Time-out status bit, $\overline{TO}$ is set. Watchdog Timer and its prescaler are cleared. The processor is put into SLEEP mode with the oscillator stopped.

SUBLW	**Subtract W from Literal**
Syntax:	[*label*] SUBLW k
Operands:	0 ≤ k ≤ 255
Operation:	k - (W) → (W)
Status Affected:	C, DC, Z
Description:	The W register is subtracted (2's complement method) from the eight-bit literal 'k'. The result is placed in the W register.

SUBWF	**Subtract W from f**
Syntax:	[*label*] SUBWF f,d
Operands:	0 ≤ f ≤ 127 d ∈ [0,1]
Operation:	(f) - (W) → (destination)
Status Affected:	C, DC, Z
Description:	Subtract (2's complement method) W register from register 'f'. If 'd' is 0, the result is stored in the W register. If 'd' is 1, the result is stored back in register 'f'.

SWAPF	**Swap Nibbles in f**
Syntax:	[*label*] SWAPF f,d
Operands:	0 ≤ f ≤ 127 d ∈ [0,1]
Operation:	(f<3:0>) → (destination<7:4>), (f<7:4>) → (destination<3:0>)
Status Affected:	None
Description:	The upper and lower nibbles of register 'f' are exchanged. If 'd' is 0, the result is placed in the W register. If 'd' is 1, the result is placed in register 'f'.

XORLW	**Exclusive OR Literal with W**
Syntax:	[*label*] XORLW k
Operands:	0 ≤ k ≤ 255
Operation:	(W) .XOR. k → (W)
Status Affected:	Z
Description:	The contents of the W register are XOR'ed with the eight-bit literal 'k'. The result is placed in the W register.

XORWF	**Exclusive OR W with f**
Syntax:	[*label*] XORWF f,d
Operands:	0 ≤ f ≤ 127 d ∈ [0,1]
Operation:	(W) .XOR. (f) → (destination)
Status Affected:	Z
Description:	Exclusive OR the contents of the W register with register 'f'. If 'd' is 0, the result is stored in the W register. If 'd' is 1, the result is stored back in register 'f'.

PIC18F458 Instruction Set Summary

Courtesy of Microchip Technology, Inc.

25.0 INSTRUCTION SET SUMMARY

The PIC18 instruction set adds many enhancements to the previous PICmicro instruction sets, while maintaining an easy migration from these PICmicro instruction sets.

Most instructions are a single program memory word (16 bits), but there are three instructions that require two program memory locations.

Each single word instruction is a 16-bit word divided into an OPCODE, which specifies the instruction type and one or more operands, which further specify the operation of the instruction.

The instruction set is highly orthogonal and is grouped into four basic categories:

- **Byte-oriented** operations
- **Bit-oriented** operations
- **Literal** operations
- **Control** operations

The PIC18 instruction set summary in Table 25-2 lists **byte-oriented**, **bit-oriented**, **literal** and **control** operations. Table 25-1 shows the opcode field descriptions.

Most **byte-oriented** instructions have three operands:

1. The file register (specified by 'f')
2. The destination of the result (specified by 'd')
3. The accessed memory (specified by 'a')

The file register designator 'f' specifies which file register is to be used by the instruction.

The destination designator 'd' specifies where the result of the operation is to be placed. If 'd' is zero, the result is placed in the WREG register. If 'd' is one, the result is placed in the file register specified in the instruction.

All **bit-oriented** instructions have three operands:

1. The file register (specified by 'f')
2. The bit in the file register (specified by 'b')
3. The accessed memory (specified by 'a')

The bit field designator 'b' selects the number of the bit affected by the operation, while the file register designator 'f' represents the number of the file in which the bit is located.

The **literal** instructions may use some of the following operands:

- A literal value to be loaded into a file register (specified by 'k')
- The desired FSR register to load the literal value into (specified by 'f')
- No operand required (specified by '—')

The **control** instructions may use some of the following operands:

- A program memory address (specified by 'n')
- The mode of the Call or Return instructions (specified by 's')
- The mode of the Table Read and Table Write instructions (specified by 'm')
- No operand required (specified by '—')

All instructions are a single word, except for three double-word instructions. These three instructions were made double-word instructions so that all the required information is available in these 32 bits. In the second word, the 4 MSbs are 1's. If this second word is executed as an instruction (by itself), it will execute as a `NOP`.

All single word instructions are executed in a single instruction cycle, unless a conditional test is true or the program counter is changed as a result of the instruction. In these cases, the execution takes two instruction cycles, with the additional instruction cycle(s) executed as a `NOP`.

The double-word instructions execute in two instruction cycles.

One instruction cycle consists of four oscillator periods. Thus, for an oscillator frequency of 4 MHz, the normal instruction execution time is 1 μs. If a conditional test is true, or the program counter is changed as a result of an instruction, the instruction execution time is 2 μs. Two-word branch instructions (if true) would take 3 μs.

Figure 25-1 shows the general formats that the instructions can have.

All examples use the format '`nnh`' to represent a hexadecimal number, where '`h`' signifies a hexadecimal digit.

The Instruction Set Summary, shown in Table 25-2, lists the instructions recognized by the Microchip Assembler (MPASM™).

Section 25.2 provides a description of each instruction.

25.1 READ-MODIFY-WRITE OPERATIONS

Any instruction that specifies a file register as part of the instruction performs a Read-Modify-Write (R-M-W) operation. The register is read, the data is modified, and the result is stored according to either the instruction or the destination designator 'd'. A read operation is performed on a register even if the instruction writes to that register.

For example, a "`clrf PORTB`" instruction will read PORTB, clear all the data bits, then write the result back to PORTB. This example would have the unintended result that the condition that sets the RBIF flag would be cleared.

TABLE 25-1: OPCODE FIELD DESCRIPTIONS

Field	Description
a	RAM access bit a = 0: RAM location in Access RAM (BSR register is ignored) a = 1: RAM bank is specified by BSR register
bbb	Bit address within an 8-bit file register (0 to 7)
BSR	Bank Select Register. Used to select the current RAM bank.
d	Destination select bit; d = 0: store result in WREG, d = 1: store result in file register f.
dest	Destination either the WREG register or the specified register file location
f	8-bit Register file address (0x00 to 0xFF)
fs	12-bit Register file address (0x000 to 0xFFF). This is the source address.
fd	12-bit Register file address (0x000 to 0xFFF). This is the destination address.
k	Literal field, constant data or label (may be either an 8-bit, 12-bit or a 20-bit value)
label	Label name
mm	The mode of the TBLPTR register for the Table Read and Table Write instructions. Only used with Table Read and Table Write instructions:
*	No change to register (such as TBLPTR with Table Reads and Writes)
*+	Post-Increment register (such as TBLPTR with Table Reads and Writes)
*-	Post-Decrement register (such as TBLPTR with Table Reads and Writes)
+*	Pre-Increment register (such as TBLPTR with Table Reads and Writes)
n	The relative address (2's complement number) for relative branch instructions, or the direct address for Call/Branch and Return instructions
PRODH	Product of Multiply high byte
PRODL	Product of Multiply low byte
s	Fast Call/Return mode select bit; s = 0: do not update into/from shadow registers s = 1: certain registers loaded into/from shadow registers (Fast mode)
u	Unused or Unchanged
WREG	Working register (accumulator)
x	Don't care (0 or 1). The assembler will generate code with x = 0. It is the recommended form of use for compatibility with all Microchip software tools.
TBLPTR	21-bit Table Pointer (points to a Program Memory location)
TABLAT	8-bit Table Latch
TOS	Top-of-Stack
PC	Program Counter
PCL	Program Counter Low Byte
PCH	Program Counter High Byte
PCLATH	Program Counter High Byte Latch
PCLATU	Program Counter Upper Byte Latch
GIE	Global Interrupt Enable bit
WDT	Watchdog Timer
$\overline{TO}$	Time-out bit
$\overline{PD}$	Power-down bit
C, DC, Z, OV, N	ALU status bits Carry, Digit Carry, Zero, Overflow, Negative
[ ]	Optional
( )	Contents
→	Assigned to
< >	Register bit field
∈	In the set of
italics	User defined term (font is courier)

FIGURE 25-1: GENERAL FORMAT FOR INSTRUCTIONS

Example Instruction

Byte-oriented file register operations

15 10	9	8	7 0
OPCODE	d	a	f (FILE #)

`ADDWF MYREG, W, B`

d = 0 for result destination to be WREG register
d = 1 for result destination to be file register (f)
a = 0 to force Access Bank
a = 1 for BSR to select bank
f = 8-bit file register address

Byte to Byte move operations (2-word)

15 12	11 0
OPCODE	f (Source FILE #)

15 12	11 0
1111	f (Destination FILE #)

`MOVFF MYREG1, MYREG2`

f = 12-bit file register address

Bit-oriented file register operations

15 12	11 9	8	7 0
OPCODE	b (BIT #)	a	f (FILE #)

`BSF MYREG, bit, B`

b = 3-bit position of bit in file register (f)
a = 0 to force Access Bank
a = 1 for BSR to select bank
f = 8-bit file register address

Literal operations

15 8	7 0
OPCODE	k (literal)

`MOVLW 0x7F`

k = 8-bit immediate value

Control operations

CALL, GOTO and Branch operations

15 8	7 0
OPCODE	n<7:0> (literal)

15 12	11 0
1111	n<19:8> (literal)

`GOTO Label`

n = 20-bit immediate value

15 8	8	7 0
OPCODE	S	n<7:0> (literal)

15 12	11 0
	n<19:8> (literal)

`CALL MYFUNC`

S = Fast bit

15 11	10 0
OPCODE	n<10:0> (literal)

`BRA MYFUNC`

15 8	7 0
OPCODE	n<7:0> (literal)

`BC MYFUNC`

TABLE 25-2: PIC18FXXX INSTRUCTION SET

Mnemonic, Operands		Description	Cycles	16-Bit Instruction Word				Status Affected	Notes
				MSb			LSb		
BYTE-ORIENTED FILE REGISTER OPERATIONS									
ADDWF	f, d, a	Add WREG and f	1	0010	01da	ffff	ffff	C, DC, Z, OV, N	1, 2
ADDWFC	f, d, a	Add WREG and Carry bit to f	1	0010	00da	ffff	ffff	C, DC, Z, OV, N	1, 2
ANDWF	f, d, a	AND WREG with f	1	0001	01da	ffff	ffff	Z, N	1,2
CLRF	f, a	Clear f	1	0110	101a	ffff	ffff	Z	2
COMF	f, d, a	Complement f	1	0001	11da	ffff	ffff	Z, N	1, 2
CPFSEQ	f, a	Compare f with WREG, skip =	1 (2 or 3)	0110	001a	ffff	ffff	None	4
CPFSGT	f, a	Compare f with WREG, skip >	1 (2 or 3)	0110	010a	ffff	ffff	None	4
CPFSLT	f, a	Compare f with WREG, skip <	1 (2 or 3)	0110	000a	ffff	ffff	None	1, 2
DECF	f, d, a	Decrement f	1	0000	01da	ffff	ffff	C, DC, Z, OV, N	1, 2, 3, 4
DECFSZ	f, d, a	Decrement f, Skip if 0	1 (2 or 3)	0010	11da	ffff	ffff	None	1, 2, 3, 4
DCFSNZ	f, d, a	Decrement f, Skip if Not 0	1 (2 or 3)	0100	11da	ffff	ffff	None	1, 2
INCF	f, d, a	Increment f	1	0010	10da	ffff	ffff	C, DC, Z, OV, N	1, 2, 3, 4
INCFSZ	f, d, a	Increment f, Skip if 0	1 (2 or 3)	0011	11da	ffff	ffff	None	4
INFSNZ	f, d, a	Increment f, Skip if Not 0	1 (2 or 3)	0100	10da	ffff	ffff	None	1, 2
IORWF	f, d, a	Inclusive OR WREG with f	1	0001	00da	ffff	ffff	Z, N	1, 2
MOVF	f, d, a	Move f	1	0101	00da	ffff	ffff	Z, N	1
MOVFF	f_s, f_d	Move f_s (source) to 1st word	2	1100	ffff	ffff	ffff	None	
		f_d (destination) 2nd word		1111	ffff	ffff	ffff		
MOVWF	f, a	Move WREG to f	1	0110	111a	ffff	ffff	None	
MULWF	f, a	Multiply WREG with f	1	0000	001a	ffff	ffff	None	
NEGF	f, a	Negate f	1	0110	110a	ffff	ffff	C, DC, Z, OV, N	1, 2
RLCF	f, d, a	Rotate Left f through Carry	1	0011	01da	ffff	ffff	C, Z, N	
RLNCF	f, d, a	Rotate Left f (No Carry)	1	0100	01da	ffff	ffff	Z, N	1, 2
RRCF	f, d, a	Rotate Right f through Carry	1	0011	00da	ffff	ffff	C, Z, N	
RRNCF	f, d, a	Rotate Right f (No Carry)	1	0100	00da	ffff	ffff	Z, N	
SETF	f, a	Set f	1	0110	100a	ffff	ffff	None	
SUBFWB	f, d, a	Subtract f from WREG with borrow	1	0101	01da	ffff	ffff	C, DC, Z, OV, N	1, 2
SUBWF	f, d, a	Subtract WREG from f	1	0101	11da	ffff	ffff	C, DC, Z, OV, N	
SUBWFB	f, d, a	Subtract WREG from f with borrow	1	0101	10da	ffff	ffff	C, DC, Z, OV, N	1, 2
SWAPF	f, d, a	Swap nibbles in f	1	0011	10da	ffff	ffff	None	4
TSTFSZ	f, a	Test f, skip if 0	1 (2 or 3)	0110	011a	ffff	ffff	None	1, 2
XORWF	f, d, a	Exclusive OR WREG with f	1	0001	10da	ffff	ffff	Z, N	
BIT-ORIENTED FILE REGISTER OPERATIONS									
BCF	f, b, a	Bit Clear f	1	1001	bbba	ffff	ffff	None	1, 2
BSF	f, b, a	Bit Set f	1	1000	bbba	ffff	ffff	None	1, 2
BTFSC	f, b, a	Bit Test f, Skip if Clear	1 (2 or 3)	1011	bbba	ffff	ffff	None	3, 4
BTFSS	f, b, a	Bit Test f, Skip if Set	1 (2 or 3)	1010	bbba	ffff	ffff	None	3, 4
BTG	f, d, a	Bit Toggle f	1	0111	bbba	ffff	ffff	None	1, 2

Note 1: When a PORT register is modified as a function of itself (e.g., `MOVF PORTB, 1, 0`), the value used will be that value present on the pins themselves. For example, if the data latch is '1' for a pin configured as input and is driven low by an external device, the data will be written back with a '0'.

2: If this instruction is executed on the TMR0 register (and, where applicable, d = `1`), the prescaler will be cleared if assigned.

3: If Program Counter (PC) is modified or a conditional test is true, the instruction requires two cycles. The second cycle is executed as a `NOP`.

4: Some instructions are 2-word instructions. The second word of these instructions will be executed as a `NOP`, unless the first word of the instruction retrieves the information embedded in these 16 bits. This ensures that all program memory locations have a valid instruction.

5: If the Table Write starts the write cycle to internal memory, the write will continue until terminated.

TABLE 25-2: PIC18FXXX INSTRUCTION SET (CONTINUED)

Mnemonic, Operands		Description	Cycles	16-Bit Instruction Word				Status Affected	Notes
				MSb			LSb		
LITERAL OPERATIONS									
ADDLW	k	Add literal and WREG	1	0000	1111	kkkk	kkkk	C, DC, Z, OV, N	
ANDLW	k	AND literal with WREG	1	0000	1011	kkkk	kkkk	Z, N	
IORLW	k	Inclusive OR literal with WREG	1	0000	1001	kkkk	kkkk	Z, N	
LFSR	f, k	Move literal (12-bit) 2nd word	2	1110	1110	00ff	kkkk	None	
		to FSRx 1st word		1111	0000	kkkk	kkkk		
MOVLB	k	Move literal to BSR<3:0>	1	0000	0001	0000	kkkk	None	
MOVLW	k	Move literal to WREG	1	0000	1110	kkkk	kkkk	None	
MULLW	k	Multiply literal with WREG	1	0000	1101	kkkk	kkkk	None	
RETLW	k	Return with literal in WREG	2	0000	1100	kkkk	kkkk	None	
SUBLW	k	Subtract WREG from literal	1	0000	1000	kkkk	kkkk	C, DC, Z, OV, N	
XORLW	k	Exclusive OR literal with WREG	1	0000	1010	kkkk	kkkk	Z, N	
DATA MEMORY ↔ PROGRAM MEMORY OPERATIONS									
TBLRD*		Table Read	2	0000	0000	0000	1000	None	
TBLRD*+		Table Read with post-increment		0000	0000	0000	1001	None	
TBLRD*-		Table Read with post-decrement		0000	0000	0000	1010	None	
TBLRD+*		Table Read with pre-increment		0000	0000	0000	1011	None	
TBLWT*		Table Write	2 (5)	0000	0000	0000	1100	None	
TBLWT*+		Table Write with post-increment		0000	0000	0000	1101	None	
TBLWT*-		Table Write with post-decrement		0000	0000	0000	1110	None	
TBLWT+*		Table Write with pre-increment		0000	0000	0000	1111	None	

Note 1: When a PORT register is modified as a function of itself (e.g., `MOVF PORTB, 1, 0`), the value used will be that value present on the pins themselves. For example, if the data latch is '1' for a pin configured as input and is driven low by an external device, the data will be written back with a '0'.

2: If this instruction is executed on the TMR0 register (and, where applicable, d = `1`), the prescaler will be cleared if assigned.

3: If Program Counter (PC) is modified or a conditional test is true, the instruction requires two cycles. The second cycle is executed as a `NOP`.

4: Some instructions are 2-word instructions. The second word of these instructions will be executed as a `NOP`, unless the first word of the instruction retrieves the information embedded in these 16 bits. This ensures that all program memory locations have a valid instruction.

5: If the Table Write starts the write cycle to internal memory, the write will continue until terminated.

TABLE 25-2: PIC18FXXX INSTRUCTION SET (CONTINUED)

Mnemonic, Operands	Description	Cycles	16-Bit Instruction Word				Status Affected	Notes
			MSb			LSb		
LITERAL OPERATIONS								
ADDLW k	Add literal and WREG	1	0000	1111	kkkk	kkkk	C, DC, Z, OV, N	
ANDLW k	AND literal with WREG	1	0000	1011	kkkk	kkkk	Z, N	
IORLW k	Inclusive OR literal with WREG	1	0000	1001	kkkk	kkkk	Z, N	
LFSR f, k	Move literal (12-bit) 2nd word	2	1110	1110	00ff	kkkk	None	
	to FSRx 1st word		1111	0000	kkkk	kkkk		
MOVLB k	Move literal to BSR<3:0>	1	0000	0001	0000	kkkk	None	
MOVLW k	Move literal to WREG	1	0000	1110	kkkk	kkkk	None	
MULLW k	Multiply literal with WREG	1	0000	1101	kkkk	kkkk	None	
RETLW k	Return with literal in WREG	2	0000	1100	kkkk	kkkk	None	
SUBLW k	Subtract WREG from literal	1	0000	1000	kkkk	kkkk	C, DC, Z, OV, N	
XORLW k	Exclusive OR literal with WREG	1	0000	1010	kkkk	kkkk	Z, N	
DATA MEMORY ↔ PROGRAM MEMORY OPERATIONS								
TBLRD*	Table Read	2	0000	0000	0000	1000	None	
TBLRD*+	Table Read with post-increment		0000	0000	0000	1001	None	
TBLRD*-	Table Read with post-decrement		0000	0000	0000	1010	None	
TBLRD+*	Table Read with pre-increment		0000	0000	0000	1011	None	
TBLWT*	Table Write	2 (5)	0000	0000	0000	1100	None	
TBLWT*+	Table Write with post-increment		0000	0000	0000	1101	None	
TBLWT*-	Table Write with post-decrement		0000	0000	0000	1110	None	
TBLWT+*	Table Write with pre-increment		0000	0000	0000	1111	None	

Note 1: When a PORT register is modified as a function of itself (e.g., `MOVF PORTB, 1, 0`), the value used will be that value present on the pins themselves. For example, if the data latch is '1' for a pin configured as input and is driven low by an external device, the data will be written back with a '0'.

2: If this instruction is executed on the TMR0 register (and, where applicable, d = `1`), the prescaler will be cleared if assigned.

3: If Program Counter (PC) is modified or a conditional test is true, the instruction requires two cycles. The second cycle is executed as a `NOP`.

4: Some instructions are 2-word instructions. The second word of these instructions will be executed as a `NOP`, unless the first word of the instruction retrieves the information embedded in these 16 bits. This ensures that all program memory locations have a valid instruction.

5: If the Table Write starts the write cycle to internal memory, the write will continue until terminated.

Answers to Selected Questions (By Chapter)

'*' by an exercise number indicates that the question is answered or partially answered in this appendix.

ANSWERS TO SELECTED CHAPTER 1 EXERCISES

*2. Create an appropriate declaration for the following (Section 1.4):

A. A constant called 'x' that will be set to 789.

Answer:

```
const int16 x = 789; // typical for CCS-PICC

        -OR-

//if #TYPE SHORT=8, INT=16, LONG=32
const int x = 789;
```

Notes: An integer (int16) value is declared in this case because 789 is a 16-bit number (+/-32767 or 0-65535) and is the next larger size variable type from a character (char or unsigned char or int8), which is only 8 bits (+/- 128 or 0-255).

B. A variable called 'fred' that will hold numbers from 3 to 456.

Answer:

```
unsigned int16 fred; // typical for CCS-PICC

        -OR-

int16 fred;

        -OR-

//if #TYPE SHORT=8, INT=16, LONG=32
unsigned int fred;
```

-OR-

int fred;

Note: In this case the numbers are always positive so either *int* or *unsigned int* will work, but the range of the number is such that it will not fit into a char or unsigned char.

C. A variable called 'sensor_out' that will contain numbers from −10 to +45.

Answer:

```
signed char sensor_out;
```

D. A variable array that will have 10 elements each holding numbers from −23 to 345.

Answer:

```
int array[10];     //if #TYPE  SHORT=8, INT=16, LONG=32
        -OR-
int16 array[10]; // typical for CCS-PICC
```

Note: In this case the numbers are such that it will not fit into a char or int8 because they can exceed 128 (8 bits).

E. A character string constant that will contain the string 'Press here to end'.

Answer:

```
const char  press_string[] = "Press here to end";
        -OR-
const int8  press_string[] = "Press here to end";
```

Notes: 'const char' and 'const int8' are considered the same type of nonvolatile memory. All three of these declarations will result in the same generated code.

F. A pointer called 'array_ptr' that will point to an array of numbers ranging from 3 to 567.

Answer:

```
//if    #TYPE    SHORT=8,   INT=16,   LONG=32
unsigned int *array_ptr;
        -OR-
unsigned int16 *array_ptr; // typical for CCS-PICC
```

Notes: When declaring the type of a pointer, it is not the size of the pointer but the size of the variable it is pointing to that is of concern. In this case the pointer is pointing to values that are larger than 8 bits but will easily fit into a 16-bit integer.

G. Use an enumeration to set 'uno', 'dos', 'tres' to 21, 22, 23, respectively.

Answer:

enum { uno=21, dos, tres };

Notes: The default for an enumeration starting label value is 0. So the first must be assigned to the desired starting value and the subsequent values will follow.

***4. Evaluate as true or false as if used in a conditional statement (Section 1.6):**

For all problems: x = 0x45; y = 0xc6

A. (x == 0x45)

Answer:

TRUE

B. (x | y)

Answer:

TRUE (because 0x45 | 0xC6 = 0xC7, and 0xC7 is not zero)

C. (x > y)

Answer:

FALSE

D. (y – 0x06 == 0xc)

Answer:

FALSE (because 0xC6 – 0x06 = 0xC0, and that is not equal to 0x0C)

***6. Evaluate the value of the variables after the fragment of code runs (Section 1.7):**

```
unsigned char cntr = 10;
unsigned int16 value = 10;
do
{
    value++;
} while (cntr < 10);
```

```
// value = ??    cntr = ??
```

Answer:

value = 11   cntr = 10

Notes: The variable *value* gets incremented, then the value of *cntr* is tested but not modified.

***10. Write a fragment of C code to declare an appropriate array and then fill the array with the powers of 2 from 2^1 to 2^6 (Section 1.7).**

One possible solution:

```
unsigned char twos[6];  // size array to hold the 6 figures.
char x,y;               // declare a couple of indexes


y = 0;
for(x = 2; x!=0x80; x<<=1)        // shift a one left until it is 2^7
        twos[y++] = x;  // store value into array and increment index
```

ANSWERS TO SELECTED CHAPTER 2 EXERCISES

***2. Describe the following memory types and delineate their uses (Section 2.4):**

A. FLASH Code Memory

The FLASH code memory is *nonvolatile* (retains its data even when power is removed) memory and it is used to store the executable code and constants because they must remain in the memory even when power is removed from the device (from Section 2.4.2).

B. Data Memory

The data memory is volatile (it loses its contents when power is removed) memory that holds the registers, the I/O registers, and the SRAM. The registers are used for short-term storage of variables, the I/O registers are used to communicate with peripheral devices, and the SRAM area is used to store variables (Section 2.4.1).

***5. Write a fragment of C language code to initialize the External Interrupt to activate on a falling edge applied to the external interrupt pin (Section 2.5).**

```
enable_interrupts(INT_EXT);  //unmask the external interrupt
enable_interrupts(global);   //enable all unmasked interrupts
```

***7. Write a fragment of C language code to initialize the Port D pins so that the upper nibble may be used for input and the lower nibble may be used for output (Section 2.6).**

```
set_tris_d(0xF0);        //upper nibble for input, lower for output (Section 2.6)
```

***10. Sketch the waveform appearing at the output of the UART when it transmits an 'H' at 9600 baud. The sketch should show voltage levels and the bit durations in addition to the waveform (Section 2.8).**

Bit time = 1/baud rate = 1/9600 = 104.2 microseconds

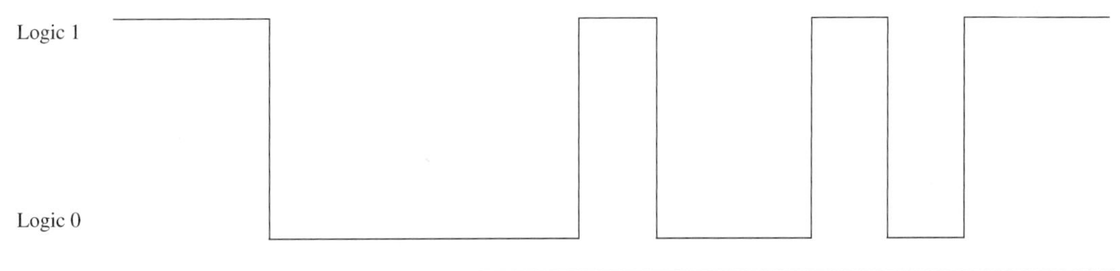

State	Idle	Idle	Active	Active	Active	Active	Active	Active	Active	Active	Active	Active	Idle
Logic Level	1	1	0	0	0	0	1	0	0	1	0	1	1
Bit Type		Start	Data	Data	Data	Data	Data	Data	Data	Data	Stop		
Data Bit Number				0	1	2	3	4	5	6	7		

***12. Compute the missing values to complete the following table that relates to analog to digital conversion (Section 2.9).**

V_{in}	$V_{fullscale}$	***Digital Out***	***# of bits***
4.2 V	10 V	107	8
1.6 V	5 V	327	10
	5 V	123	10
	10 V	223	8

$$\frac{V_{in}}{V_{fullscale}} = \frac{X}{2^n-1} \Rightarrow \frac{4.2\ (2^8-1)}{10} = 107$$

$$\frac{V_{in}}{V_{fullscale}} = \frac{X}{2^n-1} \Rightarrow \frac{1.6 \ (2^{10}-1)}{5} = 327$$

$$\frac{V_{in}}{V_{fullscale}} = \frac{X}{2^n-1} \Rightarrow \frac{123 * 5}{(2^{10}-1)} = 0.601 \text{ volt}$$

$$\frac{V_{in}}{V_{fullscale}} = \frac{X}{2^n-1} \Rightarrow \frac{223 * 10}{(2^{8}-1)} = 8.75 \text{ volts}$$

(Section 2.9.1)

SELECTED ANSWERS TO CHAPTER 3 EXERCISES

***1. What will the contents of *s* be after the following executes and "123ABX" is used as the input, and what is the total number of bytes required to hold the string? (Section 3.5)**

Answer:

7 bytes are required to hold '1','2','3','A','B','X','\0'

***4. Write a function that prints its compile date and time. Use the compiler's internal tag names to get the date and time values.**

Answer:

```
puts("\n\rCompile Date :\r");
puts(__DATE__);
puts("\rCompile Time : \r");
puts(__TIME__);
```

Note: The function *puts()* prints a string to the standard output.

***6. Write a function that inputs a 16-bit hexadecimal value and then prints the binary equivalent. Note: there is no standard output function to print binary, so it is all up to you!**

One possible solution:

```
int16 get_hex16()
{
        char s[7];          // place to put the string
        char a;
        strcpy(s,"0x");             / formatting for atol function
        for (a=0;a<4;a++)           // get the 4 nibble value
        s[a+2] = getchar();
        s[a+2] = '\0';              // terminate the string
        puts("\n\r\n\rHex Input: ");        // user feedback
        puts(s);
        puts("\n\rBinary Equivalent: ");
                                    // return the int16 val
        return (atol(s));           // converted from the string
}
void put_bin16 (int16 value)
{
        char k;                     // loop counter
        for (k=0; k<16; k++)        // do 16 digits
        {
                if (value & (0x8000>>k)) // if 2^k then print a '1'
                putchar('1');
                else
                putchar('0');   // else print a '0'
        }
}
```

Note: In this example, a 1 is shifted left, creating a mask to test each bit position for all 16 positions of the integer.

SELECTED ANSWERS TO CHAPTER 4 EXERCISES

***1. What does IDE stand for? (Section 4.3)**

Answer:

Integrated Development Environment

***4. What menu item is used to highlight all the text between a brace and its match? Also list the shortcut key (Section 4.7).**

Answer:

Menu item: Edit->Match Brace Extended

Shortcut key: Shift + Ctrl +]

***5. Figure 4.28 shows the Macro Manager window with two macros. The command line for the first macro, *send_1*, is not shown. What would be entered in the command line to send an ASCII '1'? What would be entered in the command line to send a hexadecimal value of 1? How else could the ASCII '1' be sent? (Section 4.10)**

Answer:

ASCII '1': 0x31

Hexadecimal '1': 0x01

Alternate method for ASCII '1':

Make the 'Siow' window active and type a '1'.

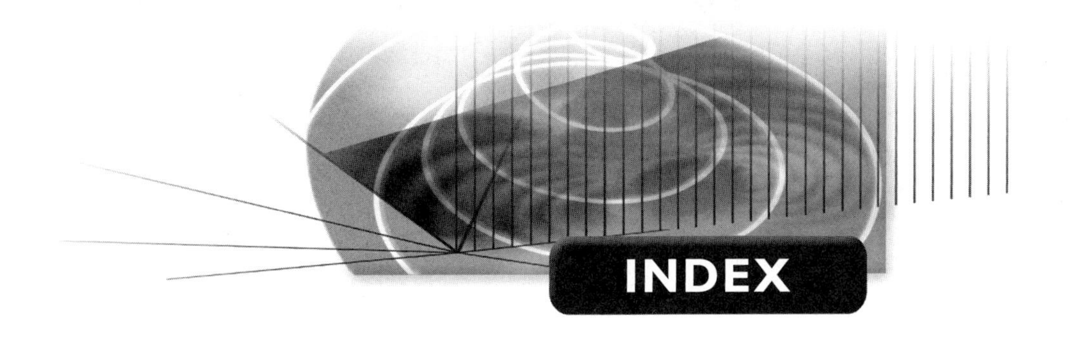

INDEX

C

D

K

L

M

N

O

Q

R

U

V

W

Z